GLOBAL STUDIES

LATIN AMERICA

ELEVENTH EDITION

Dr. Pual B. Goodwin Jr.
University of Connecticut, Storrs

OTHER BOOKS IN THE GLOBAL STUDIES SERIES
- Africa
- China
- Europe
- India and South Asia
- Japan and the Pacific Rim
- Latin America
- Russia, the Eurasian Republics, and
  Central/Eastern Europe

McGraw-Hill/Dushkin Company
2460 Kerper Blvd., Dubuque, Iowa 52001
Visit us on the Internet—http://www.dushkin.com

Staff

Larry Loeppke	*Managing Editor*
Nichole Altman	*Developmental Editor*
Lori Church	*Permissions Assistant*
Maggie Lytle	*Cover*
Tara McDermott	*Designer*
Kari Voss	*Typesetting Supervisor/Co-designer*
Jean Smith	*Typesetter*
Sandy Wille	*Typesetter*
Karen Spring	*Typesetter*

Copyright

Cataloging in Publication Data
Main entry under title: Global Studies: Latin America. 11th ed.
      1. Latin America—History. 2. Central America—History. 3. South America—History.
I. Title: Latin America. II. Goodwin, Paul, Jr., *comp*.
ISBN 0–07–286382–X   954   94–71536   ISSN 1061-2831

Eleventh Edition

Printed in the United States of America   1234567890QPDQPD0987654   Printed on Recycled Paper

LATIN AMERICA

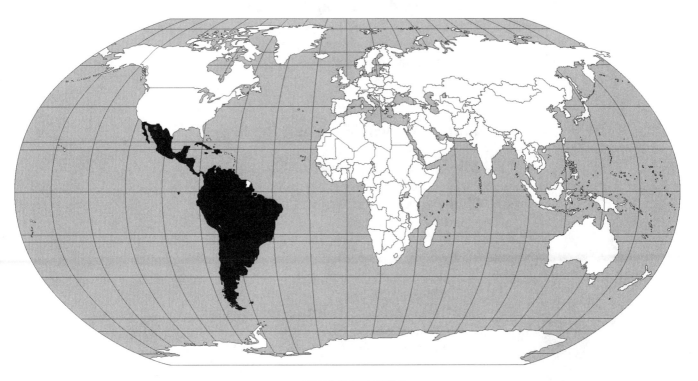

AUTHOR/EDITOR

Dr. Paul B. Goodwin Jr.

The author/editor for *Global Studies: Latin America* is a professor of Latin American history at the University of Connecticut, Storrs. Dr. Goodwin has written, reviewed, and lectured extensively at universities in the United States and many other countries. His particular areas of interest are modern Argentina and Anglo—Latin American relations. Dr. Goodwin has lectured frequently for the Smithsonian Institution, and he has authored or edited three books and numberous articles.

SERIES CONSULTANT

H. Thomas Collins
PROJECT LINKS
George Washington University

Contents

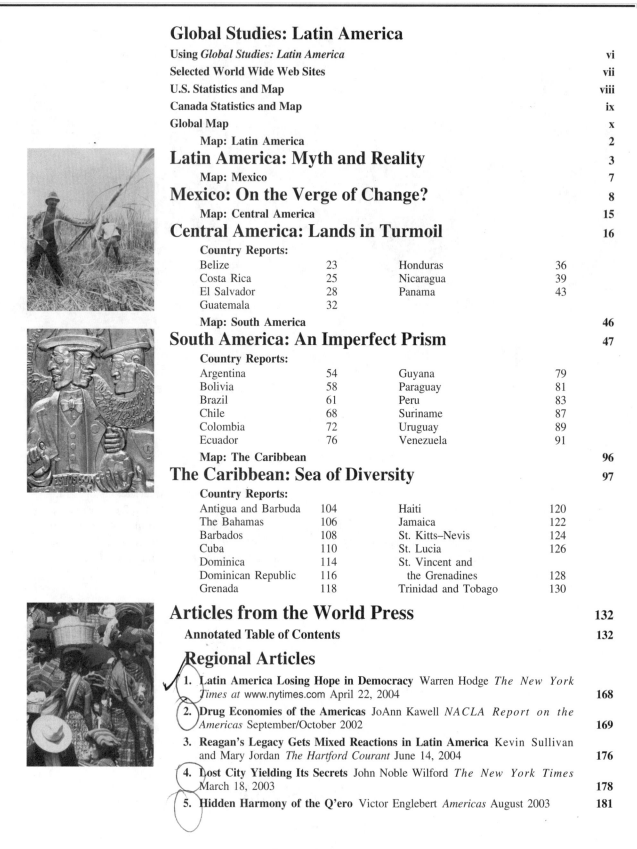

Mexico Articles

Using Global Studies: Latin America

THE GLOBAL STUDIES SERIES

The Global Studies series was created to help readers acquire a basic knowledge and understanding of the regions and countries in the world. Each volume provides a foundation of information—geographic, cultural, economic, political, historical, artistic, and religious—that will allow readers to better assess the current and future problems within these countries and regions and to comprehend how events there might affect their own well-being. In short, these volumes present the background information necessary to respond to the realities of our global age.

Each of the volumes in the Global Studies series is crafted under the careful direction of an author/editor—an expert in the area under study. The author/editors teach and conduct research and have traveled extensively through the regions about which they are writing.

In *Global Studies: Latin America*, the author/editor has written several regional essays and country reports for each of the countries included.

MAJOR FEATURES OF THE GLOBAL STUDIES SERIES

The Global Studies volumes are organized to provide concise information on the regions and countries within those areas under study. The major sections and features of the books are described here.

Regional Essays

For *Global Studies: Latin America*, the author/editor has written several essays focusing on the religious, cultural, sociopolitical, and economic differences and similarities of the countries and peoples in the various subregions of the Latin America. Regional maps accompany the essays.

Country Reports

Concise reports are written for each of the countries within the region under study. These reports are the heart of each Global Studies volume. *Global Studies: Latin America, Eleventh Edition*, contains 33 country reports, including a Mexico report, 7 reports for Central America, 12 for South America, and 13 for the Caribbean. The reports cover each *independent country* in Latin America.

The country reports are composed of five standard elements. Each report contains a detailed map visually positioning the country among its neighboring states; a summary of statistical information; a current essay providing important historical, geographical, political, cultural, and economic information; a historical timeline, offering a convenient visual survey of a few key historical events; and four "graphic indicators," with summary statements about the country in terms of development, freedom, health/welfare, and achievements.

A Note on the Statistical Reports

The statistical information provided for each country has been drawn from a wide range of sources. (The most frequently referenced are listed on page 232.) Every effort has been made to provide the most current and accurate information available. However, sometimes the information cited by these sources dif-

fers to some extent; and, all too often, the most current information available for some countries is somewhat dated. Aside from these occasional difficulties, the statistical summary of each country is generally quite complete and up to date. Care should be taken, however, in using these statistics (or, for that matter, any published statistics) in making hard comparisons among countries. We have also provided comparable statistics for the United States and Canada, which can be found on pages viii and ix.

World Press Articles

Within each Global Studies volume is reprinted a number of articles carefully selected by our editorial staff and the author/editor from a broad range of international periodicals and newspapers. The articles have been chosen for currency, interest, and their differing perspectives on the subject countries. There are 24 articles in *Global Studies: Latin America, Eleventh Edition*.

The articles section is preceded by an annotated table of contents. This resource offers a brief summary of each article.

WWW Sites

An extensive annotated list of selected World Wide Web sites can be found on the facing page (vii) in this edition of *Global Studies: Latin America*. In addition, the URL addresses for country-specific Web sites are provided on the statistics page of most countries. All of the Web site addresses were correct and operational at press time. Instructors and students alike are urged to refer to those sites often to enhance their understanding of the region and to keep up with current events.

Glossary, Bibliography, Index

At the back of each Global Studies volume, readers will find a glossary of terms and abbreviations, which provides a quick reference to the specialized vocabulary of the area under study and to the standard abbreviations used throughout the volume.

Following the glossary is a bibliography that lists general works, national histories, and current-events publications and periodicals that provide regular coverage on Latin America.

The index at the end of the volume is an accurate reference to the contents of the volume. Readers seeking specific information and citations should consult this standard index.

Currency and Usefulness

Global Studies: Latin America, like the other Global Studies volumes, is intended to provide the most current and useful information available necessary to understand the events that are shaping the cultures of the region today.

This volume is revised on a regular basis. The statistics are updated, regional essays and country reports revised, and world press articles replaced. In order to accomplish this task, we turn to our author/editor, our advisory boards, and—hopefully—to you, the users of this volume. Your comments are more than welcome. If you have an idea that you think will make the next edition more useful, an article or bit of information that will make it more current, or a general comment on its organization, content, or features that you would like to share with us, please send it in for serious consideration.

Selected World Wide Web Sites for Global Studies: Latin America

All of these Web sites are hot-linked through the *Global Studies* home page: http://www.dushkin.com/globalstudies (just click on a book).

Some Web sites are continually changing their structure and content, so the information listed may not always be available.

GENERAL SITES

CNN Online Page
http://www.cnn. com

U.S. 24-hour video news channel. News is updated every few hours.

C-SPAN Online
http://www.c-span.org

See especially C-SPAN International on the Web for International Programming Highlights and archived C-SPAN programs.

GlobalEdge
http://globaledge.msu.edu/ibrd/ibrd.asp

Connect to several international business links from this site. Included are links to a glossary of international trade terms, exporting data, international trade, current laws, and data on GATT, NAFTA, and MERCOSUR.

International Information Systems (University of Texas)
http://inic.utexas.edu

Gateway has pointers to international sites, including all Latin American countries.

Library of Congress Country Studies
http://lcweb2.loc.gov/frd/cs/cshome.html#toc

An invaluable resource for facts and analysis of 100 countries' political, economic, social, and national-security systems and installations.

Political Science Resources
http://www.psr.keele.ac.uk

Dynamic gateway to sources available via European addresses. Listed by country name, this site includes official government pages, official documents, speeches, election information, and political events.

ReliefWeb
http://www.reliefweb.int

UN's Department of Humanitarian Affairs clearinghouse for international humanitarian emergencies. It has daily updates, including Reuters and VOA, and PANA.

Social Science Information Gateway (SOSIG)
http://soig.esrc.bris.ac.uk/

Project of the Economic and Social Research Council (ESRC). It catalogs 22 subjects and lists developing countries' URL addresses..

United Nations System
http://www.sosig.ac.ulc/

The official Web site for the United Nations system of organizations. Everything is listed alphabetically, and data on UNICC and Food and Agriculture Organization are available.

UN Development Programme (UNDP)
http://www.undp.org

Publications and current information on world poverty, Mission Statement, UN Development Fund for Women, and much more. Be sure to see the Poverty Clock.

UN Environmental Programme (UNEP)
http://www.unep.org

Official site of UNEP with information on UN environmental programs, products, services, events, and a search engine.

U.S. Agency for International Development (USAID)
http://www.info.usaid.gov

Graphically presented U.S. trade statistics with Latin America and the Caribbean.

U.S. Central Intelligence Agency Home Page
http://www.odci.gov/ cia/publications/factbook/index.htm

This site includes publications of the CIA, such as the World Factbook, Factbook on Intelligence, Handbook of International Economic Statistics, CIA Maps and Publications, and much more.

U.S. Department of State Home Page
http://www.state.gov/ www/ind.html

Organized alphabetically (i.e., Country Reports, Human Rights, International Organizations, and more).

World Bank Group
http://www.worldbank.org

News (press releases, summary of new projects, speeches), publications, topics in development, and countries and regions. Links to other financial organizations are available.

World Health Organization (WHO)
http://www.who.ch

Maintained by WHO's headquarters in Geneva, Switzerland, the site uses Excite search engine to conduct keyword searches.

World Trade Organization
http://www.wto.org

Topics include foundation of world trade systems, data on textiles, intellectual property rights, legal frameworks, trade and environmental policies, and recent agreements.

MEXICO

The Mexican Government
http://world.presidencia.gob.mx

This site offers a brief overview of the organization of the Mexican Republic, including the Executive, Legislative, and Judicial Branches of the federal government.

Documents on Mexican Politics
http://www.cs.unb.ca/~alopez-o/polind.html

An archive of a large number of articles on Mexican democracy, freedom of the press, political parties, NAFTA, the economy, Chiapas, and so forth can be found on this Web site.

CENTRAL AMERICA

Central America News/Planeta
http://www.planeta.com/ecotravel/period/pubcent.html

Access to data that includes individual country reports, politics, economic news, travel, media coverage, and links to other sites are available here.

Latin World
http://www.latinworld.com

Connecting links to data on the economy and finance, businesses, culture, government, and other areas of interest are available on this site.

SOUTH AMERICA

South America Daily
http://www.southamericadaily.com

Everything you want to know about South America is available from this site—from arts and culture, to government data, to environment issues, to individual countries.

CARIBBEAN

Caribbean Studies
http://www.hist.unt.edu/09w-blk4.htm

A complete site for information about the Caribbean. Topics include general information, Caribbean religions, English Caribbean Islands, Dutch Caribbean Islands, French Caribbean Islands, Hispanic Caribbean Islands, and the U.S. Virgin Islands.

Library of Congress Report on the Islands of the Commonwealth Caribbean
http://lcweb2.loc.gov/frd/cs/extoc.html

An exteded study of the Caribbean is possible from this site.

We highly recommend that you review our Web site for expanded information and our other product lines. We are continually updating and adding links to our Web site in order to offer you the most usable and useful information that will support and expand the value of your book. You can reach us at: *http://www.dushkin.com.*

The United States (United States of America)

GEOGRAPHY

Area in Square Miles (Kilometers):
3,717,792 (9,629,091) (about 1/2 the size of Russia)

Capital (Population): Washington, DC (3,997,000)

Environmental Concerns: air and water pollution; limited freshwater resources, desertification; loss of habitat; waste disposal; acid rain

Geographical Features: vast central plain, mountains in the west, hills and low mountains in the east; rugged mountains and broad river valleys in Alaska; volcanic topography in Hawaii

Climate: mostly temperate, but ranging from tropical to arctic

PEOPLE

Population

Total: 293,000,000

Annual Growth Rate: 0.89%

Rural/Urban Population Ratio: 24/76

Major Languages: predominantly English; a sizable Spanish-speaking minority; many others

Ethnic Makeup: 77% white; 13% black; 4% Asian; 6% Amerindian and others

Religions: 56% Protestant; 28% Roman Catholic; 2% Jewish; 4% others; 10% none or unaffiliated

Health

Life Expectancy at Birth: 74 years (male); 80 years (female)

Infant Mortality: 6.69/1,000 live births

Physicians Available: 1/365 people

HIV/AIDS Rate in Adults: 0.61%

Education

Adult Literacy Rate: 97% (official)

Compulsory (Ages): 7–16; free

COMMUNICATION

Telephones: 194,000,000 main lines

Daily Newspaper Circulation: 238/1,000 people

Televisions: 776/1,000 people

Internet Users: 165,750,000 (2002)

TRANSPORTATION

Highways in Miles (Kilometers): 3,906,960 (6,261,154)

Railroads in Miles (Kilometers): 149,161 (240,000)

Usable Airfields: 14,695

Motor Vehicles in Use: 206,000,000

GOVERNMENT

Type: federal republic

Independence Date: July 4, 1776

Head of State/Government: President George W. Bush is both head of state and head of government

Political Parties: Democratic Party; Republican Party; others of relatively minor political significance

Suffrage: universal at 18

MILITARY

Military Expenditures (% of GDP): 3.2%

Current Disputes: various boundary and territorial disputes; "war on terrorism"

ECONOMY

Per Capita Income/GDP: $37,800/$10.98 trillion

GDP Growth Rate: 4%

Inflation Rate: 2.2%

Unemployment Rate: 6.2%

Population Below Poverty Line: 13%

Natural Resources: many minerals and metals; petroleum; natural gas; timber; arable land

Agriculture: food grains; feed crops; fruits and vegetables; oil-bearing crops; livestock; dairy products

Industry: diversified in both capital and consumer-goods industries

Exports: $723 billion (primary partners Canada, Mexico, Japan)

Imports: $1.148 trillion (primary partners Canada, Mexico, Japan)

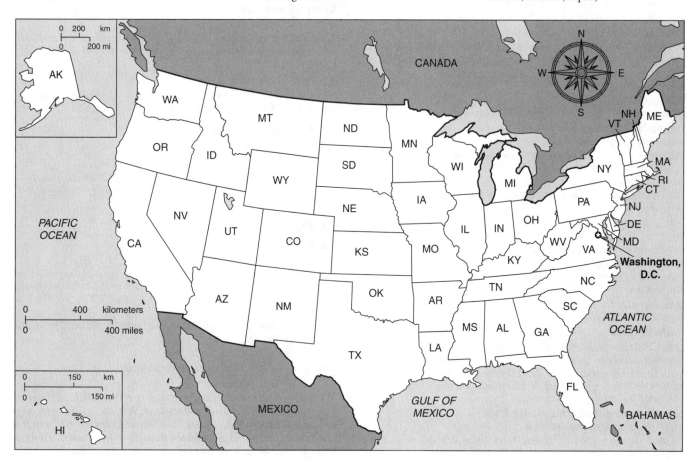

Canada

GEOGRAPHY

Area in Square Miles (Kilometers):
3,850,790 (9,976,140) (slightly larger than the United States)

Capital (Population): Ottawa (1,094,000)

Environmental Concerns: air and water pollution; acid rain; industrial damage to agriculture and forest productivity

Geographical Features: permafrost in the north; mountains in the west; central plains; lowlands in the southeast

Climate: varies from temperate to arctic

PEOPLE

Population

Total: 31,903,000

Annual Growth Rate: 0.96%

Rural/Urban Population Ratio: 23/77

Major Languages: both English and French are official

Ethnic Makeup: 28% British Isles origin; 23% French origin; 15% other European; 6% others; 2% indigenous; 26% mixed

Religions: 46% Roman Catholic; 36% Protestant; 18% others

Health

Life Expectancy at Birth: 76 years (male); 83 years (female)

Infant Mortality: 4.95/1,000 live births

Physicians Available: 1/534 people

HIV/AIDS Rate in Adults: 0.3%

Education

Adult Literacy Rate: 97%

Compulsory (Ages): primary school

COMMUNICATION

Telephones: 20,803,000 main lines

Daily Newspaper Circulation: 215/1,000 people

Televisions: 647/1,000 people

Internet Users: 16,840,000 (2002)

TRANSPORTATION

Highways in Miles (Kilometers): 559,240 (902,000)

Railroads in Miles (Kilometers): 22,320 (36,000)

Usable Airfields: 1,419

Motor Vehicles in Use: 16,800,000

GOVERNMENT

Type: confederation with parliamentary democracy

Independence Date: July 1, 1867

Head of State/Government: Queen Elizabeth II; Prime Minister Jean Chrétien

Political Parties: Progressive Conservative Party; Liberal Party; New Democratic Party; Bloc Québécois; Canadian Alliance

Suffrage: universal at 18

MILITARY

Military Expenditures (% of GDP): 1.1%

Current Disputes: maritime boundary disputes with the United States

ECONOMY

Currency ($U.S. equivalent): 1.46 Canadian dollars = $1

Per Capita Income/GDP: $27,700/$875 billion

GDP Growth Rate: 2%

Inflation Rate: 3%

Unemployment Rate: 7%

Labor Force by Occupation: 74% services; 15% manufacturing; 6% agriculture and others

Natural Resources: petroleum; natural gas; fish; minerals; cement; forestry products; wildlife; hydropower

Agriculture: grains; livestock; dairy products; potatoes; hogs; poultry and eggs; tobacco; fruits and vegetables

Industry: oil production and refining; natural-gas development; fish products; wood and paper products; chemicals; transportation equipment

Exports: $273.8 billion (primary partners United States, Japan, United Kingdom)

Imports: $238.3 billion (primary partners United States, European Union, Japan)

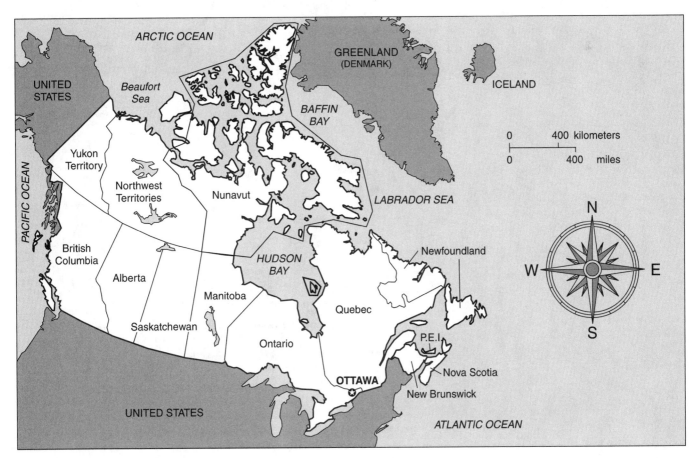

GLOBAL STUDIES

This map is provided to give you a graphic picture of where the countries of the world are located, the relationship they have with their region and neighbors, and their positions relative to major powers and power blocs. We have focused on certain areas to illustrate these crowded regions more clearly.

Latin America

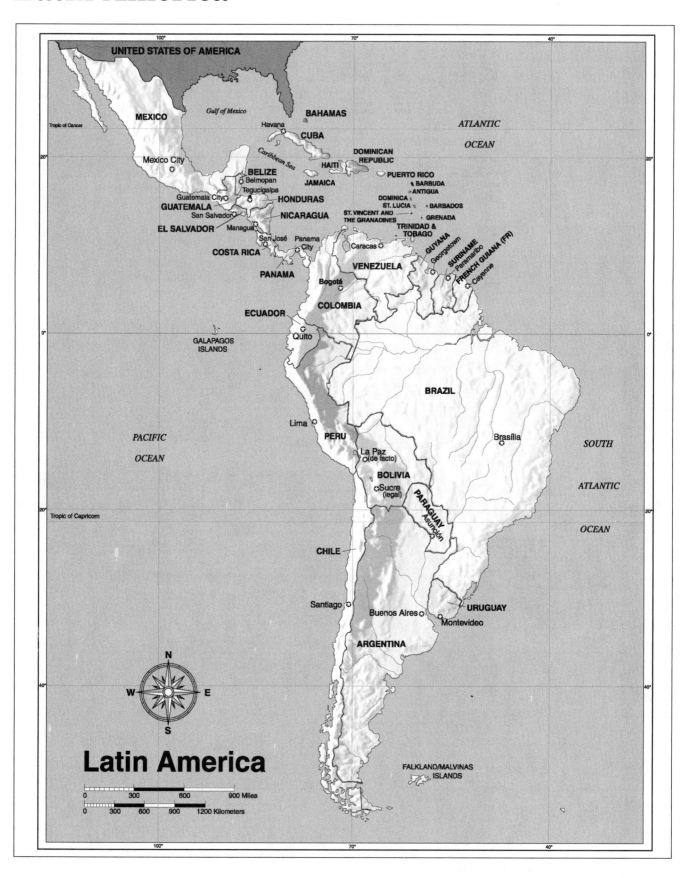

Latin America: Myth and Reality

Much of the world still tends to view Latin Americans in terms of stereotypes. The popular image of the mustachioed bandit sporting a large sombrero and draped with cartridge belts has been replaced by the figure of the modern-day guerrilla, but the same essential image, of lawlessness and violence, persists. Another common stereotype is that of the lazy Latin American who constantly puts things off until *mañana* ("tomorrow"). The implied message here is that Latin Americans lack industry and do not know how to make the best use of their time. A third widespread image is that of the Latin lover and the cult of *machismo* (manliness).

Many of those outside the culture find it difficult to conceive of Latin America as a mixture of peoples and cultures, each one distinct from the others. Indeed, it was not so long ago that then—U.S. president Ronald Reagan, after a tour of the region, remarked with some surprise that all of the countries were "different." Stereotypes spring from ignorance and bias; images are not necessarily a reflection of reality. In the words of Spanish philosopher José Ortega y Gasset: "In politics and history, if one takes accepted statements at face value, one will be sadly misled."

THE LATIN AMERICAN REALITY

The reality of Latin America's multiplicity of cultures is, in a word, complexity. Europeans, Africans, and the indigenous peoples of Latin America have all contributed substantially to these cultures. If one sets aside non-Hispanic influences for a moment, is it possible to argue, as does historian Claudio Veliz, that "the Iberian [Spanish and Portuguese] inheritance is an essential part of our lives and customs; Brazil and Spanish America [i.e., Spanish-speaking] have derived their personality from Iberia"?

Many scholars would disagree. For example, political scientist Lawrence S. Graham argues that "what is clear is that generalizations about Latin American cultural unity are no longer tenable." And that "one of the effects of nationalism has been to … lead growing numbers of individuals within the region to identify with their own nation-state before they think in terms of a more amorphous land mass called Latin America."

Granted, Argentines speak of their Argentinity and Mexicans of their *Mejicanidad.* It is true that there are profound differences that separate the nations of the region. But there exists a cultural bedrock that ties Latin America to Spain and Portugal, and beyond—to the Roman Empire and the great cultures of the Mediterranean world. African influence, too, is substantial in many parts of the region. Latin America's Indians, of course, trace their roots to indigenous sources.

To understand the nature of Latin American culture, one must remember that there exist many exceptions to the generalizations; the cultural mold is not rigid. Much of what has happened in Latin America, including the evolution of its cultures, is the result of a fortunate—and sometimes an unfortunate—combination of various factors.

(United Nations photo)

In Latin America, the family is an important element in the cultural context. These children live in a poor section of Santiago, Chile.

THE FAMILY

Let us first consider the Latin American family. The family unit has survived even Latin America's uneven economic development and the pressures of modernization. Family ties are strong and dominant. These bonds are not confined to the nuclear family of father, mother, and children. The same close ties are found in the extended family (a network of second cousins, godparents, and close friends of blood relatives). In times of difficulty, the family can be counted on to help. It is a fortress against the misery of the outside world; it is the repository of dignity, honor, and respect.

AN URBAN CIVILIZATION

In a region where the interaction of networks of families is the rule and where frequent human contact is sought out, it is not surprising to find that Latin Americans are, above all, an urban people. There are more cities of over half a million people in Latin America than in the United States.

Latin America's high percentage of urban dwellers is unusual, for urbanization is usually associated with industrialization. In Latin America, urban culture was not created by industrial growth; it actually predated it. As soon as the opportunity presented itself, the Spanish conquerors of the New World, in Veliz's words, "founded cities in which to take refuge from the barbaric, harsh, uncivilized, and rural world outside.... For those men civilization was strictly and uniquely a function of well-ordered city life."

The city, from the Spanish conquest until the present, has dominated the social and cultural horizon of Latin America. Opportunity is found in the city, not in the countryside. This cultural fact of life, in addition to economic motives, accounts for the continuing flow of population from rural to urban areas in Latin America.

A WORLD OF APPEARANCES

Because in their urban environment Latin Americans are in close contact with many people, appearances are important to them. There is a constant quest for prestige, dignity, status, and honor. People are forever trying to impress one another with their public worth. Hence, it is not unusual to see a blue-collar worker traveling to work dressed in a suit, briefcase in hand. It is not uncommon to see jungles of television antennas over shantytowns, although many are not connected to anything.

It is a society that, in the opinion of writer Octavio Paz, hides behind masks. Latin Americans convey an impression of importance, no matter how menial their position. Glen Dealy, a political scientist, writes: "And those of the lower class who must wait on tables, wash cars, and do gardening for a living can help to gain back a measure of self-respect by having their shoes shined by someone else, buying a drink for a friend ... , or concealing their occupation by wearing a tie to and from work."

MACHISMO

Closely related to appearances is *machismo*. The term is usually understood solely, and mistakenly, in terms of virility—the image of the Latin lover, for example. But machismo also connotes generosity, dignity, and honor. In many respects, macho behavior is indulged in because of social convention; it is expected of men. Machismo is also a cultural trait that cuts through class lines, for the macho is admired regardless of his social position.

THE ROLE OF WOMEN

If the complex nature of machismo is misunderstood by those outside the culture, so too is the role of women. The commonly held stereotype is that Latin American women are submissive and that the culture is dominated by males. Again, appearances mask a far more complex reality, for Latin American cultures actually allow for strong female roles. Political scientist Evelyn Stevens, for example, has found that *marianismo*—the female

(United Nations photo/Bernard P. Wolff)

The role of the indigenous woman in Latin America has been defined by centuries of tradition. This woman is spinning wool, in Chimburaso, Ecuador, just as her ancestors did.

counterpart of machismo—permeates all strata of Latin American society. Marianismo is the cult of feminine spiritual superiority that "teaches that women are semi-divine, morally superior to and spiritually stronger than men."

When Mexico's war for independence broke out in 1810, a religious symbol—the Virgin of Guadalupe—was identified with the rebels and became a rallying point for the first stirrings of Mexican nationalism. It was not uncommon in Argentine textbooks to portray Eva Perón (1919–1952), President Juan Perón's wife, in the image of the Virgin Mary, complete with a blue veil and halo. In less religious terms, one of Latin America's most popular novels, *Doña Barbara*, by Rómulo Gallegos, is the story of a female *caudillo* ("man on horseback") on the plains of Venezuela.

The Latin American woman dominates the family because of a deep-seated respect for motherhood. Personal identity is less of a problem for her because she retains her family name upon marriage and passes it on to her children. Women who work outside the home are also supposed to retain respect for their motherhood, which is sacred. In any conflict between a woman's job and the needs of her family, the employer, by custom, must grant her a leave to tend to the family's needs. Recent historical scholarship has also revealed that Latin American women have long enjoyed rights denied to women in other, more "advanced" parts of the world. For example, Latin American women were allowed to own property and to sign for mortgages in their own names even in colonial days. In the 1920s, they won the right to vote in local elections in Yucatán, Mexico, and in San Juan, Argentina.

Here again, though, appearances can be deceiving. Many Latin American constitutions guarantee equality of treatment, but reality is burdensome for women in many parts of the region. They do not have the same kinds of access to jobs that men enjoy; they seldom receive equal pay for equal work; and family life, at times, can be brutalizing.

WORK AND LEISURE

Work, leisure, and concepts of time in Latin America correspond to an entirely different cultural mindset than exists in Northern Europe and North America. The essential difference was demonstrated in a North American television commercial for a wine, in which two starry-eyed people were portrayed giving the Spanish toast *Salud, amor, y pesetas* ("Health, love, and money"). For a North American audience, the message was appropriate. But the full Spanish toast includes the tag line *y el tiempo para gozarlos* ("and the time to enjoy them").

In Latin America, leisure is viewed as a perfectly rational goal. It has nothing to do with being lazy or indolent. Indeed, in *Ariel*, by writer José Enrique Rodó, leisure is described within the context of the culture: "To think, to dream, to admire—these are the ministrants that haunt my cell. The ancients ranked them under the word *otium*, well-employed leisure, which they deemed the highest use of being truly rational, liberty of thought emancipated of all ignoble chains. Such leisure meant that use of time which they opposed to mere economic activity as the expression of a higher life. Their concept of dignity was linked closely to this lofty conception of leisure." Work, by contrast, is often perceived as a necessary evil.

CONCEPTS OF TIME

Latin American attitudes toward time also reveal the inner workings of the culture. Exasperated North American businesspeople have for years complained about the *mañana, mañana* attitude of Latin Americans. People often are late for appointments; sometimes little *appears* to get done.

For the North American who believes that time is money, such behavior appears senseless. However, Glen Dealy, in his perceptive book *The Public Man*, argues that such behavior is perfectly rational. A Latin American man who spends hours over lunch or over coffee in a café is not wasting time. For here, with his friends and relatives, he is with the source of his power. Indeed, networks of friends and families are the glue of Latin American society. "Without spending time in this fashion he would, in fact, soon have fewer friends. Additionally, he knows that to leave a café precipitously for an 'appointment' would signify to all that he must not keep someone else waiting—which further indicates his lack of importance. If he had power and position the other person would wait upon his arrival. It is the powerless who wait." Therefore, friends and power relationships are more important than rushing to keep an appointment. The North American who wants the business deal will wait. In a sense, then, the North American is the client and the Latin American is the *patrón* (the "patron," or wielder of power).

(United Nations photo/Jerry Frank)

Agriculture is the backbone of much of Latin America's cultures and economies. These workers are harvesting sugarcane on a plantation in the state of Pernambuco, Brazil.

Perceptions of time in Latin America also have a broader meaning. North American students who have been exposed to Latin American literature are almost always confused by the absence of a "logical," chronological development of the story. Time, for Latin Americans, tends to be circular rather than linear. That is, the past and the present are perceived as equally relevant—both are points on a circle. The past is as important as the present.

MYTH AND REALITY MERGE

The past that is exposed in works of Latin American literature as well as scholarly writings reflects wholly different attitudes toward what people from other cultures identify as reality. For example, in Nobel Prize—winning writer Gabriél García Márquez's classic novel *One Hundred Years of Solitude*—a fictional history of the town of Macondo and its leading family—fantasy and tall tales abound. But García Márquez drew his inspiration from stories he heard at his grandmother's knee about Aracataca, Colombia, the real town in which he grew up. The point here is that the fanciful story of the town's origins constitutes that town's memory of its past. The stories give the town a common heritage and memory.

From a North American or Northern European perspective, the historical memory is faulty. From the Latin American perspective, however, it is the perception of the past that is important, regardless of its factual accuracy. Myth and reality, appearances and substance, merge.

POLITICAL CULTURE

The generalizations drawn here about Latin American society apply also to its political culture, which is essentially authoritarian and oriented toward power and power relationships. Ideology—be it liberalism, conservatism, or communism—is little more than window dressing. It is the means by which contenders for power can be separated. As Claudio Veliz has noted, regardless of the aims of revolutionary leaders, the great upheavals in Latin America in the twentieth century, without exception, ended up by strengthening the political center, which is essentially authoritarian. This was true of the Mexican Revolution (1910), the Bolivian Revolution (1952), the Cuban Revolution (1958), and the Nicaraguan Revolution (1979).

Ideology has never been a decisive factor in the historical and social reality of Latin America. But charisma and the ability to lead are crucial ingredients. José Velasco Ibarra, five times the president of Ecuador in the twentieth century, once boasted: "Give me a balcony and I will be president!" He saw his personality, not his ideology, as the key to power.

In the realm of national and international relations, Latin America often appears to those outside the culture to be in a constant state of turmoil and chaos. It seems that every day there are reports that a prominent politician has fallen from power, border clashes have intensified, or guerrillas have taken over another section of a country. But the conclusion that chaos reigns in Latin America is most often based on the visible political and social violence, not on the general nature of a country. Political violence is often local in nature, and the social fabric of the country is bound together by the enduring social stability of the family. Again, there is the dualism of what *appears to be* and what *is*.

Much of this upheaval can be attributed to the division in Latin America between the people of Mediterranean background and the indigenous Indian populations. There may be several hundred minority groups within a single country. The problems that may arise from such intense internal differences, however, are not always necessarily detrimental, because they contribute to the texture and color of Latin American culture.

SEEING BEHIND THE MASK

In order to grasp the essence of Latin America, one must ignore the stereotypes, appreciate appearances for what they are, and attempt to see behind the mask. Latin America must be appreciated as a culture in its own right, as an essentially Mediterranean variant of Western civilization.

A Latin American world view tends to be dualistic. The family constitutes the basic unit; here one finds generosity, warmth, honor, and love. Beyond the walls of the home, in the world of business and politics, Latin Americans don their masks and enter "combat." It is a world of power relationships, of macho bravado, and of appearances. This dualism is deepseated; scholars such as Richard Morse and Glen Dealy have traced its roots to the Middle Ages. For Latin Americans, one's activities are compartmentalized into those fit for the City of God, which corresponds to religion, the home, and one's intimate circle of friends; and those appropriate for the City of Man, which is secular and often ruthless and corrupt. North Americans, who tend to measure both their public and private lives by the same yardstick, often interpret Latin American dualism as hypocrisy. Nothing could be further from the truth.

For the Latin American, life exists on several planes, has purpose, and is perfectly rational. Indeed, one is tempted to suggest that many Latin American institutions—particularly the supportive network of families and friends—are more in tune with a world that can alienate and isolate than are our own. As you will see in the following reports, the social structure and cultural diversity of Latin America add greatly to its character and, paradoxically, to its stability.

Mexico (United Mexican States)

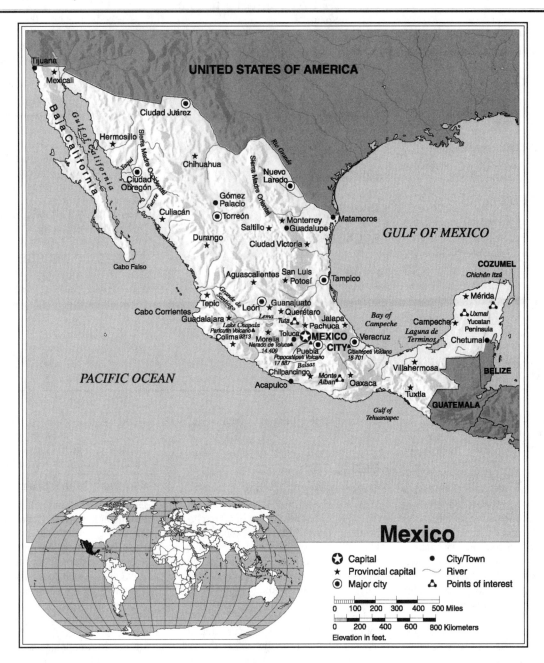

Mexico Statistics

GEOGRAPHY

Area in Square Miles (Kilometers):
764,000 (1,978,000) (about 3 times the size of Texas)

Capital (Population): Mexico City (8,500,000)

Environmental Concerns: Scarce freshwater resources; water pollution; deforestation; soil erosion; serious air pollution

Geographical Features: high, rugged mountains; low coastal plains; high plateaus; desert

Climate: varies from tropical to desert

PEOPLE

Population

Total: 104,959,594
Annual Growth Rate: 1.18%
Rural/Urban Population Ratio: 26/74

Major Languages: Spanish; various Maya, Nahuatl, and other regional indigenous languages

Ethnic Makeup: 60% Mestizo; 30% Amerindian; 9% white; 1% others

Religions: 89% Roman Catholic; 6% Protestant; 5% others

Health

Life Expectancy at Birth: 72 years (male); 77 years (female)

Infant Mortality Rate (Ratio): 21.69/1,000
Physicians Available (Ratio): 1/613

Education

Adult Literacy Rate: 92.2%
Compulsory (Ages): 6–12; free

COMMUNICATION

Telephones: 15,958,700 main lines
Daily Newspaper Circulation: 115 per 1,000 people
Televisions: 192 per 1,000
Internet Users: 10.033 million (2002)

TRANSPORTATION

Highways in Miles (Kilometers): 204,761 (329,532)
Railroads in Miles (Kilometers): 12,122 (19,510)
Usable Airfields: 1,827
Motor Vehicles in Use: 12,230,000

GOVERNMENT

Type: federal republic
Independence Date: September 16, 1810 (from Spain)
Head of State/Government: President Vicente Fox is both head of state and head of government
Political Parties: Institutional Revolutionary Party; National Action Party; Party of the Democratic Revolution; Mexican Green Ecologist Party; Workers Party
Suffrage: universal and compulsory at 18

MILITARY

Military Expenditures (% of GDP): 0.9%
Current Disputes: none

ECONOMY

Currency ($ U.S. Equivalent): 10.79 pesos = $1
Per Capita Income/GDP: $9,000/$942.2 billion

GDP Growth Rate: 1.2%
Inflation Rate: 4%
Unemployment Rate: 3.3% urban; plus considerable underemployment
Labor Force: 41.5 million
Natural Resources: petroleum; silver; copper; gold; lead; zinc; natural gas; timber
Agriculture: corn; wheat; soybeans; rice; beans; cotton; coffee; fruit; tomatoes; livestock products; wood products
Industry: food and beverages; tobacco; chemicals; iron and steel; petroleum; mining; textiles; clothing; motor vehicles; consumer durables; tourism
Exports: $164.8 billion (primary partners United States, Canada)
Imports: $168.9 billion (primary partners United States, Japan, China)

SUGGESTED WEBSITES

http://www.cia.gov/cia/
publications/factbook/geos/
mx.html#geo

Mexico Country Report

There is a story that Hernán Cortéz, the conqueror of the Aztec Empire in the sixteenth century, when asked to describe the landscape of New Spain (Mexico), took a piece of paper in his hands and crumpled it. The analogy is apt. Mexico is a tortured land of mountains and valleys, of deserts in the north and rain forests in the south. Geography has helped to create an intense regionalism in Mexico, and the existence of hundreds of *patrias chicas* ("little countries") has hindered national integration.

Much of Mexico's territory is vulnerable to earthquakes and volcanic activity. In 1943, for example, a cornfield in one of Mexico's richest agricultural zones sprouted a volcano instead of maize. In 1982, a severe volcanic eruption in the south took several hundred lives, destroyed thousands of head of livestock, and buried crops under tons of ash. Thousands of people died when a series of earthquakes struck Mexico City in 1985.

Mexico is a nation of climatic extremes. Much-needed rains often fall so hard that most of the water runs off before it can be absorbed by the soil. When rains fail to materialize, crops die in the fields. The harsh face of the land, the unavailability of water, and erosion limit the agricultural potential of Mexico. Only 10 to 15 percent of Mexico's land can be planted with crops; but because of unpredictable weather or natural disasters, good harvests can be expected

from only 6 to 8 percent of the land in any given year.

MEXICO CITY

Mexico's central region has the best cropland. It was here that the Aztecs built their capital city, the foundations of which lie beneath the current Mexican capital, Mexico City. Given their agricultural potential as well as its focus as the commercial and administrative center of the nation, Mexico City and the surrounding region have always supported a large population. For decades, Mexico City has acted as a magnet for rural poor who have given up attempts to eke out a living from the soil. In the 1940s and 1950s, the city experienced a great population surge. In that era, however, it had the capacity to absorb the tens of thousands of migrants, and so a myth of plentiful money and employment was created. Even today, that myth exercises a strong influence in the countryside; it partially accounts for the tremendous growth of the city and its greater metropolitan area, now home to approximately 18 million people.

The size and location of Mexico City have spawned awesome problems. Because it lies in a valley surrounded by mountains, air pollution is trapped. Mexico City has the worst smog in the Western Hemisphere. Traffic congestion is among the worst in the

world. And essential services—including the provision of drinkable water, electricity, and sewers—have failed to keep pace with the city's growth in population.

Social and Cultural Changes

Dramatic social and cultural changes have accompanied Mexico's population growth. These are particularly evident in Mexico City, which daily becomes less Mexican and more cosmopolitan and international.

As Mexico City has become more worldly, English words have become more common in everyday vocabulary. "Okay," "coffee break," and "happy hour" are some examples of English idioms that have slipped into popular usage. In urban centers, quick lunches and coffee breaks have replaced the traditional large meal that was once served at noon. For most people, the afternoon siesta ("nap") is a fondly remembered custom of bygone days.

Mass communication has had an incalculable impact on culture. Television commercials primarily use models who are ethnically European in appearance—preferably white, blue-eyed, and blonde. As if in defiance of the overwhelmingly Mestizo (mixed Indian and white) character of the population, Mexican newspapers and magazines carry advertisements for products guaranteed to lighten one's skin. Success has become associated with light skin.

Another symbol of success is ownership of a television. Antennas cover rooftops even in the poorest urban slums. Acute observers might note, however, that many of the antennas are not connected to anything; the residents of many hovels merely want to convey the impression that they can afford one.

Television, however, has helped to educate the illiterate. Some Mexican soap operas, for instance, incorporate educational materials. On a given day, a show's characters may attend an adult-education class that stresses basic reading and writing skills. Both the television characters and the home-viewing audience sit in on the class. Literacy is portrayed as being essential to one's success and well-being. Mexican *telenovelas*, or "soaps," have a special focus on teenagers and problems common to adolescents. Solutions are advanced within a traditional cultural context and reaffirm the central role of the family.

Cultural Survival: Compadrazgo

Despite these obvious signs of change, distinct Mexican traditions and customs have not only survived Mexico's transformation but have also flourished because of it. The chaos of city life, the hundreds of thousands of migrants uprooted from rural settings, and the sense of isolation and alienation common to city dwellers the world over are in part eased by the Hispanic institution of *compadrazgo* ("cogodparenthood" or "sponsorship").

Compadrazgo is found at all levels of Mexican society and in both rural and urban areas. It is a device for building economic

and social alliances that are more enduring than simple friendship. Furthermore, it has a religious dimension as well as a secular, or everyday, application. In addition to basic religious occasions (such as baptism, confirmation, first communion, and marriage), Mexicans seek sponsors for minor religious occasions, such as the blessing of a business, and for events as common as a graduation or a boy's first haircut.

DEVELOPMENT

President Fox's plans to privatize the state-owned electrical and petrochemical industries have been stymied by Congress and union opposition who fear a loss of jobs. It is estimated that Mexico's energy industries require tens of billions of dollars of investment that must come from the private sector.

Anthropologist Robert V. Kemper observes that the institution of compadrazgo reaches across class lines and knits the various strands of Mexican society into a whole cloth. Compadrazgo performs many functions, including providing assistance from the more powerful to the less powerful and, reciprocally, providing homage from the less powerful to the more powerful. The most common choices for *compadres* are neighbors, relatives, fellow migrants, coworkers, and employers. A remarkably flexible institution, compadrazgo is perfectly compatible with the tensions and anxieties of urban life.

Yet even compadrazgo—a form of patron/client relationship—has its limita-

tions. As Mexico City has sprawled ever wider across the landscape, multitudes of new neighborhoods have been created. Many are the result of well-planned land seizures, orchestrated by groups of people attracted by the promise of the city. Technically, such land seizures are illegal; and a primary goal of the *colonos* (inhabitants of these low-income communities) is legitimization and consequent community participation.

Beginning in the 1970s, colonos forcefully pursued their demands for legitimization through protest movements and demonstrations, some of which revealed a surprising degree of radicalism. In response, the Mexican government adopted a two-track policy: It selectively repressed the best-organized and most radical groups of colonos, and it tried to co-opt the remainder through negotiation. In the early 1980s, the government created "Citizen Representation" bodies, official channels within Mexico City through which colonos could participate, within the system, in the articulation of their demands.

From the perspective of the colonos, the establishment of the citizen organizations afforded them an additional means to advance their demands for garbage collection, street paving, provision of potable water, sewage removal, and, most critically, the regularization of land tenure—that is, legitimization. In the government's view, representation for the colonos served to win supporters for the Mexican political structure, particularly the authority of the

(AP photo/Jose Luis Magana)

A man from the state of Puebla asks for money outside a jewelry store in Mexico City, hoping to raise enough funds to supply the peasants of his community with the water and electricity that the government has been unable to provide.

official ruling party, at a time of outspoken challenge from other political sectors.

Citizens are encouraged to work within the system; potential dissidents are transformed through the process of co-optation into collaborators. In today's Mexico City, then, patronage and clientage have two faces: the traditional one of compadrazgo, the other a form of state paternalism that promotes community participation.

THE BORDER

In the past few decades, driven by poverty, unemployment, and underemployment, many Mexicans have chosen not Mexico City but the United States as the place to improve their lives. Mexican workers in the United States are not a new phenomenon. During World War II, the presidents of both nations agreed to allow Mexican workers, called *braceros*, to enter the United States as agricultural workers. They were strictly regulated. In contrast, the new wave of migrants is largely unregulated. Each year, hundreds of thousands of undocumented Mexicans illegally cross the border in search of work. It has been estimated that at any given time, between 4 million and 6 million Mexicans pursue an existence as illegal aliens in the United States.

Thousands of Mexicans are able to support families with the fruits of their labors, but, as undocumented workers, they are not protected by the law. Many are callously exploited by those who smuggle them across the border as well as by employers in the United States. For the Mexican government, however, such mass emigration has been a blessing in disguise. It has served as a kind of sociopolitical safety valve, and it has resulted in an inflow of dollars sent home by the workers.

In recent years, U.S. companies and the governments of Mexican states along the border have profited from the creation of assembly plants known as *maquiladoras*. Low wages and a docile labor force are attractive to employers, while the Mexican government reaps the benefits of employment and tax dollars. Despite the appearance of prosperity along the border, it must be emphasized that chronic unemployment in other parts of Mexico ensures the misery of millions of people. The North American Free Trade Agreement (NAFTA) hoped to alter these harsh realities, but after 10 years real wages are lower, the distribution of income has become more unequal, and Mexicans still cross the U.S. border in large numbers.

THE INDIAN "PROBLEM"

During the 1900s, urbanization and racial mixing changed the demographic face of Mexico. A government official once commented: "A country predominately Mestizo, where Indian and white are now minorities, Mexico preserves the festivity and ceremonialism of the Indian civilizations and the religiosity and legalism of the Spanish Empire." The quotation is revealing, for it clearly identifies the Indian as a marginal member of society, as an object of curiosity.

FREEDOM

Election-reform laws contributed to the victory of the Alliance for Change (National Autonomous Party, Green Party) in 2000. Drug-related corruption and violence hinder maintenance of a rule of law.

In Mexico, as is the case with indigenous peoples in most of Latin America, Indians in many quarters are viewed as obstacles to national integration and economic progress. There exist in Mexico more than 200 distinct Indian tribes or ethnic groups, who speak more than 50 languages or dialects. In the view of "progressive" Mexicans, the "sociocultural fragmentation" caused by the diversity of languages fosters political misunderstanding, insecurity, and regional incomprehension. Indians suffer from widespread discrimination. Language is not the only barrier to their economic progress. They have long endured the unequal practices of a ruling white and Mestizo elite. Indians may discover, for example, that they cannot expand a small industry, such as a furniture-making enterprise, because few financial institutions will lend a large amount of money to an Indian.

NATIONAL IDENTITY

Mexico's Mestizo face has had a profound impact on the attempts of intellectuals to understand the meaning of the term "Mexican." The question of national identity has always been an important theme in Mexican history; it became a particularly burning issue in the aftermath of the Revolution of 1910. Octavio Paz believes that most Mexicans have denied their origins: They do not want to be either Indian or Spaniard, but they also refuse to see themselves as a mixture of both. One result of this essential denial of one's ethnic roots is a collective inferiority complex. The Mexican, Paz writes, is insecure. To hide that insecurity, which stems from his sense of "inferiority," the Mexican wears a "mask." Machismo (the cult of manliness) is one example of such a mask. In Paz's estimation, aggressive behavior at a sporting event, while driving a car, or in relation-

ships with women reflects a deep-seated identity crisis.

Perhaps an analogy can be drawn from Mexican domestic architecture. Traditional Mexican homes are surrounded by high, solid walls, often topped with shards of glass and devoid of windows looking out onto the street. From the outside, these abodes appear cold and inhospitable. Once inside (once behind the mask), however, the Mexican home is warm and comfortable. Here, appearances are set aside and individuals can relax and be themselves. By contrast, many homes in the United States have vast expanses of glass that allow every passerby to see within. That whole style of open architecture, at least for homes, is jolting for many Mexicans (as well as other Latin Americans).

THE FAILURE OF THE 1910 REVOLUTION

In addition to the elusive search for Mexican identity, one of Mexican intellectuals' favorite themes is the Revolution of 1910 and what they perceive as its shortcomings. That momentous struggle (1910–1917) cost more than 1 million lives, but it offered Mexico the promise of a new society, free from the abuses of past centuries. It began with a search for truth and honesty in government; it ended with an assertion of the dignity and equality of all men and women.

The goals of the 1910 Revolution were set forth in the Constitution of 1917, a remarkable document—not only in its own era, but also today. *Article 123*, for example, which concerns labor, includes the following provisions: an eight-hour workday, a general minimum wage, and a sixweek leave with pay for pregnant women before the approximate birth date plus a six-week leave with pay following the birth. During the nursing period, the mother must be given extra rest periods each day for nursing the baby. Equal wages must be paid for equal work, regardless of sex or nationality. Workers are entitled to a participation in the profits of an enterprise (i.e., profit sharing). Overtime work must carry double pay. Employers are responsible for and must pay appropriate compensation for injuries received by workers in the performance of their duties or for occupational diseases. In 1917, such provisions were viewed as astounding and revolutionary.

Unfulfilled Promises

Unfortunately, many of the goals of 1917 have yet to be achieved. A number of writers, frustrated by the slow pace of change, concluded long ago that the Mexican Revolution was dead. Leading thinkers and

(United Nations photo)

In many ways, the Mexican people have two separate identities: one public and one private. This carved door by artist Diego Rivera, located in Chapingo, depicts the dual identity that is so much a part of Mexican culture.

writers, such as Carlos Fuentes, have bitterly criticized the failure of the Revolution to shape a more equitable society. Corruption, abuse of power, and self-serving opportunism characterize Mexico today.

One of the failed goals of the Revolution, in the eyes of critics, was an agrarian-reform program that fell short of achieving a wholesale change of land ownership or even of raising the standard of living in rural areas. Over the years, however, small-scale agriculture has sown the seeds of its own destruction. Plots of land that are barely adequate for subsistence farming have been further divided by peasant farmers anxious to satisfy the inheritance rights of their sons. More recently, government price controls on grain and corn have driven many marginal producers out of the market and off their lands.

Land Reform: One Story

Juan Rulfo, a major figure in the history of postrevolutionary literature, captured the frustration of peasants who have "benefited" from agrarian reform. "But sir," the peasant complained to the government official overseeing the land reform, "the earth is all washed away and hard. We don't think the plow will cut into the earth ... that's like a rock quarry. You'd have to make with a pick-axe to plant the seed, and even then you can't be sure that anything will come up... ." The official, cold and indifferent, responded: "You can state that in writing. And now you can go. You should be attacking the large-estate owners and not the government that is giving you the land."

More frequently, landowners have attacked peasants. During the past several years in Mexico, insistent peasant demands for a new allocation of lands have been the occasion of a number of human-rights abuses—some of a very serious character. Some impatient peasants who have occupied lands in defiance of the law have been killed or have "disappeared." In one notorious case in 1982, 26 peasants were murdered in a dispute over land in the state of Puebla. The peasants, who claimed legal title to the land, were killed by mounted gunmen, reportedly hired by local ranchers. Political parties reacted to the massacre in characteristic fashion—all attempted to manipulate the event to their own advantage rather than to address the problem of land reform. Yet years later, paramilitary bands and local police controlled by political bosses or landowners still routinely threatened and/or killed peasant activists. Indeed, access to the land was a major factor in the Maya uprising in the southern state of Chiapas that began in 1994 and, in 2004, remains unresolved.

The Promise of the Revolution

While critics of the 1910 Revolution are correct in identifying its failures, the Constitution of 1917 represents more than dashed hopes. The radical nature of the document allows governments (should they desire) to pursue aggressive egalitarian policies and still be within the law. For example, when addressing citizens, Mexican public officials often invoke the Constitution—issues tend to become less controversial if they are placed within the broad context of 1917. When President Adolfo López Mateos declared in 1960 that his government would be "extremely leftist," he quickly added that his position would be "within the Constitution." But some authorities argue that constitutional strictures can inhibit needed change. For example, the notoriously inefficient state petroleum monopoly (PEMEX) has been

(United Nations photo/Heidi Larson)

Mexican women won the right to vote in 1955. These women, at a political rally in Oaxaca, demonstrate their political consciousness.

critically short of investment capital for years. To allow private companies to invest in the oil industry would require a constitutional change that many Mexicans equate to a form of *vendepatria* (selling out the country). Indeed, in 2003–2004 Congress routinely rejected discussions of even limited private participation in a national industry.

Women's Rights

Although the Constitution made reference to the equality of women in Mexican society, it was not until World War II that the women's-rights movement gathered strength. Women won the right to vote in Mexico in 1955; by the 1970s, they had challenged laws and social customs that were prejudicial to women. Some women have served on presidential cabinets, and one became governor of the state of Colima. The most important victory for women occurred in 1974, however, when the Mexican Congress passed legislation that, in effect, asked men to grant women full equality in society—including jobs, salaries, and legal standing.

But attitudes are difficult to change with legislation, and much social behavior in Mexico still is sexist. The editor of the Mexican newspaper *Noroeste* has asserted that the most important challenge confronting president Vicente Fox is to "break the paternalistic culture." But commentator Lourdes Galaz has noted that the absence of professional women in positions of responsibility on Fox's government teams is an indication that "he lacks any commitment to the female vote."

Many Mexican men feel that there are male and female roles in society, regardless of law. Government, public corporations, private businesses, the Roman Catholic Church, and the armed forces represent important areas of male activity. The home, private religious rituals, and secondary service roles represent areas of female activity. One is clearly dominant, the other subordinate.

The Role of the Church

Under the Constitution of 1917, no religious organization is allowed to administer or teach in Mexico's primary, secondary, or normal (higher education) schools; nor may clergy participate in the education of workers or peasants. Yet between 1940 and 1979, private schools expanded to the point where they enrolled 1.5 million of the country's 17 million pupils. Significantly, more than half of the private-school population attended Roman Catholic schools. Because they exist despite the fact that they are prohibited by law, the Catholic schools demonstrate the kinds of accommodation and flexibility that are possible in Mexico. It is in the best interests of the ruling party to satisfy as many interest groups as is possible.

From the perspective of politicians, the Roman Catholic Church has increasingly tilted the balance in the direction of social justice in recent years. Some Mexican bishops have been particularly outspoken on the issue; but when liberal or radical elements in the Church embrace social change, they may cross into the jurisdiction of the state. Under the Constitution, the state is responsible for improving the welfare of its people. Some committed clergy, however, believe that religion must play an active role in the transformation of society; it must not only have compassion for the poor but must also act to relieve poverty and eliminate injustice.

In 1991, Mexican bishops openly expressed their concern about the torture and mistreatment of prisoners, political persecution, corruption, discrimination against indigenous peoples, mistreatment of Central American refugees, and electoral fraud. In previous years, the government would have reacted sharply against such charges emanating from the Church. But, in this case, there had been a significant rapprochement between the Catholic Church and the state in Mexico. The new relationship culminated with the exchange of diplomatic representatives and Pope John Paul II's successful and popular visit to Mexico in 1990. Despite better relations

at the highest level, in 1999 the bishop of Chiapas vigorously criticized the government for backing away from a 1996 accord between the state and leaders of a guerrilla insurgency and returning to a policy of violent repression.

MEXICO'S STABILITY

The stability of the Mexican state, as has been suggested, depends on the ability of the ruling elite to maintain a state of relative equilibrium among the multiplicity of interests and demands in the nation. The whole political process is characterized by bargaining among elites with various views on politics, social injustice, economic policy, and the conduct of foreign relations.

It was the Institutional Revolutionary Party (PRI), which held the presidency from 1929 until 2000, that set policy and decided what was possible or desirable. All change was generated from above, from the president and his advisers. Although the Constitution provides for a federal system, power was effectively centralized. In the words of one authority, Mexico, with its one-party rule, was not a democracy but, rather, "qualified authoritarianism." In the PRI era, Peruvian author Mario Vargas Llosa referred to Mexico as a "perfect dictatorship." Indeed, the main role of the PRI in the political system was political domination, not power sharing. Paternalistic and all-powerful, the state controlled the bureaucracies that directed the labor unions, peasant organizations, student groups, and virtually every other dimension of organized society. Even though the PRI lost the presidency in 2000, it remains the most powerful political party and retains a strong influence in Mexico's power centers.

HEALTH/WELFARE

Violence against women in Mexico first became an issue of public policy when legislation was introduced in 1990 to amend the penal code with respect to sexual crimes. Among the provisions were specialized medical and social assistance for rape victims and penalties for sexual harassment.

Historically, politicians have tended to be more interested in building their careers than in responding to the demands of their constituents. According to political scientist Peter Smith, Mexican politicians are forever bargaining with one another, seeking favors from their superiors, and communicating in a language of "exaggerated deference." They have learned how to maximize power and success within the existing political structure. By following the

"rules of the game," they move ahead. The net result is a consensus at the upper echelons of power.

In the past few decades, that consensus has been undermined. One of the great successes of the Revolution of 1910 was the rise to middle-class status of millions of people. But recent economic crises alienated that upwardly mobile sector from the PRI. People registered their dissatisfaction at the polls; in 1988, in fact, the official party finished second in Mexico City and other urban centers. In 1989, the PRI's unbroken winning streak of 60 years, facilitated by widespread electoral corruption, was broken in the state of Baja California del Norte, where the right-wing National Action Party (PAN) won the governorship. A decade of worrisome political losses prompted the PRI to consider long overdue reforms. That concern did not prevent the PRI from flagrant electoral fraud in 1988 that handed the presidency to Carlos Salinas Gortari. When it seemed apparent that the PRI would lose, the vote count was interrupted because of "computer failures". In the words of the recent autobiography of former president Miguel de la Madrid, who presided over the fraud, he was told by the PRI president: "You must proclaim the triumph of the PRI. It is a tradition that we cannot break without alarming the citizens." That "tradition" was about to end. Clearly, the PRI had lost touch with critical constituencies who were interested in fundamental change rather than party slogans and were fed up with the rampant corruption of PRI functionaries. Opposition parties continued to win elections.

In the summer of 1997, the left-of-center Party of the Democratic Revolution (PRD) scored stunning victories in legislative, gubernatorial, and municipal elections. For the first time, the PRI lost its stranglehold on the Chamber of Deputies, the lower house of Congress. Significantly, Cuauhtemoc Cardenas of the PRD was swept into power as mayor of Mexico City in the first direct vote for that position since 1928. In gubernatorial contests, the PAN won two elections and controlled an impressive seven of Mexico's 31 governorships.

Within the PRI, a new generation of leaders now perceived the need for political and economic change. President Ernesto Zedillo, worried about his party's prospects in the general elections of 2000, over the objections of old-line conservatives pushed a series of reforms in the PRI. For the first time, the party used state primaries and a national convention to choose the PRI's presidential candidate. This democratization of the party had its reflection in Zedillo's stated commitment to transform Mexican politics by giving the opposition a fair playing field. Voting was now more resistant to tampering and, as a consequence, the three major parties had to campaign for the support of the voters.

In July 2000, Vicente Fox headed a coalition of parties that adopted the name Alliance for Change and promised Mexico's electorate a "Revolution of Hope." It was a formula for success, as the PRI was swept from power. Although Fox was labeled a conservative, his platform indicated that he was above all a pragmatic politician who realized that his appeal and policies had to resonate with a wide range of sectors. Mexican voters saw in Fox someone who identified with human rights, social activism, indigenous rights, women, and the poor. He promised to be a "citizen president." Pundits described his election as a shift from an "imperial presidency" to an "entrepreneurial presidency." Indeed, Fox's economic policies, if implemented, would promote an annual growth rate of 7 percent, lower inflation, balance the budget, raise tax revenues, and improve the standard of living for Mexico's poor (who number 40 million). The private sector would drive the economy; and strategic sectors of the economy, notably electricity generation and petrochemicals, would be opened to private capital. Labor reforms would be initiated that would link salaries to productivity.

President Fox also promised a renewed dialogue with rebels in the southern state of Chiapas. There, beginning in 1994, Maya insurgents had rebelled against a government that habitually supported landowners against indigenous peoples, essentially marginalizing the latter. Led by Subcomandante Marcos, a shrewd and articulate activist who quickly became a hero not only in Chiapas but also in much of the rest of Mexico, the rebels symbolized widespread dissatisfaction with the promises of the PRI. A series of negotiations with the government from time to time interrupted the climate of violence and culminated in 1996 with the Agreements of San Andres. The government assured the Maya of their independence over issues of local governance. But lack of implementation of the agreements, in combination with attacks by the military on the Maya, doomed the accord from the outset.

ORGANIZED LABOR

Organized labor provides an excellent example of the ways in which power is wielded in Mexico and how social change occurs. Mexican trade unions have the right to organize, negotiate, and strike. Most unions historically have not been independent of the government. The major portion of the labor movement is still affiliated with the PRI through an umbrella organization known as the Confederation of Mexican Workers (CTM). The Confederation, with a membership of 3.5 million, is one of the PRI's most ardent supporters. Union bosses truck in large crowds for campaign rallies, help PRI candidates at election time, and secure from union members approval of government policies. Union bosses have been well rewarded by the system they have helped to support. Most have become moderately wealthy and acquired status and prestige. Fully one third of Mexico's senators and congressional representatives, as well as an occasional governor, come from the ranks of union leadership.

ACHIEVEMENTS

Mexican writers and artists have won world acclaim. The works of novelists such as Carlos Fuentes, Mariano Azuela, and Juan Rulfo have been translated into many languages. The graphic-art styles of Posada and the mural art of Diego Rivera, José Clemente Orozco, and David Siqueiros are distinctively Mexican.

Such a relationship must be reciprocal if it is to function properly. The CTM has used an impressive array of left-wing slogans for years to win gains for its members. It has projected an aura of radicalism when, in fact, it is not. The image is important to union members, however, for it gives them the feeling of independence from the government, and it gives a role to the true radicals in the movement. In the 1980s, cracks began to appear in the foundation of union support for the government. The economic crisis of that decade resulted in sharp cutbacks in government spending. Benefits and wage increases fell far behind the pace of inflation; layoffs and unemployment led many union members to question the value of their special relationship with the PRI. Indeed, during the 1988 elections, the Mexican newspaper *El Norte* reported that Joaquín Hernández Galicia, the powerful leader of the Oil Workers' Union, was so upset with trends within the PRI that he directed his membership to vote for opposition candidates. Not surprisingly, then, President Salinas responded by naming a new leader to the Oil Workers' Union.

Independent unions outside the Confederation of Mexican Workers capitalized on the crisis and increased their memberships. For the first time, these independent unions possessed sufficient power to challenge PRI policies. To negate the challenge from the independents, the CTM invited them to join the larger organization. Incorporation of the dissidents into the system is seen as the only way in which the system's credibility can be maintained. It illustrates the state's power to neutralize opposing forces by absorbing them into its system. The demands of labor today are strong, which presents both a challenge and an opportunity to the new government. If he is to implement successfully his economic policies, President Fox must count on the support of organized labor. While most of the leadership still identifies with its traditional patron, the PRI, if labor is to win benefits from the new government, it must cooperate. Fox is perfectly capable of using government patronage to undermine PRI influence with the workers and forge a new base of support for his Alliance for Change.

ECONOMIC CRISIS

As has been suggested, a primary threat to the consensus politics of the PRI came from the economic crisis that began to build in Mexico and other Latin American countries (notably Brazil, Venezuela, and Argentina) in the early 1980s. In the 1970s, Mexico undertook economic policies designed to foster rapid and sustained industrial growth. Credit was readily available from international lending agencies and banks at low rates of interest. Initially, the development plan seemed to work, and Mexico achieved impressive economic growth rates, in the range of 8 percent per year. The government, confident in its ability to pay back its debts from revenues generated by the vast deposits of petroleum beneath Mexico, recklessly expanded its economic infrastructure.

A glut on the petroleum market in late 1981 and 1982 led to falling prices for Mexican oil. Suddenly, there was not enough money available to pay the interest on loans that were coming due, and the government had to borrow more money—at very high interest rates—to cover the unexpected shortfall. By the end of 1982, between 35 and 45 percent of Mexico's export earnings were devoured in interest payments on a debt of $80 billion. Before additional loans could be secured, foreign banks and lending organizations, such as the International Monetary Fund, demanded that the Mexican government drastically reduce state spending. This demand translated into layoffs, inadequate funding for social-welfare programs, and a general austerity that devastated the poor and undermined the high standard of living of the middle class.

Although political reform was important to then-president Salinas, he clearly recognized that economic reform was of more compelling concern. Under Salinas, the foreign debt was renegotiated and substantially reduced.

It was hoped that the North American Free Trade Agreement (NAFTA) among Mexico, the United States, and Canada would shore up the Mexican economy and generate jobs. After a decade there is a wide range of disagreement over NAFTA's success. The Carnegie Endowment of International Peace concluded in November 2003 that the agreement failed to generate significant job growth and actually hurt hundreds of thousands of subsistence farmers who could not compete with "highly efficient and heavily subsidized American farmers." A World Bank report argued that NAFTA had "brought significant economic and social benefits to the Mexican economy," and that Mexico would have been worse off without the pact. Part of the problem lies with the globalization of the economy. Mexico has lost thousands of jobs to China as well as El Salvador, where labor is 20 percent cheaper and less strictly regulated. Five hundred of Mexico's 3,700 *maquiladoras* have closed their doors since 2001. Opposition politicians, nationalists, and those concerned with the more negative aspects of capitalism have generally fought all free-trade agreements, which they see as detrimental to Mexico's sovereignty and independence of action. Perhaps the most interesting development is not economic, but political. Analysts have noted that NAFTA has contributed to a trend toward more representative government in Mexico and that globalization of the economy undercut the state-centered regime of the PRI. Despite advances in some areas, there are still far too many Mexicans whose standard of living is below the poverty level. Of the 40 million poor, 18 million are characterized as living in "extreme poverty." Income distribution is skewed, with the richest 20 percent of the population in control of 58 percent of the nation's wealth, while the poorest 20 percent control only 4 percent.

Many of those employed workers, now estimated at 150,000 per year, will continue to make their way to the U.S. border, which remains accessible despite the passage of immigration-reform legislation and more rigorous patrolling of the border. Others will be absorbed by the so-called informal sector, or underground economy. When walking in the streets of Mexico City, one quickly becomes aware that there exists an economy that is not recognized, licensed, regulated, or "protected" by the government. Yet in the 1980s, this informal sector of the economy produced 25 to 35 percent of Mexico's gross domestic

product and served as a shield for millions of Mexicans who might otherwise have been reduced to destitution. According to George Grayson, "Extended families, which often have several members working and others hawking lottery tickets or shining shoes, establish a safety net for upward of one third of the workforce in a country where social security coverage is limited and unemployment compensation is nonexistent."

FOREIGN POLICY

The problems created by Mexico's economic policy have been balanced by a visibly successful foreign policy. Historically, Mexican foreign policy, which is noted for following an independent course of action, has been used by the government for domestic purposes. In the 1980s, President Miguel de la Madrid identified revolutionary nationalism as the historical synthesis, or melding, of the Mexican people. History, he argued, taught Mexicans to be nationalist in order to resist external aggression, and history made Mexico revolutionary in order to enable it to transform unequal social and economic structures. These beliefs, when tied to the formulation of foreign policy, have fashioned policies with a definite leftist bias. The country has often been sympathetic

to social change and has identified, at least in principle, with revolutionary causes all over the globe. The Mexican government opposed the economic and political isolation of Cuba that was so heartily endorsed by the United States. It supported the Marxist regime of Salvador Allende in Chile at a time when the United States was attempting to destabilize his government. Mexico was one of the first nations to break relations with President Anastasio Somoza of Nicaragua and to recognize the legitimacy of the struggle of the Sandinista guerrillas. In 1981, Mexico joined with France in recognizing the opposition front and guerrilla coalition in El Salvador. In the 1990s Mexico, together with several other Latin American countries, urged a negotiated solution to the armed conflict in Central America. It is also likely that Jorge Castañeda, the new foreign minister, will become very much involved in regional issues such as the internal conflict in Colombia, probably to the displeasure of the United States.

Mexico's leftist foreign policy balances conservative domestic policies. A foreign policy identified with change and social justice has the effect of softening the impact of leftist demands in Mexico for land reform or political change. Mexicans, if displeased with government domestic pol-

icies, were soothed by a vigorous foreign policy that placed Mexico in a leadership role, often in opposition to the United States. With Fox as president, there has been a more centrist position in Mexico's foreign policy, especially with regard to economic-policy formulation and the negotiation of free-trade agreements.

HARD TIMES

Mexico's future is fraught with uncertainty. In December 1994, the economy collapsed after the government could no longer sustain an overvalued peso. In just a few months, the peso fell in value by half,

Timeline: PAST

1519
Hernán Cortés lands at Vera Cruz

1521
Destruction of the Aztec Empire

1810
Mexico proclaims its independence from Spain

1846–1848
War with the United States; Mexico loses four fifths of its territory

1862–1867
The French take over the Mexican throne and install Emperor Maximillian

1876–1910
Era of dictator Porfirio Díaz: modernization

1910–1917
The Mexican Revolution

1934–1940
Land distribution under President Cárdenas

1938
Nationalization of foreign petroleum companies

1955
Women win the right to vote

1968
The Olympic Games are held in Mexico City; riots and violence

1980s
Severe economic crisis; the peso is devalued; inflation soars; the foreign-debt crisis escalates ; Maya insurgency in the state of Chiapas

1990s
NAFTA is passed; the PRI loses ground in legislative, gubernatorial, and municipal elections

PRESENT

2000s
The PRI is ousted from power
Vicente Fox is elected president

Presidential elections scheduled for 2006

while the stock market, in terms of the peso, suffered a 38 percent drop. The crash was particularly acute because the Salinas government had not invested foreign aid in factories and job creation, but had instead put most of the money into Mexico's volatile stock market. It then proceeded to spend Mexico's reserves to prop up the peso when the decline gathered momentum. Salinas's successor, President Ernesto Zedillo, had to cut public spending, sell some state-owned industries, and place strict limits on wage and price increases.

To further confound the economic crisis, the Maya insurgency in Chiapas succeeded in generating much antigovernment support in the rest of Mexico. President Zedillo claimed that the rebels, who call themselves the Zapatista Army of National Liberation (EZLN, named for Emiliano Zapata, one of the peasant leaders of the Mexican Revolution), were "neither popular, nor indigenous, nor from Chiapas." Nobel Laureate Octavio Paz condemned the uprising as an "interruption of Mexico's ongoing political and economic liberalization." The interests of the EZLN leadership, he said, were those of intellectuals rather than those of the peasantry. In other words, what happened in Chiapas was an old story of peasant Indians being used by urban intellectuals—in this instance, to challenge the PRI. Indeed, the real identity of "Subcomandante Marcos" was revealed as Rafael Sebastian Guillen Vicente, a former professor from a rich provincial family who had worked with Tzotzil and Tzeltal Maya Indians since 1984.

George Collier, however, argues that the rebellion is a response to changing governmental policies, agricultural modernization, and cultural and economic isolation. While the peasants of central Chiapas profited from PRI policies, those in the eastern part of the state were ignored. Thus, the rebellion, in essence, was a demand to be included in the largesse of the state. The demands of the EZLN were instructive: democratic reform by the state, limited autonomy for indigenous communities, an antidiscrimination law, teachers, clinics, doctors, electricity, better housing, childcare centers, and a radio station for indigenous peoples. Only vague statements were made about subdivision of large ranches.

During the presidential campaign of 2000, Fox promised to address the complaints raised by the EZLN. Legislation introduced in Congress in the spring of 2001 was designed to safeguard and promote the rights of indigenous peoples. To call attention to the debate, the Zapatistas, with government protection, embarked on a two-week-long march to Mexico City. Significantly, the marchers carried not only the flag of the EZLN but also that of Mexico. But Congress felt that the legislation could damage the nation's unity and harm the interests of local landlords in the south. When a watered-down version of the legislation was passed, Subcomandante Marcos vowed to continue the rebellion. President Fox urged that the talks continue and publicly complained about the congressional action. This was an astute move, because the EZLN could lose an important ally if it adopted an intransigent position.

In summary, the insurgency can be seen to have several roots and to serve many purposes. It is far more complex than a "simple" uprising of an oppressed people.

THE FUTURE

Journalist Igor Fuser, writing in the Brazilian newsweekly *Veja*, observed: "For pessimists, the implosion of the PRI is the final ingredient needed to set off an apocalyptic bomb composed of economic recession, guerrilla war, and the desperation of millions of Mexicans facing poverty. For optimists, the unrest is a necessary evil needed to unmask the most carefully camouflaged dictatorship on the planet."

The elections of 2000 tore away that mask, but persistent problems remain. Corruption, endemic drug-related violence, poverty, unemployment and underemployment, high debt, and inflation are daunting. President Fox admitted in his state of the nation address in September 2003 that he had failed to implement the "historic transformations our times demand." Congress has blocked many of his initiatives, the conflict with the Zapatista of Chiapas continued to simmer, jobs are being lost to the globalization of the economy, and the PRI is intent on mounting a comeback in the presidential elections scheduled for 2006. Change is critical but the policies of transformation render it problematic.

Central America

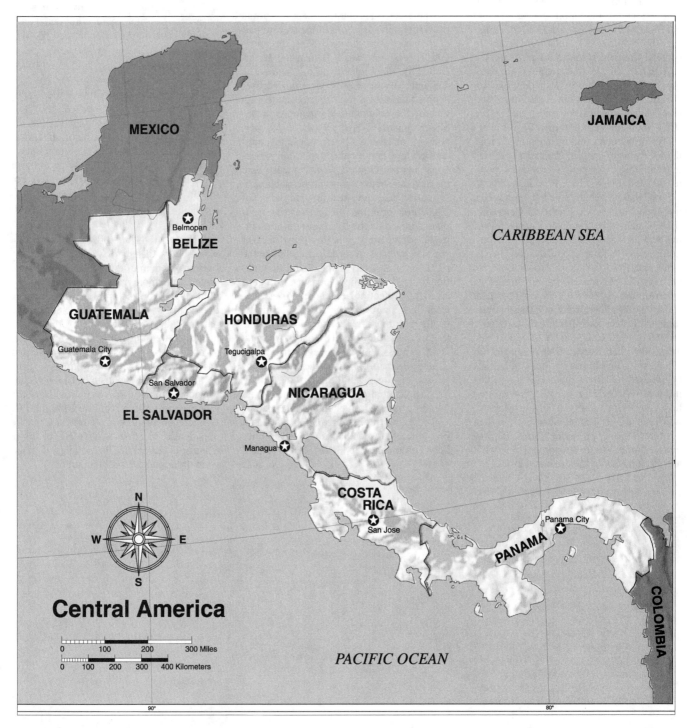

MEXICO

JAMAICA

CARIBBEAN SEA

Belmopan

BELIZE

GUATEMALA

HONDURAS

Guatemala City

Tegucigalpa

San Salvador

NICARAGUA

EL SALVADOR

Managua

COSTA RICA

San Jose

Panama City

PANAMA

COLOMBIA

N W E S

Central America

| 0 | 100 | 200 | 300 Miles |
| 0 | 100 | 200 | 300 | 400 Kilometers |

PACIFIC OCEAN

90°   80°

Much of Central America shares important historical milestones. In 1821, the states of Guatemala, Honduras, El Salvador, Costa Rica, and Nicaragua declared themselves independent of Spain. In 1822, they joined the Empire of Mexico; in 1823, they formed the United Provinces of Central America. this union lasted until 1838, when each member state severed its relations with the federation and went its own way. Since 1838, there have been more than 25 attempts to restore the union—but to no avail.

Central America: Lands in Turmoil

LIFE IN THE MOUTH OF THE VOLCANO

Sons of the Shaking Earth, a well-known study of Middle America by anthropologist Eric Wolf, captures in its title the critical interplay between people and the land in Central America. It asserts that the land is violent and that the inhabitants of the region live in an environment that is often shaken by natural disaster.

The dominant geographical feature of Central America is the impressive and forbidding range of volcanic mountains that runs from Mexico to Panama. These mountains have always been obstacles to communication, to the cultivation of the land, and to the national integration of the countries in which they lie. The volcanoes rest atop major fault lines; some are dormant, others are active, and new ones have appeared periodically. Over the centuries, eruptions and earthquakes have destroyed thousands of villages. Some have recovered, but others remain buried beneath lava and ash. Nearly every Central American city has been destroyed at one time or another; and some, such as Managua, Nicaragua, have suffered repeated devastation.

An ancient Indian philosophy speaks of five great periods of time, each doomed to end in disaster. The fifth period, which is the time in which we now live, is said to terminate with a world-destroying earthquake. "Thus," writes Wolf, "the people of Middle [Central] America live in the mouth of the volcano. Middle America … is one of the proving grounds of humanity."

Earthquakes and eruptions are not the only natural disasters that plague the region. Rains fall heavily between May and October each year, and devastating floods are common. On the Caribbean coast, hurricanes often strike in the late summer and early autumn, threatening coastal cities and leveling crops.

The constant threat of natural disaster has had a deep impact on Central Americans' views of life and development. Death and tragedy have conditioned their attitudes toward the present and the future.

GEOGRAPHY

The region is not only violent but also diverse. In political terms, Central America consists of seven independent nations: Belize, Costa Rica, El Salvador, Guatemala, Honduras, Nicaragua, and Panama. With the exception of Costa Rica and Panama, where national borders coincide with geographical and human frontiers, political boundaries are artificial and were marked out in defiance of both the lay of the land and the cultural groupings of the region's peoples.

Geographically, Central America can be divided into four broad zones: Petén–Belize; the Caribbean coasts of Guatemala, Honduras, and Nicaragua; the Pacific volcanic region; and Costa Rica–Panama.

The northern Guatemalan territory of Petén and all of Belize are an extension of Mexico's Yucatán Peninsula. The region is heavily forested with stands of mahogany, cedar, and pine, whose products are a major source of revenue for Belize.

(United Nations photo/Sygma/J. P. Laffont)

The treat of earthquakes and other natural disasters affects all Central Americans. Above, residents of Guatemala City, Guatemala, begin the long clean-up process after an earthquake.

The Caribbean lowlands, steamy and disease-ridden, are sparsely settled. The inhabitants of the Caribbean coast in Nicaragua include Miskito Indians and the descendants of English-speaking blacks who first settled the area in the seventeenth century. The Hispanic population there was small until recently. Coastal Honduras, however, presents a different picture. Because of heavy investments by foreign companies in the region's banana industry, it is a pocket of relative prosperity in the midst of a very poor country whose economy is based on agricultural production and textiles.

The Pacific volcanic highlands are the cultural heartland of Central America. Here, in highland valleys noted for their springlike climate, live more than 80 percent of the population of Central America; here are the largest cities. In cultural terms, the highlands are home to the whites, mixed bloods, Hispanicized Indians known as Ladinos, and pure-blooded Indians who are descended from the Maya. These highland groups form a striking ethnic contrast to the Indians (such as the Miskito), mulattos, and blacks of the coastlands. The entire country of El Salvador falls within this geographical zone. Unlike its neighbors, there is a uniformity to the land and people of El Salvador.

(United Nations photo/Jerry Frank)

Central American Indians are firmly tied to their traditional beliefs and have strongly resisted the influence of European culture, as evidenced by this Cuna woman of Panama's Tubala Island.

The fourth region, divided between the nations of Costa Rica and Panama, constitutes a single geographical unit. Mountains form the spine of the isthmus. In Costa Rica, the Central Mesa has attracted 70 percent of the nation's population, because of its agreeable climate.

CLIMATE AND CULTURE

The geographic and biological diversity of Central America— with its cool highlands and steaming lowlands, its incredible variety of microclimates and environments, its seemingly infinite types of flora and fauna, and its mineral wealth—has been a major factor in setting the course of the cultural history of Central America. Before the Spanish conquest, the environmental diversity favored the cultural cohesion of peoples. The products of one environmental niche could easily be exchanged for the products of another. In a sense, valley people and those living higher up in the mountains depended on one another. Here was one of the bases for the establishment of the advanced culture of the Maya.

The cultural history of Central America has focused on the densely populated highlands and Pacific plains—those areas most favorable for human occupation. Spaniards settled in the same regions, and centers of national life are located there today. But if geography has been a factor in bringing peoples to-

gether on a local level, it has also contributed to the formation of regional differences, loyalties, interests, and jealousies. Neither Maya rulers nor Spanish bureaucrats could triumph over the natural obstacles presented by the region's harsh geography. The mountains and rain forests have mocked numerous attempts to create a single Central American state.

CULTURES IN CONFLICT

Although physical geography has interacted with culture, the contact between Indians and Spaniards since the sixteenth century has profoundly shaped the cultural face of today's Central America. According to historian Ralph Woodward, the religious traditions of the indigenous peoples, with Christianity imperfectly superimposed over them, "together with the violence of the Conquest and the centuries of slavery or serfdom which followed, left clear impressions on the personality and mentality of the Central American Indian."

To outsiders, the Indians often appear docile and obedient to authority, but beneath this mask may lie intense emotions, including distrust and bitterness. The Indians' vision is usually local and oriented toward the village and family; they do not identify themselves as Guatemalan or Nicaraguan. When challenged, Indians have fought to defend their rights, and a long succession of rebellions from colonial days until the present attests to their sense of what is just and what is not. The Indians, firmly tied to their traditional beliefs and values, have tried to resist modernization, despite government programs and policies designed to counter what urbanized whites perceive as backwardness and superstition.

Population growth, rather than government programs and policies, has had a great impact on the region's Indian peoples and has already resulted in the recasting of cultural traditions. Peasant villages in much of Central America have traditionally organized their ritual life around the principle of *mayordomía,* or sponsorship. Waldemar Smith, an anthropologist who has explored the relationship between the *fiesta* (ceremony) system and economic change, has shown the impact of changing circumstances on traditional systems. In any Central American community in any given year, certain families are appointed *mayordomos,* or stewards, of the village saints; they are responsible for organizing and paying for the celebrations in their names. This responsibility ordinarily lasts for a year. One of the outstanding features of the fiesta system is the phenomenal costs that the designated family must bear. An individual might have to expend the equivalent of a year's earnings or more to act as a sponsor in a community fiesta. Psychological and social burdens must also be borne by the mayordomos, for they represent their community before its saints. Mayordomos, who in essence are priests for a year, are commonly expected to refrain from sexual activity for long periods as well as to devote much time to ritual forms.

The office, while highly prestigious, can also be dangerous. Maya Indians, for example, believe that the saints use the weather as a weapon to punish transgressions, and extreme weather is often traced to ritual error or sins on the part of the mayordomo, who might on such occasions actually be jailed.

Since the late 1960s, the socioeconomic structure of much of the area heavily populated by Indians has changed, forcing changes in traditional cultural forms, including the fiesta system. Expansion of markets and educational opportunity, the absorption of much of the workforce in seasonal plantation labor, more efficient transportation systems, and population growth have precipitated change. Traditional festivals in honor of a community's saints have significantly diminished in importance in a number of towns. Costs have been reduced or several families have been made responsible for fiesta sponsorship. This reflects not only modernization but also crisis. Some communities have become too poor to support themselves—and the expensive fiestas have, naturally, suffered.

This increasing poverty is driven in part by population growth, which has exerted tremendous pressure on people's access to land. Families that cannot be sustained on traditional lands must now seek seasonal wage labor on sugarcane, coffee, or cotton plantations. Others emigrate. The net result is a culture under siege. Thus, while the fiestas may not vanish, they are surely in the process of change.

The Ladino World

The word *Ladino* can be traced back to the Roman occupation of Spain. It referred to someone who had been "Latinized" and was therefore wise in the ways of the world. The word has several meanings in Central America. In Guatemala, it refers to a person of mixed blood, or *Mestizo*. In most of the rest of Central America, however, it refers to an Indian who has adopted white culture.

The Ladinos are caught between two cultures, both of which initially rejected them. The Ladinos attempted to compensate for their lack of cultural roots and cultural identity by aggressively carving out a place in Central American society. Often acutely status-conscious, Ladinos typically contrast sharply with the Indians they physically resemble. Ladinos congregate in the larger towns and cities, speak Spanish, and seek a livelihood as shopkeepers or landowners. They compose the local elite in Guatemala, Nicaragua, Honduras, and El Salvador (the latter country was almost entirely Ladinoized by the end of the nineteenth century), and they usually control regional politics. They are often the most aggressive members of the community, driven by the desire for self-advancement. Their vision is frequently much broader than that of the Indian; they have a perspective that includes the capital city and the nation. The vast majority of the population speak Spanish; few villages retain the use of their original, native tongues.

The Elite

For the elite, who are culturally "white," the city dominates their social and cultural horizons. For them, the world of the Indian is unimportant—save for the difficult questions of social integration and modernization. Businesspeople and bureaucrats, absentee landlords, and the professional class of doctors, lawyers, and engineers constitute an urban elite who are cosmopolitan and sophisticated. Wealth, status, and "good blood" are the keys to elite membership.

The Disadvantaged

The cities have also attracted disadvantaged people who have migrated from poverty-stricken rural regions in search of eco-

(United Nations photo)

Many Central Americans migrate from rural areas to the urban centers, but it is frequently beyond the capacity of urban areas to support them. The child pictured above has to get the family's water from a single, unsanitary community tap.

nomic opportunity. Many are self-employed as peddlers, small-scale traders, or independent craftspeople. Others seek low-paying, unskilled positions in industry, construction work, and transportation. Most live on the edge of poverty and are the first

19

(World Bank photo/Jaime Martin-Escobal)

The migration of poor rural people to Central American urban centers has caused large numbers of squatters to take up residence in slums. The crowded conditions in urban El Salvador, as shown in this photograph, are typical results of this phenomenon.

to suffer in times of economic recession. But there exist Hispanic institutions in this harsh world that help people of all classes to adjust. In each of the capital cities of Central America, lower-sector people seek help and sustenance from the more advantaged elements in society. They form economic and social alliances that are mutually beneficial. For example, a tradesman might approach a well-to-do merchant and seek advice or a small loan. In return, he can offer guaranteed service, a steady supply of crafts for the wholesaler, and a price that is right. It is a world built on mutual exchanges.

These networks, when they function, bind societies together and ease the alienation and isolation of the less advantaged inhabitants. Of course, networks that cut through class lines can effectively limit class action in pursuit of reforms; and, in many instances, the networks do not exist or are exploitive.

POPULATION MOVEMENT

For many years, Central Americans have been peoples in motion. Migrants who have moved from rural areas into the cities have often been driven from lands they once owned, either because of the expansion of landed estates at the expense of the smaller landholdings, population pressure, or division of the land into plots so small that subsistence farming is no longer possible. Others have moved to the cities in search of a better life.

Population pressure on the land is most intense in El Salvador. No other Latin American state utilizes the whole of its

territory to the extent that El Salvador does. Most of the land is still privately owned and is devoted to cattle farming or to raising cotton and coffee for the export market. There is not enough land to provide crops for a population that has grown at one of the most rapid rates in the Western Hemisphere. There are no unpopulated lands left to occupy. Agrarian reform, even if successful, will still leave hundreds of thousands of peasants without land.

Many Salvadorans have moved to the capital city of San Salvador in search of employment. Others have crossed into neighboring countries. In the 1960s, thousands moved to Honduras, where they settled on the land or were attracted to commerce and industry. By the end of that decade, more than 75 percent of all foreigners living in Honduras had crossed the border from El Salvador. Hondurans, increasingly concerned by the growing presence of Salvadorans, acted to stem the flow and passed restrictive and discriminatory legislation against the immigrants. The tension, an ill-defined border, and festering animosity ultimately brought about a brief war between Honduras and El Salvador in 1969.

Honduras, with a low population density (about 139 persons per square mile, as compared to El Salvador's 721), has attracted population not only from neighboring countries but also from the Caribbean. Black migrants from the "West Indian" Caribbean islands known as the Antilles have been particularly attracted to Honduras's north coast, where they have been able to find employment on banana plantations or in the light industry that has increasingly been established in the

(UN Photo 135/228/Syoma/J. P. Laffont)

Guatemala, City, Guatemala, has attracted numerous people from poverty-stricken rural regions in search of economic opportunities. Many are self-employed as peddlers, small-scale traders, or independent craftspeople. These Guatemalan Indians are buying household supplies in a makeshift outdoor market.

area. The presence of these Caribbean peoples in moderate numbers has more sharply focused regional differences in Honduras. The coast, in many respects, is Caribbean in its peoples' identity and outlook; while peoples of the highlands of the interior identify with the capital city of Tegucigalpa, which is Hispanic in culture.

THE REFUGEE PROBLEM

Recent turmoil in Central America created yet another group of people on the move—refugees from the fighting in their own countries or from the persecution by extremists of the political left and right. For example, thousands of Salvadorans crowded into Honduras's western province. In the south, Miskito Indians, fleeing from Nicaragua's Sandinista government, crossed the Río Coco in large numbers. Additional thousands of armed Nicaraguan counterrevolutionaries camped along the border. Only in 1990–1991 did significant numbers of Salvadorans move back to their homeland. With the declared truce between Sandinistas and Contras and the election victory of Violeta Chamorro, Nicaraguan refugees were gradually repatriated. Guatemalan Indians sought refuge in southern Mexico, and Central Americans of all nationalities resettled in Costa Rica and Belize.

El Salvadorans, who began to emigrate to the United States in the 1960s, did so in much greater numbers with the onset of the El Salvadoran Civil War, which killed approximately 70,000 people and displaced about 25 percent of the nation's population. The Urban Institute, a Washington, D.C.–based research group, estimated in 1986 that there were then about ¾ million El Salvadorans—of a total population of just over 5 million—living in the United States. Those emigrants became a major source of dollars for El Salvador; it is estimated that they now send home about $500 million a year.

While that money has undoubtedly helped to keep the nation's economy above water, it has also generated, paradoxically, a good deal of anti–U.S. sentiment in El Salvador. Lindsey Gruson, a reporter for *The New York Times*, studied the impact of expatriate dollars in Intipuca, a town 100 miles southwest of the capital, and concluded that they had a profound impact on Intipuqueño culture. The influx of money was an incentive not to work, and townspeople said that the "free" dollars "perverted cherished values" and were "breaking up many families."

THE ROOTS OF VIOLENCE

Central America still feels the effects of civil war and violence. Armies, guerrillas, and terrorists of the political left and right have exacted a high toll on human lives and property. The civil wars and guerrilla movements that spread violence to the region sprang from each of the societies in question.

A critical societal factor was (and remains) the emergence of a middle class in Central America. In some respects, people of the middle class resemble the Mestizos or Ladinos, in that their wealth and position have placed them above the masses. But, like the Mestizos and Ladinos, they have been denied access to the upper reaches of power, which is the special preserve of the elite. Since World War II, it has been members of the middle class who have called for reform and a more equitable distribution of the national wealth. They have also attempted to forge alliances of opportunity with workers and peasants.

Nationalistic, assertive, restless, ambitious, and, to an extent, ruthless, people of the middle class (professionals, intellectuals, junior officers in the armed forces, office workers, businesspeople, teachers, students, and skilled workers) demand a greater voice in the political world. They want governments that are responsive to their interests and needs; and, when governments have proven unresponsive or hostile, elements of the middle class have chosen confrontation.

In the civil war that removed the Somoza family from power in Nicaragua in 1979, for example, the middle class played a critical leadership role. Guerrilla leaders in El Salvador were middle class in terms of their social origins, and there was significant middle-class participation in the unrest in Guatemala.

Indeed, Central America's middle class is among the most revolutionary groups in the region. Although middle-class people are well represented in antigovernment forces, they also resist changes that would tend to elevate those below them on the social scale. They are also significantly represented among right-wing groups, whose reputation for conservative views is accompanied by systematic terror.

Other societal factors also figure prominently in the violence in Central America. The rapid growth of population since the 1960s has severely strained each nation's resources. Many rural areas have become overpopulated, poor agricultural practices have caused extensive soil erosion, the amount of land available to subsistence farmers is inadequate, and poverty and misery are pervasive. These problems have combined to compel rural peoples to migrate to the cities or to whatever frontier lands are still available. In Guatemala, government policy drove Indians from ancestral villages in the highlands to "resettlement" villages in the low-lying, forested Petén to the north. Indians displaced in this manner often—not surprisingly—joined guerrilla movements. They were not attracted to insurgency by the allure of socialist or communist ideology; they simply responded to violence and the loss of their lands with violence against the governments that pursued such policies.

The conflict in this region does not always pit landless, impoverished peasants against an unyielding elite. Some members of the elite see the need for change. Most peasants have not taken up arms, and the vast majority wish to be left in peace. Others who desire change may be found in the ranks of the military or within the hierarchy of the Roman Catholic Church. Reformers are drawn from all sectors of society. It is thus more appropriate to view the conflict in Central America as a civil war rather than a class struggle, as civil wars cut through the entire fabric of a nation.

Much of today's criminal violence in urban areas of Central America, and particularly in El Salvador and Honduras, is a direct consequence of the years of civil war. Young children of refugees, who relocated to large United States cities as adolescents, often imitated the gang culture to which they were exposed. When they returned to Central America and encountered a society that they did not recognize, they could not find jobs and the gang culture was replicated.

ECONOMIC PROBLEMS

Central American economies, always fragile, have in recent years been plagued by a combination of vexing problems. Foreign debt, inflation, currency devaluations, recession, and, in some instances, outside interference have had deleterious effects on the standard of living in all the countries. Civil war, insurgency, corruption and mismanagement, and population growth have added fuel to the crisis—not only in the region's economies but also in their societies. Nature, too has played an important contributory role in the region's economic and social malaise. Hurricane Mitch, which struck Central American in 1998, killed thousands, destroyed crops and property, and disrupted the infrastructure of roads and bridges in Honduras, Nicaragua, Guatemala, and El Salvador.

Civil war in El Salvador brought unprecedented death and destruction and was largely responsible for economic deterioration and a decline of well over one third of per capita income from 1980 to 1992. Today, fully two thirds of the working-age population are either unemployed or underemployed. The struggle of the Sandinista government of Nicaragua against U.S.-sponsored rebels routinely consumed 60 percent of government spending; even with peace, much of the budget was earmarked for economic recovery. In Guatemala, a savage civil war lasted more than a generation; took more than 140,000 lives; strained the economy; depressed wages; and left unaddressed pressing social problems in education, housing, and welfare. Although the violence has subsided, the lingering fears conditioned by that violence have not. U.S. efforts to force the ouster of Panamanian strongman Manuel Antonio Noriega through the application of economic sanctions probably harmed middle-class businesspeople in Panama more than Noriega.

Against this backdrop of economic malaise there have been some creative attempts to solve, or at least to confront, pressing problems. In 1987, the Costa Rican government proposed a series of debt-for-nature swaps to international conservation groups, such as the Nature Conservancy. The first of the transactions took place in 1988, when several organizations purchased more than $3 million of Costa Rica's foreign debt at 17 percent of face value. The plan called for the government to exchange with the organizations part of Costa Rica's external debt for government bonds; the conservation groups would then invest the earnings of the bonds in the management and protection of Costa Rican national parks. According to the National Wildlife Federation, while debt-for-nature swaps are not a cure-all for the Latin American debt crisis, at least the swaps can go some distance toward protecting natural resources and encouraging ecologically sound, long-term economic development.

INTERNAL AND EXTERNAL DIMENSIONS OF CONFLICT

The continuing violence in much of Central America suggests that internal dynamics are perhaps more important than the overweening roles formerly ascribed to Havana, Moscow, and Washington. The removal of foreign "actors" from the stage lays bare the real reasons for violence in the region: injustice, power, greed, revenge, and racial and ethnic discrimination. Havana, Moscow, and Washington, among others, merely used

Central American violence in pursuit of larger policy goals. And Central American governments and guerrilla groups were equally adept at using foreign powers to advance their own interests, be they revolutionary or reactionary.

Panama offers an interesting scenario in this regard. It, like the rest of Central America, is a poor nation comprised of subsistence farmers, rural laborers, urban workers, and unemployed and underemployed people dwelling in the shantytowns ringing the larger cities. For years, the pressures for reform in Panama were skillfully rechanneled by the ruling elite toward the issue of the Panama Canal. Frustration and anger were deflected from the government, and an outdated social structure was attributed to the presence of a foreign power—the United States—in what Panamanians regarded as their territory.

Central America, in summary, is a region of diverse geography and is home to peoples of many cultures. It is a region of strong local loyalties; its problems are profound and perplexing. The violence of the land is matched by the violence of its peoples as they fight for something as noble as justice or human rights, or as ignoble as political power or self-promotion.

Belize

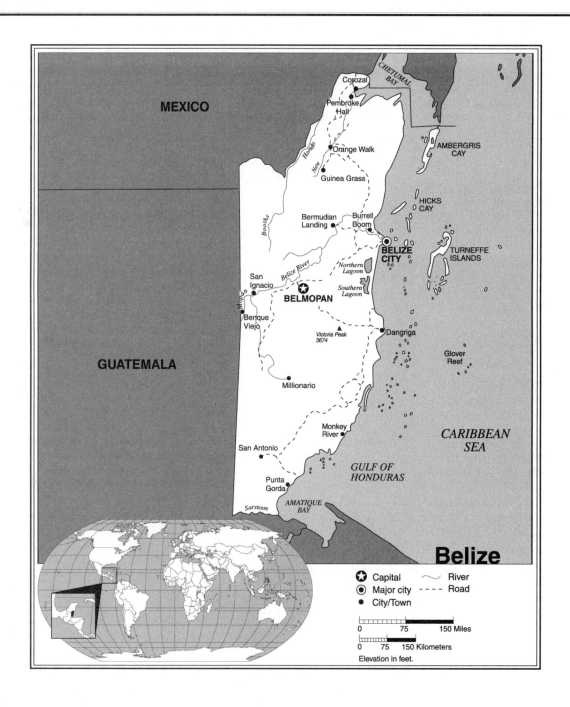

Belize Statistics

GEOGRAPHY

Area in Square Miles (Kilometers): 8,866 (22,963) (about the size of Massachusetts)

Capital (Population): Belmopan (6,800)

Environmental Concerns: deforestation; water pollution

Geographical Features: flat, swampy coastal plain; low mountains in south

Climate: tropical; very hot and humid

PEOPLE

Population

Total: 272,945

Annual Growth Rate: 2.39%

Rural/Urban Population Ratio: 54/46

Ethnic Makeup: 48.7% Mestizo; 24.9% Creole; 10.6% Maya; 6.1% Garifuna; 9.7% others

Major Languages: English; Spanish; Maya; Garifuna

Religions: 49.6% Roman Catholic; 27% Protestant; 14% others; 9.4% unaffiliated

Health

Life Expectancy at Birth: 65 years (male); 69 years (female)
Infant Mortality Rate (Ratio): 26.37/1,000
Physicians Available (Ratio): 1/1,546

Education

Adult Literacy Rate: 70.3%
Compulsory (Ages): 5–14

COMMUNICATION

Telephones: 31,300 main lines
Televisions: 109 per 1,000 people
Internet Users: 30,000

TRANSPORTATION

Highways in Miles (Kilometers): 1,723 (2,872)
Railroads in Miles (Kilometers): none
Usable Airfields: 44
Motor Vehicles in Use: 5,600

GOVERNMENT

Type: parliamentary democracy
Independence Date: September 21, 1981 (from the United Kingdom)
Head of State/Government: Governor General Sir Colville Young (represents Queen Elizabeth II); Prime Minister Said Musa
Political Parties: People's United Party; United Democratic Party; National Alliance for Belizean Rights
Suffrage: universal at 18

MILITARY

Military Expenditures (% of GDP): 2%
Current Disputes: border dispute with Guatemala

ECONOMY

Currency ($ U.S. Equivalent): 2.00 Belize dollars = $1
Per Capita Income/GDP: $4,900/$1.28 billion

GDP Growth Rate: 3.7%

Inflation Rate: 1.9%

Unemployment Rate: 9.1%

Labor Force: 90,000

Natural Resources: arable land; timber; fish; hydropower

Agriculture: bananas; cocoa; citrus fruits; sugarcane; lumber; fish; cultured shrimp

Industry: garment production; food processing; tourism; construction

Exports: $207.8 million (primary partners Mexico, United States, European Union)

Imports: $500.6 million (primary partners United States, Mexico, United Kingdom)

SUGGESTED WEBSITE

http://www.cia.gov/cia/
publications/factbook/geos/
bh.html#Geo

Belize Country Report

THE "HISPANICIZATION" OF A COUNTRY

Belize was settled in the late 1630s by English woodcutters who also indulged in occasional piracy at the expense of the Spanish crown. The loggers were interested primarily in dye-woods, which, in the days before chemical dyes, were essential to British textile industries. The country's name is derived from Peter Wallace, a notorious buccaneer who, from his base there, haunted the coast in search of Spanish shipping. The natives shortened and mispronounced Wallace's name until he became known as "Belize."

As a British colony (called British Honduras), Belize enjoyed relative prosperity as an important entrepôt, or storage depot for merchandise, until the completion of the Panama Railway in 1855. With the opening of a rail route to the Pacific, commerce shifted south, away from Caribbean ports. Belize entered an economic tailspin (from which it has never entirely recovered). Colonial governments attempted to diversify the colony's agricultural base and to attract foreign immigration to develop the land. But, except for some Mexican settlers and a few former Confederate soldiers who came to the colony after the U.S. Civil War, the immigration policy failed. Economically depressed, its population exposed to the ravages of yellow fever, malaria, and dengue (a tropical fever), Be-

lize was once described by British novelist Aldous Huxley in the following terms: "If the world had ends, Belize would be one of them."

Living conditions improved markedly by the 1950s, and the colony began to move toward independence from Great Britain. Although self-governing by 1964, Belize did not become fully independent until 1981, because of Guatemalan threats to invade what it even today considers a lost province, stolen by Britain. British policy calls for a termination of its military presence, even though Guatemalan intentions toward Belize are ambivalent.

DEVELOPMENT

Belize has combined its tourism and environmental-protection offices into one ministry, which holds great promise for ecotourism. Large tracts of land have been set aside to protect jaguars and other endangered species. But there is also pressure on the land from rapid population growth.

For most of it's history Belize has been culturally British with Caribbean overtones. English common law is practiced in the courts, and politics are patterned on the English parliamentary system. A large percent of the people are Protestants. The Belizeans are primarily working-class poor

and middle-class shopkeepers and merchants. There is no great difference between the well-to-do and the poor in Belize, and few people fall below the absolute poverty line.

Thirty percent of the population are Creole (black and English mixture), 6 percent Garifuna (black and Indian mixture). The Garifuna originally inhabited the Caribbean island of St. Vincent. In the eighteenth century, they joined with native Indians in an uprising against the English authorities. As punishment, virtually all the Garifuna were deported to Belize.

Despite a pervasive myth of racial democracy in Belize, discrimination exists. Belize is not a harmonious, multiethnic island in a sea of violence. For example, sociologist Bruce Ergood notes that in Belize it "is not uncommon to hear a light Creole bad-mouth 'blacks,' even though both are considered Creole. This reflects a vestige of English colonial attitude summed up in the saying, 'Best to be white, less good to be mulatto, worst to be black....' "

FREEDOM

Legislation was passed in October 2000 that calls on trade unions and employers to negotiate over unionization of the workplace if that is the desire of the majority of workers.

A shift in population occurred in the 1980s because of the turmoil in neighboring Central American states. For years, well-educated, English-speaking Creoles had been leaving Belize in search of better economic opportunities in other countries; but this was more than made up for by the inflow of perhaps as many as 40,000 Latin American refugees fleeing the fighting in the region. Spanish is now the primary language of a significant percentage of the population, and some Belizeans are concerned about the "Hispanicization" of the country.

HEALTH/WELFARE

In a speech to the Christian Workers Union, Prime Minister Said Musa noted: "Higher wages will not mean much if families cannot obtain quality and affordable health care services. What good are higher wages if there are not enough classrooms in which to place the children? What good are higher wages if we are forced to live in fear of the criminal elements in society? A workers' movement must … concern itself not only with wages but also with the overall quality of life of its members."

Women in Belize suffer discrimination that is deeply rooted in the cultural, social, and economic structures of the society, even though the government promotes their par-

ticipation in the nation's politics and development process. Great emphasis is placed on education and health care. Tropical diseases, once the primary cause of death in Belize, were brought under control by a government program of insect spraying. Better health and nutritional awareness are emphasized in campaigns to encourage breastfeeding and the selection and preparation of meals using local produce.

ACHIEVEMENTS

Recent digging by archaeologists has uncovered several Maya sites that have convinced scholars that the indigenous civilization in the region was more extensive and refined than experts had previously believed.

With the new millenium, Belize has increasingly turned its attention to the impact of globalization. Concern was expressed by the government about job security and the need for education and training in the skills necessary to compete in a global marketplace. National Trade Union Congress president Dorene Quiros noted that "global institutions are not meeting the basic needs of people," and promises by international organizations to do better have produced only modest results. Worrying,

too, is the rising incidence of violent urban crime and growing involvement in the South American drug trade.

Timeline: PAST

1638
Belize is settled by English Logwood cutters

1884
Belize is declared an independent Crown colony

1972
Guatemala threatens to invade

1981
Independence from Great Britain

1990s
Belize becomes an ecotourism destination
Said Musa is elected prime minister

PRESENT

2000s
Guatemala continues territorial claims to Belize

Negotiations with Guatemala over border issues continue

Costa Rica (Republic of Costa Rica)

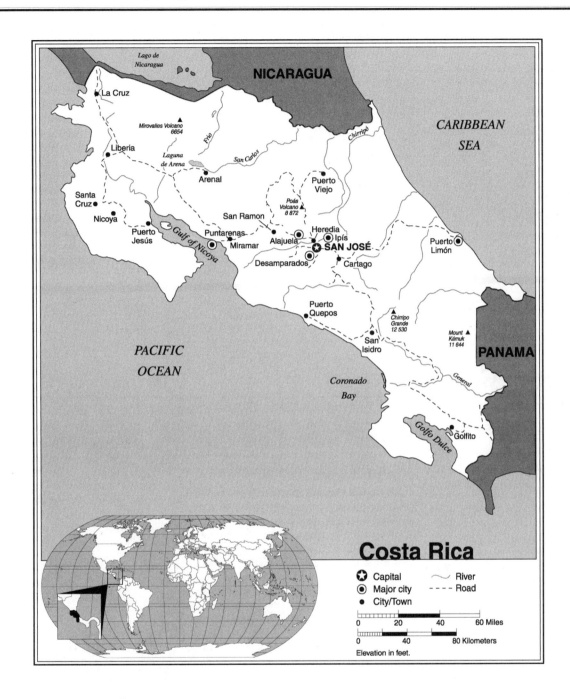

Costa Rica Statistics

GEOGRAPHY

Area in Square Miles (Kilometers):
19,700 (51,022) (about the size of West Virginia)

Capital (Population): San José (325,000)

Environmental Concerns: deforestation; soil erosion

Geographical Features: coastal plains separated by rugged mountains

PEOPLE

Population

Total: 3,956,507
Annual Growth Rate: 1.52%

Rural/Urban Population Ratio: 50/50

Major Language: Spanish

Ethnic Makeup: 96% white (including a few Mestizos); 2% black; 1% Indian; 1% Chinese

Religions: 76.3% Roman Catholic; 13.7% Evangelical; 5% others

Health

Life Expectancy at Birth: 74 years (male);
79 years (female)
Infant Mortality Rate (Ratio): 10.26/1,000
Physicians Available (Ratio): 1/763

Education

Adult Literacy Rate: 95%
Compulsory (Ages): 6–15; free

COMMUNICATION

Telephones: 1.132 million main lines
Daily Newspaper Circulation: 102 per
1,000 people
Televisions: 102 per 1,000 people
Internet Users: 800,000 (2000)

TRANSPORTATION

Highways in Miles (Kilometers): 21,363
(37,273)
Railroads in Miles (Kilometers): 593
(950)
Usable Airfields: 155
Motor Vehicles in Use: 119,000

GOVERNMENT

Type: democratic republic
Independence Date: September 15, 1821
(from Spain)
Head of State/Government: President
Abel Pacheco is both head of state and
head of government
Political Parties: Social Christian Unity
Party; National Liberation Party;
National Integration Party; People
United Party; Democratic Party;
National Independent Party; others
Suffrage: universal and compulsory at 18

MILITARY

*Military Expenditures (% of Central Gov-
ernment Expenditures):* 0.4%
Current Hostilities: none

ECONOMY

Currency ($ U.S. Equivalent): 365 colons
= $1

Per Capita Income/GDP: $7,100/$26
billion
GDP Growth Rate: 7%
Inflation Rate: 10.8%
Unemployment Rate: 5.7%; 7.5%
underemployment
Labor Force: 1,377,000
Natural Resources: hydropower
Agriculture: coffee; bananas; sugar; corn;
rice; beans; potatoes; beef; timber
Industry: microprocessors; food
processing; textiles and clothing;
construction materials; fertilizer; plastic
products; tourism
Exports: $6.4 billion (primary partners
United States, European Union, Central
America)
Imports: $7.5 billion (primary partners
United States, Japan, Mexico)

SUGGESTED WEBSITE

http://www.cia.gov/cia/
publications/factbook/geos/
cs.html

Costa Rica Country Report

COSTA RICA: A DIFFERENT TRADITION?

Costa Rica has often been singled out as politically and socially unique in Latin America. It is true that the nation's historical
development has not been as directly influenced by Spain as its neighbors' have, but
this must not obscure the essential Hispanic
character of the Costa Rican people and
their institutions. Historian Ralph Woodward observes that historically, Costa
Rica's "uniqueness was the product of her
relative remoteness from the remainder of
Central America, her slight economic importance to Spain, and her lack of a non-
white subservient class and corresponding
lack of a class of large landholders to exploit
its labors." Indeed, in 1900, Costa Rica had
a higher percentage of farmers with small-
and medium-range operations than any
other Latin American country.

The nature of Costa Rica's economy allowed a wider participation in politics and
fostered the development of political institutions dedicated to the equality of all people, which existed only in theory in other
Latin American countries. Costa Rican
politicians, since the late nineteenth century, have endorsed programs that have
been largely middle class in content. The
government has consistently demonstrated

a commitment to the social welfare of its
citizens.

AN INTEGRATED SOCIETY

Despite the recent atmosphere of crisis and
disintegration in Central America, Costa
Rica's durable democracy has avoided the
twin evils of oppressive authoritarianism
and class warfare. But what might be con-
strued as good luck is actually a reflection
of Costa Rica's history. In social, racial,
linguistic, and educational terms, Costa
Rica is an integrated country without the
fractures and cleavages that typify the rest
of the region.

Despite its apparent uniqueness, Costa
Rica is culturally an integral part of Latin
America and embodies what is most positive about Hispanic political culture. The
government has long played the role of benevolent patron to the majority of its citizens. Opposition and antagonism have
historically been defused by a process of
accommodation, mutual cooperation, and
participation. In the early 1940s, for example, modernizers who wanted to create a
dynamic capitalist economy took care to
pacify the emerging labor movement with
appropriate social legislation and benefits.
Moreover, to assure that development did
not sacrifice social welfare, the state assumed a traditional role with respect to the

economy—that is, it took an active role in
the production and distribution of income.
After much discussion, in 1993, the Costa
Rican Congress authorized the privatization of the state-owned cement and fertilizer companies. In both cases, according to
Latin American Regional Reports, "a 30%
stake [would] be reserved for employees,
20% [would] be offered to private investors, and the remainder [would] be shared
out between trade unions … and cooperatives." Tight controls were retained on
banking, insurance, oil refining, and public
utilities.

DEVELOPMENT

Economic goals of Costa Rica's
government have included
streamlining or eliminating
restrictive government
regulations, in order to encourage more
foreign investment and the introduction of
high-tech training centers to rural areas to
integrate the most underdeveloped areas of
the country into the process of national
development.

Women, who were granted the right to
vote in the 1940s, have participated freely
in Costa Rica's elections. Women have
served as a vice president, minister of foreign commerce, and president of the Legis-

lative Assembly. Although in broader terms the role of women is primarily domestic, they are legally unrestricted. Equal work, in general, is rewarded by equal pay for men and women. But women also hold, as a rule, lower-paying jobs.

POLITICS OF CONSENSUS

Costa Rica's political stability is assured by the politics of consensus. Deals and compacts are the order of the day among various competing elites. Political competition is open, and participation by labor and peasants is expanding. Election campaigns provide a forum to air differing viewpoints, to educate the voting public, and to keep politicians in touch with the population at large.

Costa Rica frequently has had strong, charismatic leaders who have been committed to social democracy and have rejected a brand of politics grounded in class differences. The country's democracy has always reflected the paternalism and personalities of its presidents.

FREEDOM

Despite Costa Rica's generally enviable human-rights record, there is some de facto discrimination against blacks, Indians, and women (domestic violence against women is a serious problem). The press is free. A stringent libel law, however, makes the media cautious in reporting of personalities.

This tradition was again endorsed when José María Figueres Olsen won the presidential election on February 6, 1994. Figueres was the son of the founder of the modern Costa Rican democracy, and he promised to return to a reduced version of the welfare state. But, by 1996, in the face of a sluggish economy, the populist champion adopted policies that were markedly pro-business. As a result, opinion polls rapidly turned against him. In the 1998 presidential election, an unprecedented 13 political parties ran candidates, which indicated to the three leading parties that citizens no longer believed in them and that political reforms were in order.

Other oft-given reasons for Costa Rica's stability are the high levels of tolerance exhibited by its people and the absence of a military establishment. Costa Rica has had no military establishment since a brief civil war in 1948. Government officials have long boasted that they rule over a country that has more teachers than soldiers. There is also a strong public tradition that favors demilitarization. Costa Rica's auxiliary

forces, however, could form the nucleus of an army in a time of emergency.

The Costa Rican press is among the most unrestricted in Latin America, and differing opinions are openly expressed. Humanrights abuses are virtually nonexistent in the country, but there is a general suspicion of Communists in this overwhelmingly middle-class, white society. And some citizens are concerned about the antidemocratic ideas expressed by ultra-conservatives.

The aftermath of Central America's civil wars is still being felt. Although thousands of refugees returned to Nicaragua with the advent of peace, many thousands more remained in Costa Rica. Economic malaise in Nicaragua combined with the devastation of Hurricane Mitch in 1998 sent thousands of economic migrants across the border into Costa Rica. "Ticos" are worried by the additional strain placed on government resources in a country where more than 80 percent of the population are covered by social-security programs, and approximately 60 percent are provided with pensions and medical benefits.

The economy has been under stress since 1994, and President Figueres was forced to reconsider many of his statist policies. While the export sector remained healthy, domestic industry languished and the internal debt ballooned. The Costa Rican–American Chamber of Commerce observed that "Costa Rica, with its tiny $8.6 billion GDP and 3.5 million people, can not afford a government that consistently overspends its budget by 5 percent or more and then sells short-term bonds, mostly to state institutions, to finance the deficit." In 1997, there was a vigorous debate over the possible privatization of many state entities in an effort to reduce the debt quickly. But opponents of privatization noted that state institutions were important contributors to the high standard of living in the country.

HEALTH/WELFARE

Costa Ricans enjoy the highest standard of living in Central America. But Costa Rica's indigenous peoples, in part because of their remote location, have inadequate schools, health care, and access to potable water. Fully 20% of the population live in poverty.

Acknowledging that the world had entered a new phase of development, President Miguel Angel Rodríguez introduced a new economic program in January 2001. Called *Impulso* ("Impulse"), the plan, as reported in *The Tico Times*, noted that for

Modern growing and distribution techniques are making more and better coffee available, both for export and for the domestic market. This worker is shown picking coffee on a plantation in Costa Rica.

Costa Rica to compete in the new global economy, "knowledge, technology, quality of human resources and the development of telecommunication and transportation infrastructures are fundamental determinants of national prosperity." The old model of economic development, which, according to the president, was characterized by "a diversification of exports, liberalized markets and high levels of foreign investment," must be replaced with a fresh approach "rooted in advanced technological development, a highly qualified labor force, and exports of greater value." President Abel Pacheco, elected in April 2002, has essentially embraced a similar approach to economic development.

THE ENVIRONMENT

At a time when tropical rain forests globally are under assault by developers, cattle barons, and land-hungry peasants, Costa Rica has taken concrete action to protect its environment. Minister of Natural Resources Álvaro Umana was one of those responsible for engineering an imaginative debt-for-nature swap. In his words: "We would like to see debt relief support conservation … a policy that everybody agrees is good." Since 1986, the Costa Rican government has authorized the conversion of $75 million in commercial debt into bonds. Interest generated by those bonds has supported a variety of projects, such as the enlargement and protection of La Amistad, a 1.7 million-acre reserve of tropical rain forest.

ACHIEVEMENTS

 In a region torn by civil war and political chaos, Costa Rica's years of free and democratic elections stand as a remarkable achievement in political stability and civil rights. President Óscar Arias was awarded the Nobel Peace Prize in 1987; he remains a respected world leader.

About 13 percent of Costa Rica's land is protected currently in a number of national parks. It is hoped that very soon about 25 percent of the country will be designated as national parkland in order to protect tropical rain forests as well as the even more endangered tropical dry forests.

Much of the assault on the forests typically has been dictated by economic necessity and/or greed. In one all-too-common scenario, a small- or middle-size cacao grower discovers that his crop has been decimated by a blight. Confronted by disaster, he will usually farm the forest surrounding his property for timber and then torch the remainder. Ultimately, he will likely sell his land to a cattle rancher, who will transform what had once been rain forest or dry forest into pasture.

In an effort to break this devastating pattern, at least one Costa Rican environmental organization has devised a workable plan to save the forests. Farmers are introduced to a variety of cash crops so that they will not be totally dependent on a single crop. Also, in the case of cacao, for example, the farmer will be provided with a disease- or blight-resistant strain to lessen further the chances of crop losses and subsequent conversion of land to cattle pasture.

Scientists in Costa Rica are concerned that tropical forests are being destroyed before their usefulness to humankind can be fully appreciated. Such forests contain a treasure-trove of medicinal herbs. In Costa Rica, for example, there is at least one plant common to the rain forests that might be beneficial in the struggle against AIDS.

Timeline: PAST

1522
Spain establishes its first settlements in Costa Rica

1821
Independence from Spain

1823
Costa Rica is part of the United Provinces of Central America

1838
Costa Rica Becomes independent as a separate state

1948
Civil war; reforms; abolition of the army

1980s
Costa Rica takes steps to protect its tropical rain forests and dry forests

1990s
Ecotourism to Costa Rica increases; Miguel Angel Rodríguez is elected president

PRESENT

2000s
Abel Pacheco is elected president in April 2002

El Salvador (Republic of El Salvador)

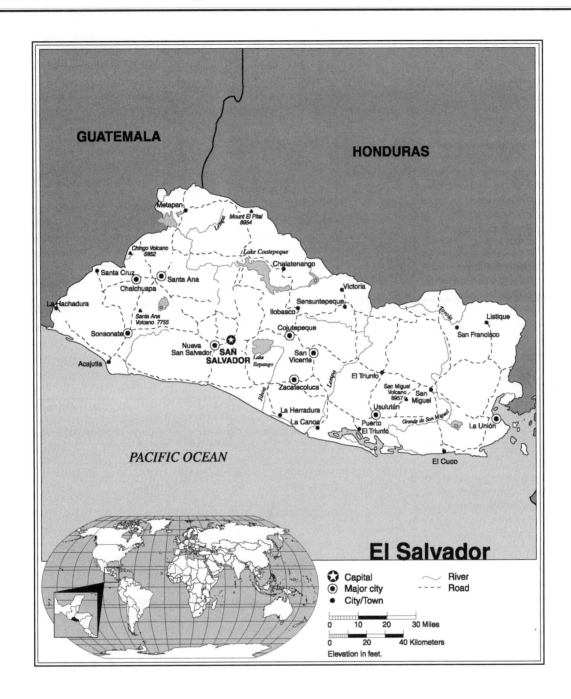

El Salvador Satistics

GEOGRAPHY

Area in Square Miles (Kilometers): 8,292 (21,476) (about the size of Massachusetts)

Capital (Population): San Salvador (1,214,000)

Environmental Concerns: deforestation; soil erosion; water pollution; soil contamination

Geographical Features: a hot coastal plain in south rises to a cooler plateau and valley region; mountainous in north, including many volcanoes

Climate: tropical; distinct wet and dry seasons

PEOPLE

Population

Total: 6,587,541

Annual Growth Rate: 1.78%
Rural/Urban Population Ratio: 55/45
Ethnic Makeup: 94% Mestizo;
   Amerindian 1% white; 9%
Major Language: Spanish
Religions: 75% Roman Catholic; 25%
   Protestant groups

Health

Life Expectancy at Birth: 67 years (male);
   74 years (female)
Infant Mortality Rate (Ratio): 25.93/1,000
Physicians Available (Ratio): 1/1,219

Education

Adult Literacy Rate: 80.2%
Compulsory (Ages): 7–16; free

COMMUNICATION

Telephones: 483,000 main lines
Daily Newspaper Circulation: 53 per
   1,000 people
Televisions: 91 per 1,000
Internet Users: 4 (2000)

TRANSPORTATION

Highways in Miles (Kilometers): 6,196
   (9,977)
Railroads in Miles (Kilometers): 374 (602)
Usable Airfields: 85
Motor Vehicles in Use: 80,000

GOVERNMENT

Type: republic
Independence Date: September 15, 1821
   (from Spain)
Head of State/Government: President
   Francisco Flores is both head of state and
   head of government
Political Parties: Farabundo Martí
   National Liberation Front; National
   Republican Alliance; National
   Conciliation Party; Christian Democratic
   Party; Democratic Convergence; others
Suffrage: universal at 18

MILITARY

Military Expenditures (% of GDP): 1.1%
Current Disputes: border disputes

ECONOMY

Currency ($ U.S. Equivalent): U.S. Dollar
Per Capita Income/GDP: $4,800/$30.99
   billion
GDP Growth Rate: 1.4%
Inflation Rate: 2.1%
Unemployment Rate: 6.5%
Labor Force: 2,350,000
Natural Resources: hydropower;
   geothermal power; petroleum; arable land
Agriculture: coffee; sugarcane; corn; rice;
   beans; oilseed; cotton; sorghum; beef;
   dairy products; shrimp
Industry: food processing; beverages;
   petroleum; chemicals; fertilizer; textiles;
   furniture; light metals
Exports: $3.2 billion (primary partners
   United States, Guatemala, Germany)
Imports: $5.5 billion (primary partners
   United States, Guatemala, Mexico)

SUGGESTED WEBSITE

http://cia.gov/cia/publications/
   factbook/index.html

El Salvador Country Report

EL SALVADOR: A TROUBLED LAND

El Salvador, a small country, was engaged until 1992 in a civil war that cut through class lines, divided the military and the Roman Catholic Church, and severely damaged the social and economic fabric of the nation. It was the latest in a long series of violent sociopolitical eruptions that have plagued the country since its independence in 1821.

In the last quarter of the nineteenth century, large plantation owners—spurred by the sharp increase in the world demand for coffee and other products of tropical agriculture—expanded their lands and estates. Most of the new land was purchased or taken from Indians and Mestizos (those of mixed white and Indian blood), who, on five occasions between 1872 and 1898, took up arms in futile attempts to preserve their land. The once-independent Indians and Mestizos were reduced to becoming tenant farmers, sharecroppers, day laborers, or peons on the large estates. Indians, when deprived of their lands, also lost much of their cultural and ethnic distinctiveness. Today, El Salvador is an overwhelmingly Mestizo society.

The uprooted peasantry was controlled in a variety of ways. Some landowners played the role of *patrón* and assured workers the basic necessities of life in return for their labor. Laws against "vagabonds" (those who, when stopped by rural police, did not have a certain amount of money in their pockets) assured plantation owners a workforce and discouraged peasant mobility.

To enforce order further, a series of security organizations—the National Guard, the National Police, and the Treasury Police— were created by the central government. Many of these security personnel actually lived on the plantations and estates and followed the orders of the owner. Although protection of the economic system was their primary function, over time elements of these organizations became private armies.

DEVELOPMENT

Since 1991, the government has been able to attract substantial investment in a new industry of low-wage, duty-free assembly plants patterned after the maquiladora industries along Mexico's border with the United States. Advantageous tax laws and a free-market climate favorable to business are central to the government's development policy.

This phenomenon lay at the heart of much of the "unofficial" violence in El Salvador in recent years. In Salvadoran society, personal loyalties to relatives or local strongmen competed with and often superseded loyalty to government officials. Because of this, the government was unable to control some elements within its security forces.

In an analysis of the Salvadoran Civil War, it is tempting to place the rich, right-wing landowners and their military allies on one side; and the poor, the peasantry, and the guerrillas on the other. Such a division is artificial, however, and fails to reflect the complexities of the conflict. Granted, the military and landowners had enjoyed a mutually beneficial partnership since 1945. But there were liberal and conservative factions within the armed forces, and, since the 1940s, there had been some movement toward needed social and economic reforms. It was a military regime in 1949 that put into effect the country's first social-security legislation. In 1950, a Constitution was established that provided for public-health programs, women's suffrage, and extended social-security coverage. The reformist impulse continued in the 1960s, when it became legal to organize opposition political parties.

A TIME FOR CHANGE

Food production increased in the 1970s by 44 percent, a growth that was second in Latin America only to Brazil's. Although much of the food grown was exported to world markets, some of the revenue generated was used for social programs in El Salvador. Life expectancy increased, the death rate fell, illiteracy declined, and the percentage of government expenditures on public health, housing, and education was among the highest in Latin America.

The programs and reforms, in classic Hispanic form, were generated by the upper classes. The elite believed that state-sponsored changes could be controlled in such a way that traditional balances in society would remain intact and elite domination of the government would be assured.

The origin of El Salvador's Civil War may be traced to 1972, when the Christian Democratic candidate for president, José Napoleón Duarte, is believed to have won the popular vote but was deprived of his victory when the army declared the results false and handed the victory to its own candidate. Impatient and frustrated, middle-class politicians and student leaders from the opposition began to consider more forceful ways to oust the ruling class.

By 1979, guerrilla groups had become well established in rural El Salvador, and some younger army officers grew concerned that a successful left-wing popular revolt was a distinct possibility. Rather than wait for revolution from below, which might result in the destruction of the military as an institution, the officers chose to seize power in a coup and manipulate change from above. Once in power, this *junta*, or ruling body, moved quickly to transform the structure of Salvadoran society. A land-reform program, originally developed by civilian reformers and Roman Catholic clergy, was adopted by the military. It would give the campesinos ("peasants") not only land but also status, dignity, and respect.

FREEDOM

The end of the Civil War brought an overall improvement in human rights in El Salvador. News from across the political spectrum, often critical of the government, is reported in El Salvador, although foreign journalists seem to be the target of an unusually high level of muggings, robberies, and burglaries. Violence against women is widespread. Judges often dismiss rape cases on the pretext that the victim provoked the crime.

In its first year, 1980, the land-reform program had a tremendous impact on the landowning elite—37 percent of the lands producing cotton and 34 percent of the coffee-growing lands were confiscated by the government and redistributed. The junta also nationalized the banks and assumed control of the sale of coffee and sugar. Within months, however, several peasant members of the new cooperatives and the government agricultural advisers sent to help them were gunned down. The violence spread. Some of the killings were attributed to government security men in the pay of dispossessed landowners, but most of the killings may have been committed by the army.

In the opinion of a land-reform program official, the army was corrupt and had returned to the cooperatives that it had helped to establish in order to demand money for protection and bribes. When the peasants refused, elements within the army initiated a reign of terror against them.

In 1989, further deterioration of the land-reform program was brought about by Supreme Court decisions and by policies adopted by the newly elected rightwing government of President Alfredo Cristiani. Former landowners who had had property taken for redistribution to peasants successfully argued that seizures under the land reform were illegal. Subsequently, five successive land-reform cases were decided by the Supreme Court in favor of former property owners.

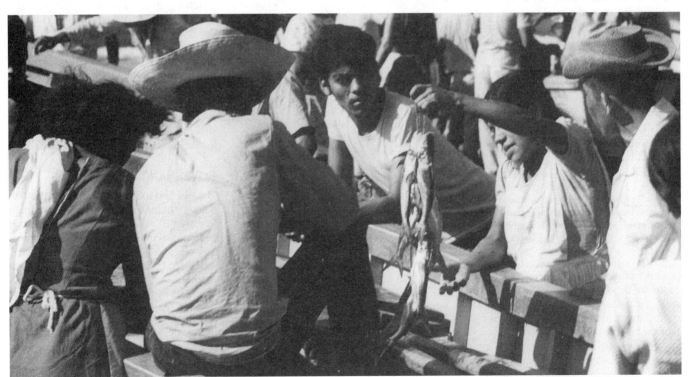

(Y. Nagata/PAS United Nations photo)

Civil strife disrupted much of El Salvador's agrarian production, and a lack of fishery planning necessitated importing from other parts of the world. With a new and efficient program to take advantage of fish in domestic waters, El Salvador has been able to develop an effective food industry from the sea.

Cristiani, whose right-wing National Republican Alliance Party (ARENA) fought hard against land reform, would not directly attack the land-reform program—only because such a move would further alienate rural peasants and drive them into the arms of left-wing guerrillas. Instead, Cristiani favored the reconstitution of collective farms as private plots. Such a move, according to the government, would improve productivity and put an end to what authorities perceived as a form of U.S.–imposed "socialism." Critics of the government's policy charged that the privatization plan would ultimately result in the demise of land reform altogether.

Yet another problem was that many of the collectives established under the reform were (and remain) badly in debt. A 1986 study by the U.S. Agency for International Development reported that 95 percent of the cooperatives could not pay interest on the debt they were forced to acquire to compensate the landlords. *New York Times* reporter Lindsey Gruson noted that the world surplus of agricultural products as well as mismanagement by peasants who suddenly found themselves in the unfamiliar role of owners were a large part of the reason for the failures. But the government did not help. Technical assistance was not provided, and the tremendous debt gave the cooperatives a poor credit rating, which made it difficult for them to secure needed fertilizer and pesticides.

Declining yields and, for many families, lives of increasing desperation have been the result. Some peasants must leave the land and sell their plots to the highest bidder. This will ultimately bring about a re-concentration of land in the hands of former landlords.

HEALTH/WELFARE

Many Salvadorans suffer from parasites and malnutrition. El Salvador has one of the highest infant mortality rates in the Western Hemisphere, largely because of polluted water. Potable water is readily available to only 10% of the population.

Other prime farmland lay untended because of the Civil War. Violence drove many peasants from the land to the slums of the larger cities. And free-fire zones established by the military (in an effort to destroy the guerrillas' popular base) and guerrilla attacks against cooperatives (in an effort to sabotage the economy and further destabilize the country) had a common victim: the peasantry.

Some cooperatives and individual families failed to bring the land to flower be-cause of the poor quality of the soil they inherited. Reporter Gruson told the story of one family, which was, unfortunately, all too common:

> José ... received 1.7 acres on a rock-pocked slope an hour's walk from his small shack. José ... used to sell some of his beans and rice to raise a little cash. But year after year his yields have declined. Since he cannot afford fertilizers or insecticides, the corn that survives the torrential rainy season produces pest-infested ears the size of a baby's foot. Now, he has trouble feeding his wife and seven children.
>
> "The land is no good," he said. "I've been working it for 12 years and my life has gotten worse every year. I don't have anywhere to go, but I'll have to leave soon."

After the coup, several governments came and went. The original reformers retired, went into exile, or went over to the guerrillas. The Civil War continued into 1992, when a United Nations–mediated cease-fire took effect. The extreme right and left regularly utilized assassination to eliminate or terrorize both each other and the voices of moderation who still dared to speak out. The death squads and guerrillas claimed their victims from all social classes. Some leaders, such as former president Duarte, described a culture of violence in El Salvador that had become part of the national character.

HUMAN-RIGHTS ISSUES

Through 1992, human-rights abuses still occurred on a wide scale in El Salvador. Public order was constantly disrupted by military operations, guerrilla raids, factional hatreds, acts of revenge, personal grudges, pervasive fear, and a sense of uncertainty about the future. State-of-siege decrees suspended all constitutional rights to freedom of speech and press. However, self-censorship, both in the media and by individuals, out of fear of violent reprisals, was the leading constraint on free expression in El Salvador.

Release of the report in 1993 by the UN's "Truth Commission," a special body entrusted with the investigation of human-rights violations in El Salvador, prompted the right wing–dominated Congress to approve an amnesty for those named. But progress has been made in other areas. The National Police have been separated from the Defense Ministry; and the National Guard, Civil Defense forces, and the notorious Treasury Police have been abolished. A new National Civilian Police, comprised of 20 percent of National Police, 20 percent former Farabundo Martí National Liberation Front (FMLN) guerrillas, and 60 percent with no involvement on either side in the Civil War, was instituted in 1994.

In El Salvador, as elsewhere in Latin America, the Roman Catholic Church was divided. The majority of Church officials backed government policy and supported the United States' contention that the violence in El Salvador was due to Cuban-backed subversion. Other clergy strongly disagreed and argued convincingly that the violence was deeply rooted in historical social injustice.

ACHIEVEMENTS

Despite the violence of war, political power has been transferred via elections at both the municipal and national levels. Elections have helped to establish the legitimacy of civilian leaders in a region usually dominated by military regimes.

Another endemic problem that confronts postwar El Salvador is widespread corruption. It is a human-rights issue because corruption and its attendant misuse of scarce resources contribute to persistent or increased poverty and undermine the credibility and stability of government at all levels. According to the nonprofit watchdog group *Probidad*, "El Salvador has a long history of corruption.... Before the first of many devastating earthquakes on January 13, 2001, El Salvador was the third poorest country in Latin America.... Influence peddling between construction companies and their friends and families in government and other corrupt practices resulted in many unnecessary deaths, infrastructure damage, and irregularities in humanitarian assistance distribution."

GOVERNANCE

The election to the presidency of José Napoleón Duarte in 1984 was an important first step in establishing the legitimacy of government in El Salvador, as were municipal elections in 1985. The United States supported Duarte as a representative of the "democratic" middle ground between the guerrillas of the FMLN and the right-wing ARENA party. Ironically, U.S. policy in fact undermined Duarte's claims to legitimacy and created a widespread impression that he was little more than a tool for U.S. interests.

Yet while the transfer of power to President Cristiani via the electoral process in 1989 reflected the will of those who voted, it did not augur well for the lessening of human-rights abuses. With respect to the guerrillas of the FMLN, Cristiani made it clear that the government would set the terms for any talks about ending the Civil War. For its part, the FMLN warned that it would make the country "ungovernable." In effect, then, the 1989 election results polarized the country's political life even more.

After several unsuccessful efforts to bring the government and the guerrillas to the negotiating table, the two sides reached a tentative agreement on constitutional reforms in April 1991 at a UN–sponsored meeting in Mexico City. The military, judicial system, and electoral process were all singled out for sweeping changes. By October, the FMLN had promised to lay down its arms; and near midnight on December 31, the final points of a peace accord were agreed upon. Final refinements of the agreement were drawn up in New York, and a formal signing ceremony was staged in Mexico City on January 16, 1992. The official cease-fire took effect February 1, ending the 12-year Civil War that had claimed more than 70,000 lives and given El Salvador the reputation of a bloody and abusive country.

Implementation of the agreement reached between the government and the FMLN has proven contentious. "But," according to *Boston Globe* correspondent Pamela Constable, "a combination of war-weariness and growing pragmatism among leaders of all persuasions suggests that once-bitter adversaries have begun to develop a modus vivendi."

President Cristiani reduced the strength of the army from 63,000 to 31,500 by February 1993, earlier than provided for by the agreement; and the class of officers known as the *tondona*, who had long dominated the military and were likely responsible for human-rights abuses, were forcibly retired by the president on June 30, 1993. Land,

judicial, and electoral reforms followed. Despite perhaps inevitable setbacks because of the legacy of violence and bitterness, editor Juan Comas wrote that "most analysts are inclined to believe that El Salvador's hour of madness has passed and the country is now on the road to hope."

In 1998, President Armando Calderón Sol surprised both supporters and opponents when he launched a bold program of reforms. The first three years of his administration had been characterized by indecision. Political scientist Tommie Sue Montgomery noted that his "reputation for espousing as policy the last viewpoint he has heard has produced in civil society both heartburn and black humor." But a combination of factors created new opportunities for Calderón. The former guerrillas of the FMLN were divided and failed to take advantage of ARENA's apparent weak leadership; a UN–sponsored program of reconstruction and reconciliation was short of funds and, by 1995, had lost momentum; and presidential elections were looming in 1999. A dozen years of war had left the economic infrastructure in disarray. The economy had, at best, remained static, and while the war raged, there had been no attempt to modernize. During his final year in office, Calderón developed reform policies of modernization, privatization, and free-market competition. Interestingly, his reforms generated opposition from former guerrillas, who are now represented in the Legislature by the FMLN, as well as from some members of the traditional conservative economic elite.

Perhaps one result of Calderón's reforms was the decisive victory of ARENA at the polls in 1999, and again in 2004. The FMLN, on the other hand, won municipal and legislative elections in 2003, which gave them the largest voting bloc in congress.

Serious problems remain. Half of the population live in poverty and the devastating earthquake of 2001 left a million homeless. A BBC report notes: "Poverty, civil war, natural disaster, and consequent dislocations have left their mark on … society, which is among the most violent and crime ridden in the Americas."

Timeline: PAST

1524
Present-day El Salvador is occupied by Spanish settlers from Mexico

1821
Independence from Spain is declared

1822
El Salvador is part of the United Provinces of Central America

1838
El Salvador becomes independent as a separate state

1969
A brief war between El Salvador and Honduras

1970
Guerrilla warfare in El Salvador

1979
Army officers seize power in a coup; Civil War

1990s
A cease-fire takes effect on February 1, 1992, officially ending the Civil War

PRESENT

2000s
Earthquakes devastate towns and cities, with a heavy loss of life and extensive infrastructure damage

Anthony Saca wins 2004 presidential election

Guatemala (Republic of Guatemala)

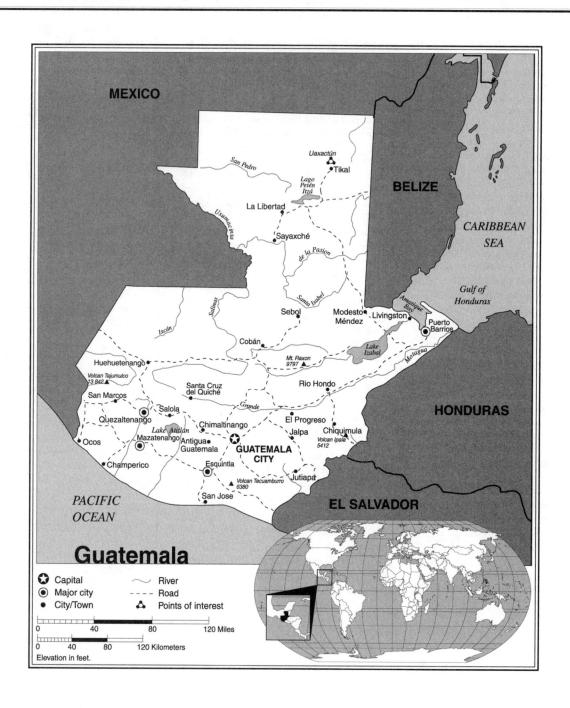

Guatemala

- ✪ Capital
- ◉ Major city
- ● City/Town
- ◬ Points of interest
- 〰 River
- --- Road

0   40   80   120 Miles

0   40   80   120 Kilometers

Elevation in feet.

Guatemala Statistics

GEOGRAPHY

Area in Square Miles (Kilometers): 42,000 (108,780) (about the size of Tennessee)

Capital (Population): Guatemala City (2,205,000)

Environmental Concerns: deforestation; soil erosion; water pollution

Geographical Features: mostly mountains, with narrow coastal plains and a rolling limestone plateau (Peten)

Climate: temperate in highlands; tropical on coasts

PEOPLE

Population

Total: 14,280,596

Annual Growth Rate: 2.63%

Rural/Urban Population Ratio: 61/39

Ethnic Makeup: 56% Ladino (Mestizo

and Westernized Indian); 44%
Amerindian

Major Languages: Spanish; Maya
languages

Religions: predominantly Roman
Catholic; Protestant and Maya

Health

Life Expectancy at Birth: 64 years (male);
66 years (female); 44 years (Indian
population)

Infant Mortality Rate (Ratio): 36.91/1,000

Physicians Available (Ratio): 1/2,356

Education

Adult Literacy Rate: 70.6%

Compulsory (Ages): 7–14; free

COMMUNICATION

Telephones: 846,000 main lines

Daily Newspaper Circulation: 29 per
1,000 people

Televisions: 45 per 1,000

Internet Users: 400,000

TRANSPORTATION

Highways in Miles (Kilometers): 8,135
(13,100)

Railroads in Miles (Kilometers): 552 (884)

Usable Airfields: 477

Motor Vehicles in Use: 199,000

GOVERNMENT

Type: Constitutional democratic republic

Independence Date: September 15, 1821
(from Spain)

Head of State/Government: President
Oscar Jose Rafael Berger (January 2004)
is both head of state and head of
government

Political Parties: National Centrist Union;
Christian Democratic Party; National
Advancement Party; National Liberation
Movement; Social Democratic Party;
Revolutionary Party; Guatemalan
Republican Front; Democratic Union;
New Guatemalan Democratic Front

Suffrage: universal at 18

MILITARY

Military Expenditures (% of GDP): 0.8%

Current Disputes: border dispute with
Belize

ECONOMY

Currency ($ U.S. Equivalent): 7.94
quetzals = $1

Per Capita Income/GDP: $4,100/$56.53
billion

GDP Growth Rate: 2.2%

Inflation Rate: 5.6%

Unemployment Rate: 7.5%

Labor Force: 4.2 million

Natural Resources: petroleum; nickel; rare
woods; fish; chicle; hydropower

Agriculture: sugarcane; corn; bananas;
coffee; beans; cardamom; livestock

Industry: sugar; textiles and clothing;
furniture; chemicals; petroleum; metals;
rubber; tourism

Exports: $2.7 billion (primary partners
United States, El Salvador, Honduras)

Imports: $5.7 billion (primary partners
United States, Mexico, South Korea, El
Salvador)

SUGGESTED WEBSITE

http://www.cia.gov/cia/
publications/factbook/index.html

Guatemala Country Report

GUATEMALA: PEOPLES IN CONFLICT

Ethnic relations between the descendants of Maya Indians, who comprise 44 percent of Guatemala's population, and whites and Ladinos (Hispanicized Indians) have always been unfriendly and have contributed significantly to the nation's turbulent history. During the colonial period and since independence, Spaniards, Creoles (in Guatemala, whites born in the New World—as opposed to in Nicaragua, where Creoles are defined as native-born blacks), and Ladinos have repeatedly sought to dominate the Guatemalan Indian population, largely contained in the highlands, by controlling the Indians' land and their labor.

The process of domination was accelerated between 1870 and 1920, as Guatemala's entry into world markets hungry for tropical produce such as coffee resulted in the purchase or extensive seizures of land from Indians. Denied sufficient lands of their own, Indians were forced onto the expanding plantations as debt peons. Others were forced to labor as seasonal workers on coastal plantations; many died there because of the sharp climatic differences.

THE INDIAN AND INTEGRATION

Assaulted by the Ladino world, highland Indians withdrew into their own culture and built social barriers between themselves and the changing world outside their villages. Those barriers have persisted until the present.

DEVELOPMENT

Unchecked spending by the government in 2000–2001 led to rapid swelling of the national debt as revenues failed to keep pace. Without economic reform, the IMF will refuse further loans to Guatemala, and the country's peace process will be further endangered.

For the Guatemalan governments that have thought in terms of economic progress and national unity, the Indians have always presented a problem. A 2003 presidential candidate stated: "Indigenous groups do not speak of a 'political system'; they speak of community consensus, and their conception of community is very local.... How do you have a functioning nation state, one where indigenous groups participate actively in protecting their political interests, and yet still respect the cultural practices of other indigenous groups for whom participation in Western political institutions is deemed undesireable?"

According to anthropologist Leslie Dow, Jr., Guatemalan governments too easily explain the Indian's lack of material prosperity in terms of the "deficiencies" of Indian culture. Indian "backwardness" is better explained by elite policies calculated to keep Indians subordinate. Social, political, and economic deprivations have consistently and consciously been utilized by governments anxious to maintain the Indian in an inferior status.

Between 1945 and 1954, however, there was a period of remarkable social reform in Guatemala. Before the reforms were cut short by the resistance of landowners, factions within the military, and a U.S. Central Intelligence Agency–sponsored invasion, Guatemalan governments made a concerted effort to integrate the Indian into national life. Some Indians who lived in close proximity to large urban centers such as the capital, Guatemala

(United Nations photo/152/271/Antoinette Jongen)

This elderly Indian woman of San Mateo looks back on a life experience of economic and social prejudice. In recent years, Indians in Guatemala have pursued their rights by exercising their voting power. On occasion, they have resorted to violence, which has been repressed swiftly and mercilessly by the government. But the power of the ballot box has finally begun to reap gains.

City, learned that their vote had the power to effect changes to their benefit. They also realized that they were unequal not because of their illiteracy, "backwardness," poverty, or inability to converse in Spanish, but because of governments that refused to reform their political, social, and economic structures.

In theory, indigenous peoples in Guatemala enjoy equal legal rights under the Constitution. In fact, however, they remain largely outside the national culture, do not speak Spanish, and are not integrated into the national economy. Indian males are far more likely to be impressed into the army or guerrilla units. Indigenous peoples in Guatemala have suffered most of the combat-related casualties and repeated abuses of their basic human rights. There remains a pervasive discrimination against Indians in white society. Indians have on occasion challenged state policies that they have considered inequitable and repressive. But if they become too insistent on change, threaten violence or societal upheaval, or support and/or join guerrilla groups, government repression is usually swift and merciless.

GUERRILLA WARFARE

A civil war, which was to last for 36 years, developed in 1960. Guatemala was plagued by violence, attributed both to left-wing insurgencies in rural areas and to armed forces' counterinsurgency operations. Led by youthful middle-class rebels, guerrillas gained strength because of several factors: the radical beliefs of some Roman Catholic priests in rural areas; the ability of the guerrillas to mobilize Indians for the first time; and the "demonstration effect" of events elsewhere in Central America. Some of the success is explained by the guerrilla leaders' ability to converse in Indian languages. Radical clergy increased the recruitment of Indians into the guerrilla forces by suggesting that revolution was an acceptable path to social justice. The excesses of the armed forces in their search for subversives drove other Indians into the arms of the guerrillas. In some parts of the highlands, the loss of ancestral lands to speculators or army officers was sufficient to inspire the Indians to join the radical cause.

According to the *Latin American Regional Report* for Mexico and Central America, government massacres of guerrillas and their actual or suspected supporters were frequent and "characterized by clinical savagery." At times, the killing was selective, with community leaders and their families singled out. In other instances, entire villages were destroyed and all the inhabitants slaughtered. "Everything depends on the army's perception of the local level of support for the guerrillas," according to the report.

FREEDOM

Former president Ramiro de León Carpio warned those who would violate human rights, saying that the law would punish those guilty of abuses, "whether or not they are civilians or members of the armed forces." The moment has come, he continued, "to change things and improve the image of the army and of Guatemala."

To counterbalance the violence, once guerrillas were cleared from an area, the government implemented an "Aid Program to Areas in Conflict." Credit was offered to small farmers to boost food production in order to meet local demand, and displaced and jobless people were enrolled in food-for-work units to build roads or other public projects.

By the mid-1980s, most of the guerrillas' military organizations had been destroyed. This was the result not only of successful counterinsurgency tactics by the Guatemalan military but also of serious errors of judgment by guerrilla leaders. Impatient and anxious for change, the guerrillas had overestimated the willingness of the Guatemalan people to rebel. They also had underestimated the power of the military establishment. Surviving guerrilla units maintained an essentially defensive posture for the remainder of the decade. In 1989, however, the guerrillas regrouped. The subsequent in-

tensification of human-rights abuses and the climate of violence were indicative of the military's response.

There was some hope for improvement in 1993, in the wake of the ouster of President Jorge Serrano, whose attempt to emulate the "self-coup" of Peru's Alberto Fujimori failed. Guatemala's next president, Ramiro de León Carpio, was a human-rights activist who was sharply critical of security forces in their war against the guerrillas of the Guatemalan National Revolutionary Unity (URNG). Peace talks between the government and guerrillas had been pursued with the Roman Catholic Church as intermediary for several years, with sparks of promise but no real change. In July 1993, de León announced a new set of proposals to bring to an end the decades of bloodshed that had resulted in 140,000 deaths. Those proposals were the basis for the realization of a peace agreement worked out under the auspices of the United Nations in December 1996.

But the underlying causes of the violence still must be addressed. Colin Woodard, writing in *The Chronicle of Higher Education*, reported that the peace accords promised to "reshape Guatemala as a democratic, multicultural society." But an estimated 70 percent of the Maya Indians still live in poverty, and more than 80 percent are illiterate. Estuardo Zapeta, Guatemala's first Maya newspaper columnist, writes: "This is a multicultural, multilingual society.... As long as we leave the Maya illiterate, we're condemning them to being peasants. And if that happens, their need to acquire farmland will lead us to another civil war." This, however, is only one facet of a multifaceted set of issues. The very complexity of Guatemalan society, according to political scientist Rachel McCleary, "make[s] it extremely difficult to attain a consensus at the national level on the nature of the problems confronting society." But the new ability of leaders from many sectors of society to work together to shape a meaningful peace is a hopeful sign.

Although the fighting has ended, fear persists. Journalist Woodard wrote in July 1997: "In many neighborhoods [in Guatemala City] private property is protected by razor wire and patrolled by guards with pump-action shotguns." One professor at the University of San Carlos observed, "It is good that the war is over, but I am pessimistic about the peace.... There is intellectual freedom now, but we are very unsure of the permanence of that freedom. It makes us very cautious."

URBAN VIOLENCE

Although most of the violence occurred in rural areas, urban Guatemala did not es-

cape the horrors of the Civil War. The following characterization of Guatemalan politics, written by an English traveler in 1839, is still relevant today: "There is but one side to the politics in Guatemala. Both parties have a beautiful way of producing unanimity of opinion, by driving out of the country all who do not agree with them."

During the Civil War, right-wing killers murdered dozens of leaders of the moderate political left to prevent them from organizing viable political parties that might challenge the ruling elite. These killers also assassinated labor leaders if their unions were considered leftist or antigovernment. Leaders among university students and professors "disappeared" because the national university had a reputation as a center of leftist subversion. Media people were gunned down if they were critical of the government or the right wing. Left-wing extremists also assassinated political leaders associated with "repressive" policies, civil servants (whose only "crime" was government employment), military personnel and police, foreign diplomats, peasant informers, and businesspeople and industrialists associated with the government.

Common crime rose to epidemic proportions in Guatemala City (as well as in the capitals of other Central American republics). Many of the weapons that once armed the Nicaraguan militias and El Salvador's civil-defense patrols found their way onto the black market, where, according to the Managua newspaper *Pensamiento Propio*, they were purchased by the Guatemalan Army, the guerrillas of the URNG, and criminals.

HEALTH/WELFARE

While constitutional bars on child labor in the industrial sector are not difficult to enforce, in the informal and agricultural sectors such labor is common. It is estimated that 5,000 Guatemalan children live on the streets and survive as best they can. They are often targeted for elimination by police and death squads.

The fear of official or unofficial violence has always inhibited freedom of the press in Guatemala. Early in the 1980s, the Conference on Hemispheric Affairs noted that restrictions on the print media and the indiscriminate brutality of the death squads "turned Guatemala into a virtual no-man's land for journalists." Lingering fears and memories of past violence tend to limit the exercise of press freedoms guaranteed by the Constitution. The U.S. State Department's Country Reports notes that "the media continues to exercise a degree of self-censorship on certain topics.... The lack of aggressive investigative reporting dealing with the military and human rights violations apparently is due to self-censorship."

HEALTH CARE AND NUTRITION

In rural Guatemala, half the people have a diet that is well below the minimum daily caloric intake established by the Food and Agricultural Organization. Growth in the staple food crops (corn, rice, beans, wheat) has failed to keep pace with population growth. Marginal malnutrition is endemic.

Health services vary, depending on location, but are uniformly poor in rural Guatemala. The government has begun pilot programs in three departments to provide basic primary health care on a wide scale. But some of these well-intentioned policies have failed because of a lack of sensitivity to cultural differences. Anthropologist Linda Greenberg has observed that the Ministry of Health, as part of its campaign to bring basic health-care services to the hinterlands, introduced midwives who were ignorant of Indian traditions. For Guatemalan Indians, pregnancy is considered an illness that demands specific care, calling for certain foods, herbs, body positions, and interpersonal relations between expectant mother and Indian midwife. In Maya culture, traditional medicine has spiritual, psychological, physical, social, and symbolic dimensions. Ministry of Health workers too often dismiss traditional practices as superstitious and unscientific. Their insensitivity and ignorance create ineffectual health-care programs.

THE FUTURE

In February 1999, a UN–sponsored Commission for Historical Clarification, in a harsh nine-volume report, blamed the Guatemalan government for acts of genocide against the Maya during the long Civil War. The purpose of the report was not to set the stage for criminal prosecutions but to examine the root causes of the Civil War and explain how the conflict developed over time. It was hoped that the report signaled the first steps toward national reconciliation and the addressing of human-rights issues, long ignored by those in power.

ACHIEVEMENTS

 Guatemalan novelist Miguel Ángel Asturias gained an international reputation for his works about political oppression. In 1967, he was awarded the Nobel Prize for Literature. Rigoberta Menchú Tum won the Nobel Peace Prize in 1992 for her passionate support of the Maya peoples of Guatemala.

But the high command of the military and its civilian allies, accused of planning and executing a broad range of atrocities against the Maya, may perceive the report as a threat to their position and their future. In fact, the government has done little to implement the recommendations called for in the 1996 peace accords that ended the Civil War. Former President Efraín Rios Montt, who engineered the assault against the Maya during the civil war, lost his congressional seat—and his immunity to prosecution—in 2004. Although he now faces charges of genocide for his scorched earth policy, Guatemala's current president, Oscar Berger, a wealthy farmer backed by the nation's traditional power brokers, has remained non-committal on Ríos Montt's fate.

Not surprisingly, the poor and disadvantaged are increasingly frustrated. Illiteracy, infant mortality and malnutrition are among the highest in Central America while life expectancy is among the lowest. Two-thirds of Guatemala's children live in poverty. Violence remains endemic.

Timeline: PAST

1523
Guatemala is conquered by Spanish forces from Mexico

1821
Independence

1822–1838
Guatemala is part of the United Provinces of Central America

1838
Guatemala becomes independent as a separate state

1944
Revolution; many reforms

1954
A CIA–sponsored coup deposes the reformist government

1976
An earthquake leaves 22,000 dead

1977
Human-rights abuses lead to the termination of U.S. aid

1990s
Talks between the government and guerrillas end 36 years of violence

PRESENT

2000s
Economic problems multiply

Oscar Berger elected president in November 2003

Honduras (Republic of Honduras)

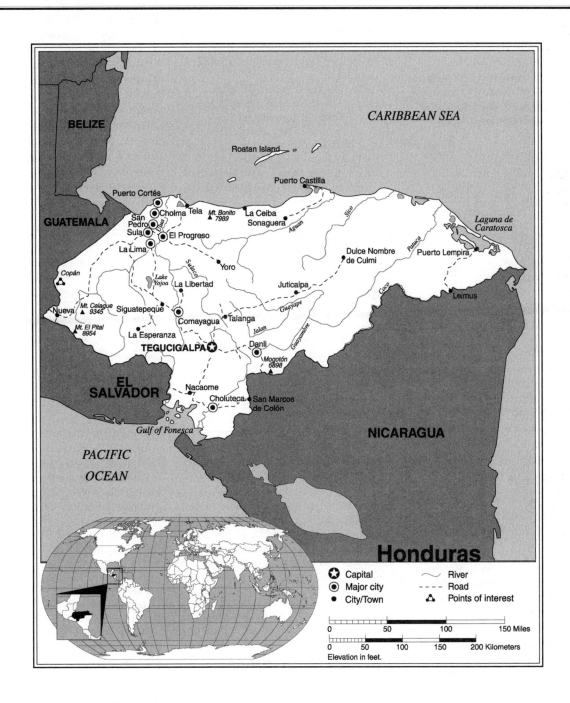

Honduras Statistics

GEOGRAPHY

Area in Square Miles (Kilometers): 43,267 (112,090) (slightly larger than Tennessee)

Capital (Population): Tegucigalpa (995,000)

Environmental Concerns: urbanization; deforestation; land degradation and soil erosion; mining pollution

Geographical Features: mostly mountainous in the interior; narrow coastal plains

Climate: subtropical, but varies with elevation (temperate highlands)

PEOPLE*

Population

Total: 6,823,568

Annual Growth Rate: 2.2%

Rural/Urban Population Ratio: 56/44

Ethnic Makeup: 90% Mestizo (European and Indian mix); 7% Indian; 2% African; 1% European, Arab, and Asian
Major Language: Spanish
Religions: 97% Roman Catholic; a small Protestant minority

Health

Life Expectancy at Birth: 65 years (male); 67 years (female)
Infant Mortality Rate (Ratio): 29.64/1,000
Physicians Available (Ratio): 1/1,586

Education

Adult Literacy Rate: 76.2%
Compulsory (Ages): 7–13; free

COMMUNICATION

Telephones: 322,500 main lines
Daily Newspaper Circulation: 45 per 1,000 people
Televisions: 29 per 1,000
Internet Users: 168,600

TRANSPORTATION

Highways in Miles (Kilometers): 9,563 (15,400)

Railroads in Miles (Kilometers): 369 (595)
Usable Airfields: 119
Motor Vehicles in Use: 185,000

GOVERNMENT

Type: democratic constitutional republic
Independence Date: September 15, 1821 (from Spain)
Head of State/Government: President Ricardo (Joest) Maduro is both head of state and head of government
Political Parties: Liberal Party; National Party of Honduras; National Innovation and Unity Party–Social Democratic Party; Christian Democratic Party; others
Suffrage: universal and compulsory at 18

MILITARY

Military Expenditures (% of GDP): 1.5%
Current Disputes: boundary disputes with El Salvador and Nicaragua

ECONOMY

Currency ($ U.S. Equivalent): 17.35 lempiras = $1 *Per Capita Income/GDP:* $2,600/$17.46 billion

GDP Growth Rate: 2.5%
Inflation Rate: 7.7%
Unemployment Rate: 27.5%
Labor Force: 2,300,000
Natural Resources: timber; gold; silver; copper; lead; zinc; iron ore; antimony; coal; fish; hydropower
Agriculture: bananas; coffee; citrus fruits; beef; timber; shrimp
Industry: sugar; coffee; textiles and clothing; wood products
Exports: $1.4 billion (primary partners United States, Japan, Germany)
Imports: $3.1 billion (primary partners United States, Guatemala, Netherland Antilles)

SUGGESTED WEBSITE

http://www.cia.gov/cia/
publications/factbook/geos/
ho.html

---

*Note: Estimates for Honduras explicitly take into account the effects of excess mortality due to AIDS.

Honduras Country Report

HONDURAS:
THE CONTAGION OF VIOLENCE

In political terms, Honduras resembles much of the rest of Central America. Frequent changes of government, numerous constitutions, authoritarian leaders, widespread corruption, and an inability to solve basic problems are common to Honduras and to the region. A historian of Honduras once wrote that his country's history could be "written in a tear."

DEVELOPMENT

The Central American Free Trade Zone, of which Honduras is a member, will reduce tariffs by 5% to 20% on more than 5,000 products traded within the region. In the coming years, more products will be included and tariffs will be progressively lowered.

In terms of social policy, however, Honduras stands somewhat apart from its neighbors. It was slower to modernize, there were no great extremes of wealth between landowners and the rest of the population, and society appeared more paternalistic and less exploitive than was the case in other Central American states.

"Ironically," notes journalist Loren Jenkins, "the land's precarious existence as a poor and unstable backwater has proven almost as much a blessing as a curse." Honduras lacks the sharp social divisions that helped to plunge Nicaragua, El Salvador, and Guatemala into rebellion and civil war. And Honduran governments have seemed somewhat more responsive to demands for change. Still, Honduras is a poor country. Its people have serious problems—widespread illiteracy, malnutrition, and inadequate health care and housing.

A WILLINGNESS TO CHANGE

In 1962 and 1975, agrarian-reform laws were passed and put into effect with relative success. The Honduran government, with the aid of peasant organizations and organized labor, was able to resettle 30,000 families on their own land. Today, two thirds of the people who use the land either own it or have the legal right to its use. Labor legislation and social-security laws were enacted in the early 1960s. Even the Honduran military, usually corrupt, has at times brought about reform. An alliance of the military and organized labor in the early 1970s produced a series of reforms in

response to pressure from the less advantaged sectors of the population; in 1974, the military government developed a five-year plan to integrate the rural poor into the national economy and to increase social services in the area. The state has often shown a paternalistic face rather than a brutal, repressive one. The capacity for reform led one candidate in the 1981 presidential campaign to comment: "We Hondurans are different. There is no room for violence here."

There are now many signs of change. Agrarian reform slowed after 1976, prompting a peasant-association leader to remark: "In order to maintain social peace in the countryside, the peasants' needs will have to be satisfied to avoid revolt." In 1984, the Honduran government initiated a land-titling program and issued about 1,000 titles per month to landless peasants. The government's agrarian-reform program, which is under the control of the National Agrarian Institute, has always been characterized by the carrot and the stick. While some *campesinos* ("peasants") have been granted titles to land, others have been jailed or killed. Former military and security personnel apparently murdered several indigenous minority rights leaders in 2004.

Honduran campesinos, according to *Central America Report*, "have had a long and combative history of struggling for land rights." In 1987, hundreds of peasants were jailed as "terrorists" as a result of land invasions. Occupation of privately owned lands has become increasingly common in Honduras and reflects both population pressure on and land hunger of the peasantry. Land seizures by squatters are sometimes recognized by the National Agrarian Institute. In other cases, the government has promoted the relocation of people to sparsely populated regions of the country. Unfortunately, the chosen relocation sites are in tropical rain forests, which are already endangered throughout the region. The government wishes to transform the forests into rubber and citrus plantations or into farms to raise rice, corn, and other crops.

Peasants who fail to gain access to land usually migrate to urban centers in search of a better life. What they find in cities such as the capital, Tegucigalpa, are inadequate social services, a miserable standard of living, and a municipal government without the resources to help. In 1989, Tegucigalpa was deeply in debt, mortgaged to the limit, months behind in wage payments to city workers, and plagued by garbage piling up in the streets. In 2004 the capital city was plagued by a crime wave conducted by youth gangs, drug trafficking, police implication in high profile crimes, and the murder of street children by death squads.

FREEDOM

Former president Reina reduced the power of the military. Constitutional reforms in 1994–1995 replaced obligatory military service and the press-gang recruitment system with voluntary service. As a result, the size of the army declined.

The nation's economy as a whole fared badly in the late 1980s. But by 1992, following painful adjustments occasioned by the reforms of the government of President Rafael Callejas, the economy again showed signs of growth. Real gross domestic product reached 3.5 percent, and inflation was held in check. Still, unemployment remained a persistent problem; some agencies calculated that two thirds of the workforce lacked steady employment. A union leader warned: "Unemployment leads to desperation and becomes a time bomb that could explode at any moment."

In addition to internal problems, pressure has been put on Honduras by the International Monetary Fund. According to the *Caribbean & Central America Report*, the

first phase of a reform program agreed to with the IMF succeeded in stabilizing the economy through devaluation of the lempira (the Honduran currency), public spending cuts, and increased taxes. But economic growth declined, and international agencies urged a reduction in the number of state employees as well as an accelerated campaign to privatize state-owned enterprises. The government admitted that there was much room for reform, but one official complained: "As far as they [the IMF] are concerned, the Honduran state should make gigantic strides, but our position is that this country cannot turn into General Motors overnight."

Opposition to the demands of international agencies was quick to materialize. One newspaper warned that cuts in social programs would result in violence. Trade-union and Catholic Church leaders condemned the social costs of the stabilization program despite the gains recorded in the credit-worthiness of Honduras.

HUMAN AND CIVIL RIGHTS

In theory, despite the continuing violence in the region, basic freedoms in Honduras are still intact. The press is privately owned and free of government censorship. There is, however, a quietly expressed concern about offending the government, and self-censorship is considered prudent. Moreover, it is an accepted practice in Honduras for government ministries and other agencies to have journalists on their payrolls.

Honduran labor unions are free to organize and have a tradition of providing their rank-and-file certain benefits. Unions are allowed to bargain, but labor laws guard against "excessive" activity. A complex procedure of negotiation and arbitration must be followed before a legal strike can be called. If a government proves unyielding, labor will likely pass into the ranks of the opposition.

In 1992, Honduras's three major workers' confederations convinced the private sector to raise the minimum wage by 13.7 percent, the third consecutive year of increases. Nevertheless, the minimum wage, which varies by occupation and location, is not adequate to provide a decent standard of living, especially in view of inflation. One labor leader pointed out that the minimum wage will "not even buy tortillas." To compound workers' problems, the labor minister admitted that about 30 percent of the enterprises under the supervision of his office paid wages *below* the minimum. To survive, families must pool the resources of all their working members. Predictably, health and safety laws are usually ignored. As is the case in the rural sector, the gov-

ernment has listened to the complaints of workers—but union leaders have also on occasion been jailed.

The government is also confronted with the problem of an increasing flow of rural poor into the cities. Employment opportunities in rural areas have declined as landowners have converted cropland into pasture for beef cattle. Because livestock raising requires less labor than growing crops, the surplus rural workers seek to better their opportunities in the cities. But the new migrants have discovered that Honduras's commercial and industrial sectors are deep in recession and cannot provide adequate jobs.

HEALTH/WELFARE

Honduras remains one of the region's poorest countries. Serious shortcomings are evident in education and health care, and economic growth is essentially erased by population growth. Approximately half of the population live in poverty.

Fortunately, many of the 300,000 refugees from Nicaragua and El Salvador have returned home. With the election of President Violeta Chamorro in Nicaragua, most of the 20,000 rebel Contras laid down their arms and went home, thus eliminating—from the perspective of the Honduran government—a source of much violence in its border regions.

To the credit of the Honduran government, which is under strong pressure from conservative politicians and businesspeople as well as elements within the armed forces for tough policies against dissent, allegations vis-à-vis human-rights abuses are taken seriously. (In one celebrated case, the Inter-American Court of Human Rights, established in 1979, found the government culpable in at least one person's "disappearance" and ordered the payment of an indemnification to the man's family. While not accepting any premise of guilt, the government agreed to pay. More important, according to the COHA *Washington Report*, the decision sharply criticized "prolonged isolation" and "incommunicado detention" of prisoners and equated such abuses with "cruel and inhuman punishment.") Former president Carlos Roberto Reina was a strong advocate of human rights as part of his "moral revolution." In 1995, he took three steps in this direction: a special prosecutor was created to investigate human-rights violations, human-rights inquiries were taken out of the hands of the military and given to a new civilian Department of Criminal Investigation, and promises were made to follow up on cases of disappearances during previous administrations. While Honduras may no longer be characterized as "the peaceable kingdom," the government has not lost touch with its people and still acts out a traditional role of patron.

From the mid-1980s to the mid-1990s, the most serious threat to civilian government came from the military. The United States' Central American policy boosted the prestige, status, and power of the Honduran military, which grew confident in its ability to forge the nation's destiny. With the end of the Contra–Sandinista armed struggle in Nicaragua, there was a dramatic decline in military assistance from the United States. This allowed President Reina to assert civilian control over the military establishment.

ACHIEVEMENTS

The small size of Honduras, in terms of territory and population, has produced a distinctive literary style that is a combination of folklore and legend.

The sharp drop in U.S. economic assistance to Honduras—it fell from $229 million in 1985 to about $50 million in 1997—has revealed deep problems with the character of that aid. *Wall Street Journal* reporter Carla Anne Robbins writes that "Honduras's experience suggests that massive, politically motivated cash transfers … can buy social peace, at least temporarily, but can't guarantee lasting economic growth or social development." Rather, such unconditional aid "may have slowed development by making it possible for the government to put off economic reform." On the other hand, some of the aid that found its way to programs that were not politically motivated has also been lost. One program provided access to potable water and was credited with cutting the infant mortality rate by half. Other programs funded vaccinations and primary-education projects. In the words of newspaperman and development expert Juan Ramón Martínez: "Just when you [the United States] started getting it right, you walked away."

President Reina's "moral revolution" also moved to confront the problem of endemic official corruption. In June 1995, Reina alluded to the enormity of the task when he said that if the government went after all of the guilty, "there would not be enough room for them in the prisons." In 1998, just as the Honduran economy was beginning to recover from economic setbacks occasioned by turmoil in the influential Asian financial markets, Hurricane Mitch wreaked havoc on the nation's infrastructure. Roads, bridges, schools, clinics, and homes were destroyed, and thousands of lives were lost. Freshwater wells had to be reconstructed. Banana plantations were severely damaged. Recovery from this natural disaster will be prolonged and costly.

Although the press in Honduras is legally "free," *Honduras This Week* notes that many journalists have close ties to the business community and allow these contacts "rather than impartial journalism" to determine the substance of their articles. Moreover, there is a disturbing tendency for the media to praise the government in power as a "means of gaining favor with that administration."

President Ricardo Maduro has made a determined effort to crack down on a rampant crime wave. Undoubtedly his focus has been sharpened by the loss of his son to criminal violence in 1998. But crime in part springs from more basic social problems. These, too, must be addressed.

Timeline: PAST

1524
Honduras is settled by Spaniards from Guatemala

1821
Independence from Spain

1822–1838
Honduras is part of the United Provinces of Central America

1838
Honduras becomes independent as a separate country

1969
Brief border war with El Salvador

1980s
Tensions with Nicaragua grow

1990s
Carlos Flores Facusse is elected president; Hurricane Mitch causes enormous death and destruction

PRESENT

2000s
AIDS is an increasing problem

Ricardo Maduro elected president in 2002

Nicaragua (Republic of Nicaragua)

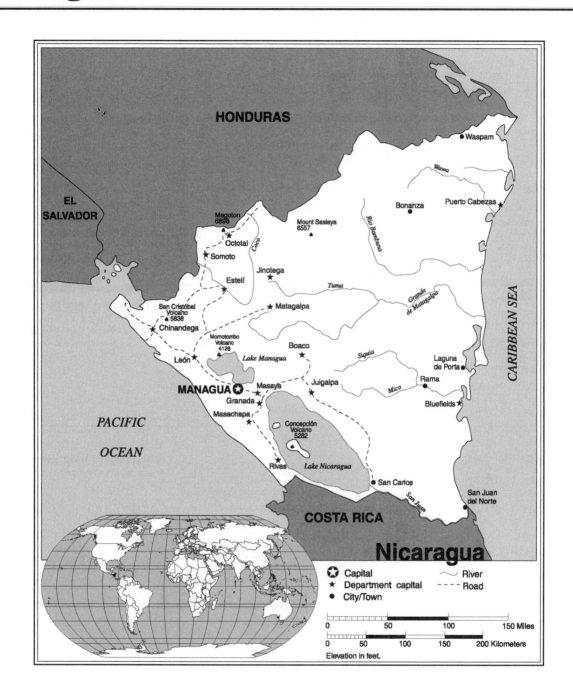

Nicaragua Statistics

GEOGRAPHY

Area in Square Miles (Kilometers): 49,985 (129,494) (about the size of New York)

Capital (Population): Managua (1,124,000)

Environmental Concerns: deforestation; soil erosion; water pollution

Geographical Features: extensive Atlantic coastal plains rising to central interior mountains; narrow Pacific coastal plain interrupted by volcanoes

Climate: tropical, but varies with elevation (temperate highlands)

PEOPLE

Population

Total: 5,359,759

Annual Growth Rate: 1.97%

Rural/Urban Population Ratio: 37/63

Ethnic Makeup: 69% Mestizo; 17% white; 9% black; 5% Amerindian
Major Language: Spanish
Religions: 95% Roman Catholic; 5% Protestant

Health

Life Expectancy at Birth: 67 years (male); 72 years (female)
Infant Mortality Rate (Ratio): 34.7/1,000
Physicians Available (Ratio): 1/1,566

Education

Adult Literacy Rate: 67.5%
Compulsory (Ages): 7–13; free

COMMUNICATION

Telephones: 171,600 main lines
Daily Newspaper Circulation: 31 per 1,000 people
Televisions: 48 per 1,000
Internet Users: 90,000

TRANSPORTATION

Highways in Miles (Kilometers): 9,829 (16,382)
Railroads in Miles (Kilometers): none

Usable Airfields: 182
Motor Vehicles in Use: 145,000

GOVERNMENT

Type: republic
Independence Date: September 15, 1821 (from Spain)
Head of State/Government: President Enrique Bolanos Geyer is both head of state and head of government
Political Parties: Independent Liberal Party; Liberal Alliance; Neoliberal Party; Conservative Party; National Action Party; Sandinista National Liberation Front; many others
Suffrage: universal at 16

MILITARY

Military Expenditures (% of GDP): 1.2%
Current Disputes: territorial or boundary disputes with Colombia, Honduras, and El Salvador

ECONOMY

Currency ($ U.S. Equivalent): 14.06 córdobas oros = $1

Per Capita Income/GDP: $2,200/$11.49 billion
GDP Growth Rate: 1.4%
Inflation Rate: 5.3%
Unemployment Rate: 22% (official rate); substantial underemployment
Labor Force: 1,700,000
Natural Resources: gold; silver; copper; tungsten; lead; zinc; timber; fish
Agriculture: coffee; bananas; sugarcane; cotton; rice; corn; tobacco; soya; beans; livestock
Industry: food processing; chemicals; machinery; metals products; textiles and clothing; petroleum; beverages; footwear; wood
Exports: $632 million (primary partners United States, Germany, El Salvador)
Imports: $1.6 billion (primary partners United States, Costa Rica, Guatemala)

SUGGESTED WEBSITE

http://www.cia.gov/cia/
publications/factbook/index.html

Nicaragua Country Report

NICARAGUA: A NATION IN RECOVERY

Nicaraguan society, culture, and history have been molded to a great extent by the country's geography. A land of volcanoes and earthquakes, the frequency of natural disasters in Nicaragua has profoundly influenced its peoples' perceptions of life, death, and fate. What historian Ralph Woodward has written about Central America is particularly apt for Nicaraguans: Fatalism may be said to be a "part of their national mentality, tempering their attitudes toward the future. Death and tragedy always seem close in Central America. The primitive states of communication, transportation, and production, and the insecurity of human life, have been the major determinants in the region's history…."

Nicaragua is a divided land, with distinct geographic, cultural, racial, ethnic, and religious zones. The west-coast region, which contains about 90 percent of the total population, is overwhelmingly white or Mestizo (mixed blood), Catholic, and Hispanic. The east coast is a sharp contrast, with its scattered population and multiplicity of Indian, Creole (in Nicaragua, native-born blacks), and Hispanic ethnic groups.

The east coast's geography, economy, and isolation from Managua, the nation's

DEVELOPMENT

The possibility of the construction of a "dry canal" across Nicaragua has raised the hopes of thousands for a better future. A group of Asian investors is investigating the construction of a 234-mile-long rail link between the oceans to carry container cargo.

capital city, have created a distinct identity among its people. Many east-coast citizens think of themselves as *costeños* ("coast dwellers") rather than Nicaraguans. Religion reinforces this common identity. About 70 percent of the east-coast population, regardless of ethnic group, are members of the Protestant Moravian Church. After a century and a half of missionary work, the Moravian Church has become "native," with locally recruited clergy. Among the Miskito Indians, Moravian pastors commonly replace tribal elders as community leaders. The Creoles speak English and originally arrived either as shipwrecked or escaped slaves or as slave labor introduced by the British to work in the lumber camps and plantations in the seventeenth century. Many Creoles and Miskitos feel a greater sense of allegiance to the

British than to Nicaraguans from the west coast, who are regarded as foreigners.

SANDINISTA POLICIES

Before the successful 1979 Revolution that drove the dictator Anastasio Somoza from power, Nicaraguan governments generally ignored the east coast. Revolutionary Sandinistas—who took their name from a guerrilla, Augusto César Sandino, who fought against occupying U.S. forces in the late 1920s and early 1930s—adopted a new policy toward the neglected region. The Sandinistas were concerned with the east coast's history of rebelliousness and separatism, and they were attracted by the economic potential of the region (palm oil and rubber). Accordingly, they hastily devised a bold campaign to unify the region with the rest of the nation. Roads, communications, health clinics, economic development, and a literacy campaign for local inhabitants were planned. The Sandinistas, in defiance of local customs, also tried to organize the local population into mass formations—that is, organizations for youth, peasants, women, wage earners, and the like. It was believed in Managua that such groups would unite the people behind the

(United Nations photo/Jerry Frank)

This lakeside section of Managua, Nicaragua, was destroyed by an earthquake in 1974. The region is often shaken by both large and small earthquakes.

government and the Revolution and facilitate the economic, political, and social unification of the region.

In general, the attempt failed, and regional tensions within Nicaragua persist to this day. Historically, costeños were unimpressed with the exploits of the guerrilla Sandino, who raided U.S. companies along the east coast in the 1930s. When the companies left or cut back on operations, workers who lost their jobs blamed Sandino rather than the worldwide economic crisis of the 1930s. Consequently, there was a reluctance to accept Sandino as the national hero of the new Nicaragua. Race and class differences increased due to an influx of Sandinistas from the west. Many of the new arrivals exhibited old attitudes and looked down on the east-coast peoples as "uncivilized" or "second class."

The Miskito Question

In 1982, the government forced 10,000 Indians from their ancestral homes along the Río Coco because of concern with border security. As a result, many Indians joined the Contras, U.S.–supported guerrillas who fought against the Sandinista regime.

In an attempt to win back the Miskito and associated Indian groups, the government decided on a plan of regional autonomy. The significance of the Sandinista policy was that the government finally appreciated how crucial regional differences are in Nicaragua. Cultural and ethnic differences must be respected if Managua expects to rule its peoples effectively. The lesson learned by the Sandinistas was taken to heart by the subsequent Chamorro government, which was the first in history to appoint a Nicaraguan of Indian background to a ministerial-level position. A limited self-government granted to the east-coast region by the Sandinistas in 1987 has been maintained; local leaders were elected to office in 1990.

A Mixed Record

The record of the Sandinista government was mixed. When the rebels seized power in 1979, they were confronted by an economy in shambles. Nineteen years of civil war had taken an estimated 50,000 lives and destroyed half a billion dollars' worth of factories, businesses, medical facilities, and dwellings. Living standards had tumbled to 1962 levels, and unemployment had reached an estimated 25 percent.

Despite such economic difficulties, the government made great strides in the areas of health and nutrition. A central goal of official policy was to provide equal access to health services. The plan had more success in urban areas than in rural ones. The government emphasized preventive, rather than curative, medicine. Preventive medi-

cine included the provision of clean water, sanitation, immunization, nutrition, and maternal and child care. People were also taught basic preventive medical techniques. National campaigns to wipe out malaria, measles, and polio had reasonable success. But because of restricted budgets, the health system was overloaded, and there was a shortage of medical supplies. In the area of nutrition, basic foodstuffs such as grains, oil, eggs, and milk were paid for in part by the government in an effort to improve the general nutritional level of Nicaraguans.

By 1987, the Sandinista government was experiencing severe economic problems that badly affected all social programs. In 1989, the economy, for all intents and purposes, collapsed. Hyperinflation ran well over 100 percent a month; and in June 1989, following a series of mini-devaluations, the nation's currency was devalued by an incredible 100 percent. Commerce was virtually paralyzed.

The revolutionary Sandinista government, in an attempt to explain the economic debacle, with some justice argued that the Nicaragua that it had inherited in 1979 had been savaged and looted by former dictator Somoza. The long-term costs of economic reconstruction; the restructuring of the economy to redistribute wealth; the trade embargo erected by the United States and North American diplomatic pressure, designed to discourage lending or aid from international institutions such as the International Monetary Fund; and the high cost of fighting a war against the U.S.–supported Contra rebels formed the backdrop to the crisis. Opposition leaders added to this list various Sandinista economic policies that discouraged private business.

FREEDOM

Diverse points of view have been freely and openly discussed in the media. Radio, the most important medium for news distribution in Nicaragua, has conveyed a broad range of opinion.

The impact of the economic crisis on average Nicaraguans was devastating. Overnight, prices of basic consumer goods such as meat, rice, beans, milk, sugar, and cooking oil were increased 40 to 80 percent. Gasoline prices doubled. School-teachers engaged in work stoppages in an effort to increase their monthly wages of about $15, equal to the pay of a domestic servant. (To put the teachers' plight into perspective, note that the cost of a liter of milk absorbed fully 36 percent of a day's pay.)

As a hedge against inflation, other Nicaraguans purchased U.S. dollars on the black market. *Regionews*, published in Managua, noted that conversion of córdobas into dollars was "seen as a better proposition than depositing them in savings accounts."

Economic travail inevitably produces dissatisfaction; opinion polls taken in July 1989 signaled political trouble for the Sandinistas. The surveys reflected an electorate with mixed feelings. While nearly 30 percent favored the Sandinistas, 57 percent indicated that they would not vote for President Daniel Ortega.

The results of the election of 1990 were not surprising, for the Sandinistas had lost control of the economy. They failed to survive a strong challenge from the opposition, led by the popular Violeta Chamorro.

Sandinista land reform, for the most part, consisted of the government's confiscation of the huge estates of the ousted Somoza family. These lands amounted to more than 2 million acres, including about 40 percent of the nation's best farmland. Some peasants were given land, but the government preferred to create cooperatives. This policy prompted the criticism that the state had simply become an old-style landowner. The Sandinistas replied that "the state is not the same state as be-

fore; it is a state of producers; we organized production and placed it at the disposal of the people." In 1990, there were several reports of violence between Sandinista security forces and peasants and former Contras who petitioned for private ownership of state land.

The Role of the Church

The Revolution created a sharp division within the Roman Catholic Church in Nicaragua. Radical priests, who believed that Christianity and Marxism share similar goals and that the Church should play a leading role in social change and revolution, were at odds with traditional priests fearful of "godless communism." Since 1979, many radical Catholics had become involved in social and political projects; several held high posts in the Sandinista government.

One priest of the theology of liberation was interviewed by *Regionews*. The interviewer stated that an "atheist could say, 'These Catholics found a just revolution opposed by the Church hierarchy. They can't renounce their religion and are searching for a more convenient theology. But it's their sense of natural justice that motivates them.'" The priest replied: "I think that's evident and that Jesus was also an 'atheist,' an atheist of the religion as practiced in his time. He didn't believe in the God of the priests in the temples who were allied with Caesar. Jesus told of a new life. And the 'atheist' that exists in our people doesn't believe in the God that the hierarchy often offers us. He believes in life, in man, in development. God manifests Himself there. A person who believes in life and justice in favor of the poor is not an atheist." The movement, he noted, would continue "with or without approval from the hierarchy."

The Drift to the Left

As has historically been the case in revolutions, after a brief period of unity and excitement, the victors begin to disagree over policies and power. For a while in Nicaragua, there was a perceptible drift to the left, and the Revolution lost its image of moderation. While radicalization was a dynamic inherent in the Revolution, it was also pushed in a leftward direction by a hostile U.S. foreign policy that attempted to bring down the Sandinista regime through its support of the Contras. In 1987, however, following the peace initiatives of Latin American governments, the Sandinista government made significant efforts to project a more moderate image. *La Prensa*, the main opposition newspaper, which the Sandinistas had shut down in 1986, was again allowed to publish. Radio Católica,

another source of opposition to the government, was given permission to broadcast after its closure the year before. And anti-government demonstrations were permitted in the streets of Managua.

Significantly, President Ortega proposed reforms in the country's election laws in April 1989, to take effect before the national elections in 1990. The new Nicaraguan legislation was based on Costa Rican and Venezuelan models, and in some instances was even more forward-looking.

An important result of the laws was the enhancement of political pluralism, which allowed for the National Opposition Union (UNO) victory in 1990. Rules for organizing political parties, once stringent, were loosened; opposition parties were granted access to the media; foreign funding of political parties was allowed; the system of proportional representation permitted minority parties to maintain a presence; and the opposition was allowed to monitor the elections closely.

HEALTH/WELFARE

Nicaragua's deep debt and the austerity demands of the IMF have had a strongly negative effect on citizens' health. As people have been driven from the health service by sharp cuts in government spending, the incidence of malnutrition in children has risen. Reported deaths from diarrhea and respiratory problems are also on the increase.

The Sandinistas realized that to survive, they had to make compromises. In need of breathing space, the government embraced the Central American Peace Plan designed by Costa Rican president Óscar Arias and designed moderate policies toisolate the United States.

On the battlefield, the cease-fire unilaterally declared by the Sandinistas was eventually embraced by the Contras as well, and both sides moved toward a political solution of their differences. Armed conflict formally ended on June 27, 1990, although sporadic violence continued in rural areas.

A PEACEFUL TRANSITION

It was the critical state of the Nicaraguan economy that in large measure brought the Sandinistas down in the elections of 1990. Even though the government of Violeta Chamorro made great progress in the demilitarization of the country and national reconciliation, the economy remained a time bomb.

The continuing economic crisis and disagreements over policy directions destroyed the original base of Chamorro's political support. Battles between the legislative and executive branches of government virtually paralyzed the country. At the end of 1992, President Chamorro closed the Assembly building and called for new elections. But by July 1995, an accord had been reached between the two contending branches of government. Congress passed a "framework law" that created the language necessary to implement changes in the Sandinista Constitution of 1987. The Legislative Assembly, together with the executive branch, are pledged to the passage of laws on matters such as property rights, agrarian reform, consumer protection, and taxation. The July agreement also provided for the election of the five-member Consejo Supremo Electoral (Supreme Electoral Council), which oversaw the presidential elections in November 1996.

The election marked something of a watershed in Nicaraguan political history. Outgoing president Chamorro told reporters at the inauguration of Arnoldo Alemán Lâcayo: "For the first time in more than 100 years … one civilian, democratically elected president will hand over power to another." But the election did not mask the fact that Nicaragua was still deeply polarized and that the Sandinistas only grudgingly accepted their defeat.

President Alemán sought a dialogue with the Sandinistas, and both sides agreed to participate in discussions to study poverty, property disputes occasioned by the Sandinista policy of confiscation, and the need to attract foreign investment.

The Alemán administration confronted a host of difficult problems. In the Western Hemisphere, only Haiti is poorer. Perhaps 80 percent of the population are unemployed or underemployed, and an equal percentage live below the poverty line. Just as the economy began to show some signs of recovery from years of war, Hurricane Mitch devastated the country in 1998 and profoundly set back development efforts, as all available resources had to be husbanded to reconstruct much of Nicaragua's infrastructure.

Economic malaise compounded by allegations of corruption and illegal enrichment undermined the credibility of the Alemán government. Dissatisfaction among voters was registered at the polls, resulting in Sandinista victories in Managua and nine of 17 provincial capitals in municipal elections in November 2000. A contributing factor was the emergence of the Conservative Party, which split the anti-Sandinista vote. Interestingly, the Sandinista victor in Managua, Herty Lewites, has styled himself as a "revolutionary businessman and defender of social justice"— that is, a popular pragmatist.

Organized labor has shown a similar pragmatic dimension in Nicaragua. Labor leaders have quietly supported both globalization and the policies of the World Trade Organization because of the jobs that would be created. Another effect of globalization, not only in Nicaragua but also throughout the region, has been the further erosion of the *siesta* (nap) tradition. In the words of a Nicaraguan-government official, the emerging world economy demands that "we stay open all day."

The administration of President Enrique Bolano Geyer, elected in 2002, has seen some improvement. There was some economic growth in 2003, private investment has increased, and exports have risen. For the foreseeable future, however, Nicaragua will remain poor.

Timeline: PAST

1522
Nicaragua is explored by Gil González

1821
Independence from Spain

1823
Nicaragua joins the United Provinces of Central America

1838
Nicaragua becomes independent as a separate state

1855
William Walker and filibusters (U.S. insurgents) invade Nicaragua

1928–1934
Augusto César Sandino leads guerrillas against occupying U.S. forces

1934–1979
Domination of Nicaragua by the Somoza family

1979
Sandinista guerrillas oust the Somoza family

1990s
A cease-fire allows an opening for political dialogue; Hurricane Mitch devastates the country

PRESENT

2000s
Sandinistas win municipal elections in November 2000

Enrique Bolano Geyer elected president in 2002

Panama (Republic of Panama)

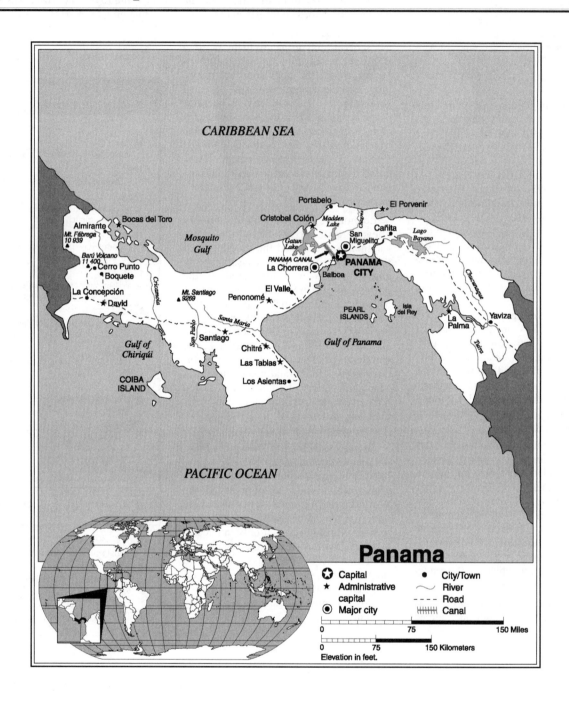

Panama

- ⊛ Capital
- ★ Administrative capital
- ⊙ Major city
- • City/Town
- ∿ River
- - - - Road
- ⊦⊦⊦⊦ Canal

0   75   150 Miles
0   75   150 Kilometers
Elevation in feet.

Panama Statistics

GEOGRAPHY

Area in Square Miles (Kilometers): 30,185 (78,200) (about the size of South Carolina)

Capital (Population): Panama City (465,000)

Environmental Concerns: water pollution; deforestation; land degradation; soil erosion

Geographical Features: interior mostly steep, rugged mountains and dissected upland plains; coastal areas largely plains and rolling hills

Climate: tropical marine

PEOPLE

Population

Total: 3,000,463
Annual Growth Rate: 1.34%

Rural/Urban Population Ratio: 44/56

Major Languages: Spanish; English

Ethnic Makeup: 70% Mestizo; 14% West Indian; 10% white; 6% Indian and others

Religions: 85% Roman Catholic; 15% Protestant and others

Health

Life Expectancy at Birth: 69 years (male); 74 years (female)

Infant Mortality Rate (Ratio): 20.8/1,000

Physicians Available (Ratio): 1/856

Education

Adult Literacy Rate: 92.6%

Compulsory (Ages): for 6 years between 6–15; free

COMMUNICATION

Telephones: 366,000 main lines

Daily Newspaper Circulation: 62 per 1,000 people

Televisions: 13 per 1,000

Internet Users 120,000

TRANSPORTATION

Highways in Miles (Kilometers): 6,893 (11,100)

Railroads in Miles (Kilometers): 208 (355)

Usable Airfields: 105

Motor Vehicles in Use: 227,000

GOVERNMENT

Type: constitutional democracy

Independence Date: November 3, 1903 (from Colombia)

Head of State/Government: President Mireya Elisa Moscoso Rodriguez is both head of state and head of government

Political Parties: Nationalist Republican Liberal Movement; Solidarity Party; Authentic Liberal Party; Arnulfista Party; Christian Democratic Party; Papa Egoro Movement; Democratic Revolutionary Party; Independent Democratic Union; National Liberal Party; Labor Party; others

Suffrage: universal and compulsory at 18

MILITARY

Military Expenditures (% of GDP): 1.2%

Current Disputes: none

ECONOMY

Currency ($ U.S. Equivalent): 1.00 balboa = $1

Per Capita Income/GDP: $6,300/$18.62 billion

GDP Growth Rate: 3.24%

Inflation Rate: 1.3%

Unemployment Rate: 14.5%

Labor Force: 1,044,000

Natural Resources: copper; mahogany forests; shrimp; hydropower

Agriculture: bananas; rice; corn; coffee; sugarcane; vegetables; livestock; fishing

Industry: construction; petroleum; brewing; sugar; canal traffic/tourism

Exports: $5.2 billion (primary partners United States, Sweden, Costa Rica)

Imports: $6.6 billion (primary partners United States, Central America and Caribbean)

SUGGESTED WEBSITE

http://www.cia.gov/cia/
publications/factbook/index.html

Panama Country Report

PANAMA: A NATION AND A CANAL

The Panama Canal, opened to shipping in 1914, has had a sharp impact on Panamanian political life, foreign policy, economy, and society. Panama is a country of minorities and includes blacks, Mestizos (mixed Indian and white), Indians, and Chinese. Many of the blacks and Chinese are the children or grandchildren of the thousands of workers who were brought to Panama to build the canal. Unable to return home, they remained behind, an impoverished people, ignored for decades by a succession of Panamanian governments.

DEVELOPMENT

Because many new ships can no longer transit the Panama Canal it has become imperative to add new, larger locks, and to widen the existing channel. This will require billions of dollars in new investment.

The government has usually been dominated by whites, although all of the country's minorities are politically active. In areas where Indians comprise a majority of the population, they play significant roles in provincial political life. Some, such as the

San Blas islanders—famous for the art form known as Mola, which consists of different colored fabrics that are cut away to make designs—live in self-governing districts. Although Indians are not restricted to tribal areas, most remain by choice, reflecting a long tradition of resistance to assimilation and defense of their cultural integrity.

Panama's economy has both profited and suffered from the presence of the canal. Because governments traditionally placed too much reliance on the direct and indirect revenues generated by the canal tolls, they tended to ignore other types of national development. Much of Panama's economic success in the 1980s, however, was the result of a strong service sector associated with the presence of a large number of banks, the Panama Canal, and the Colón Free Zone. Agriculture and industry, on the other hand, usually experienced slow growth rates.

Because of U.S. control of the canal and the Canal Zone, this path between the seas continuously stoked the fires of Panamanian nationalism. The high standard of living and the privileges enjoyed by U.S. citizens residing in the Canal Zone contrasted sharply with the poverty of Panamanians. President Omar Torrijos became

a national hero in 1977 when he signed the Panama Canal Treaties with U.S. president Jimmy Carter. The treaties provided for full Panamanian control over the canal and its revenues by 1999.

FREEDOM

Panama's indigenous population of 194,000 have the same political rights as other citizens. In 1992, Cuna Indians asked for the creation of an additional reserve to prohibit incursions by squatters into areas traditionally considered their own.

Panamanian officials spoke optimistically of their plans for the bases they would soon inherit, citing universities, modern container ports, luxury resorts, and retirement communities. But there was much concern over the loss of an estimated $500 million that tens of thousands of American troops, civilians, and their dependents had long pumped into the Panamanian economy. Moreover, while all agreed that the canal itself would be well run, because Panamanians had been phased into its operation, there was pessimism about the lack of planning for ancillary facilities.

(Courtesy Dr. Paul Goodwin)

The Panama Canal has been of continuing importance to the country since it opened in 1914. Full control of the canal was turned over to Panama in 1999, marking the end of U.S. involvement and representing a source of Panamanian nationalism.

In 1995, more than 300 poor, landless people a day were moving into the Zone and were clearing forest for crops. The rain forest in the Canal Basin supplies not only the water essential to the canal but also the drinking water for about 40 percent of Panama's population. Loss of the rain forest could prove catastrophic. One official noted: "If we lose the Canal Basin we do not lose only our water supply, it will also be the end of the Canal itself."

A RETURN TO CIVILIAN GOVERNMENT

President Torrijos, who died in a suspicious plane crash in 1981, left behind a legacy that included much more than the treaties. He elevated the National Guard to a position of supreme power in the state and ruled through a National Assembly of community representatives.

The 1984 elections appeared to bring to fruition the process of political liberalization initiated in 1978. But even though civilian rule was officially restored, the armed forces remained the real power behind the throne. Indeed, spectacular revelations in 1987 strongly suggested that Defense Forces chief general Manuel Antonio Noriega had rigged the 1984 elections. He was also accused of drug trafficking, gun running, and money laundering.

In February 1988, Noriega was indicted by two U.S. grand juries and charged with using his position to turn Panama into a center for the money-laundering activities of the Medellín, Colombia, drug cartel and providing protection for cartel members living temporarily in Panama.

Attempts by Panamanian president Eric Arturo Delvalle to oust the military strongman failed, and Delvalle himself was forced into hiding. Concerted efforts by the United States to remove Noriega from power—including an economic boycott, plans to kidnap the general and have the CIA engineer a coup, and saber-rattling by the dispatch of thousands of U.S. troops to the Canal Zone—proved fruitless.

The fraud and violence that accompanied an election called by Noriega in 1989 to legitimize his government and the failure of a coup attempt in October ultimately resulted in the invasion of Panama by U.S. troops in December. Noriega was arrested, brought to the United States for trial, and eventually was convicted on drug-trafficking charges.

HEALTH/WELFARE

The Care Group, which is affiliated with Harvard Medical School, Beth Israel Hospital, and Panama's excellent Hospital Nacionál, reached agreement to create the region's first teaching hospital in the area of emergency care. Physicians from all of Latin America would be welcomed to the facility.

The U.S. economic sanctions succeeded in harming the wrong people. Noriega and his cronies were shielded from the economic crisis by their profits from money laundering. But many other Panamanians were devastated by the U.S. policy.

Nearly a decade after the invasion by U.S. troops to restore democracy and halt drug trafficking, the situation in Panama remains problematic. The country is characterized by extremes of wealth and poverty, and corruption is pervasive. The economy is still closely tied to drugmoney laundering, which has reached levels higher than during the Noriega years.

Elections in 1994 reflected the depth of popular dissatisfaction. Three quarters of the voters supported political movements that had risen in opposition to the policies and politics imposed on Panama by the U.S. invasion. The new president, Ernesto Pérez Balladares, a 48-year-old economist and businessman and a former supporter of Noriega, promised "to close the Noriega chapter" in Panama's history. During his term, he pushed ahead with privatization, the development of the Panama Canal Zone, a restructuring of the foreign debt, and initiatives designed to enhance tourism.

Unfortunately, Pérez seemed to have inherited some of the personalist tendencies of his predecessors. In 1998, he pushed for a constitutional change that would have allowed him to run for reelection in 1999. When put to a referendum in August 1998, Panamanians resoundingly defeated the ambitions of the president.

The 1999 elections, without the participation of Pérez, produced a close campaign between Martín Torrijos, the son of Omar, and Mireya Moscoso, the widow of the president who had been ousted by Omar Torrijos. Moscoso emerged as a winner, with 44 percent of the vote, and became Panama's first woman president.

Moscoso opposed many of Pérez's freemarket policies and was especially critical of any further plans to privatize state-owned industries. Moscoso identified her administration with the inauguration of a "new era" for Panama's poor. Her social policies stood in direct contrast to the more economically pragmatic approach of her predecessors. Continued domination of the Legislature by the opposition render social reform difficult, but the president felt that she had to intercede on behalf of the poor, who constitute one third of the population. Diversification of the economy remains a need, as Panama is still overly dependent on canal revenues and traditional agricultural exports. As supplement to the income produced by the canal, the Panama Canal Railway has been refurbished so that it will be able to transport container cargo in less time than it takes for a ship to transit the canal.

SOCIAL POLICIES

As is the case in most Latin American nations, Panama's Constitution authorizes the state to direct, regulate, replace, or create economic activities designed to increase the nation's wealth and to distribute the benefits of the economy to the greatest number of people. The harsh reality is that the income of one third of Panama's population frequently fails to provide for families' basic needs.

Women, who won the right to vote in the 1940s, are accorded equal political rights under the law and hold a number of important government positions, including the presidency. But as in all of Latin America, women do not enjoy the same opportunities for advancement as men. There are also profound domestic constraints to their freedom. Panamanian law, for example, does not recognize community property; divorced or deserted women have no protection and can be left destitute, if that is the will of their former spouses. Many female heads-of-household from poor areas are obliged to work for the government, often as street cleaners, in order to receive support funds from the authorities.

ACHIEVEMENTS

The Panama Canal, which passed wholly to Panamanian control in 1999, is one of the greatest engineering achievements of the twentieth century. A maze of locks and gates, it cuts through 50 miles of the most difficult terrain on Earth.

With respect to human rights, Panama's record is mixed. The press and electronic media, while theoretically free, have experienced some harassment. In 1983, the Supreme Court ruled that journalists need not be licensed by the government. Nevertheless, both reporters and editors still exercise a calculated self-censorship, and press conduct in general is regulated by an official "Morality and Ethics Commission," whose powers are broad and vague. In 2001, some journalists complained that the Moscoso government used criminal antidefamation laws to intimidate the press in general, and its critics in particular.

In May 2004 Martín Torrijos was elected president with about 47 percent of the vote. Although he is the flag bearer of a political party built by military strongmen, including his father and Noriega, he has promised change. Cloaking himself in the garb of a populist, Torrijos has presented an image in both the cities and the countryside as the defender of the poor. His economic policies will likely embrace a significant reconstruction of the Panama Canal to allow the passage of larger ships. The need for huge amounts of private investment will tend to temper his populism.

Timeline: PAST

1518
Panama City is established

1821–1903
Panama is a department of Colombia

1903
Independence from Colombia

1977
The signing of the Panama Canal Treaties

1980s
The death of President Omar Torrijos creates a political vacuum American troops invade Panama; Manuel Noriega surrenders to face drug charges in the United States

1990s
Mireya Moscoso is elected as Panama's first woman president The last U.S. troops leave Panama; the Panama Canal passes to wholly local control

PRESENT

2000s
Climatic changes have been accelerated by deforestation

Martín Torrijos elected president in 2004

South America

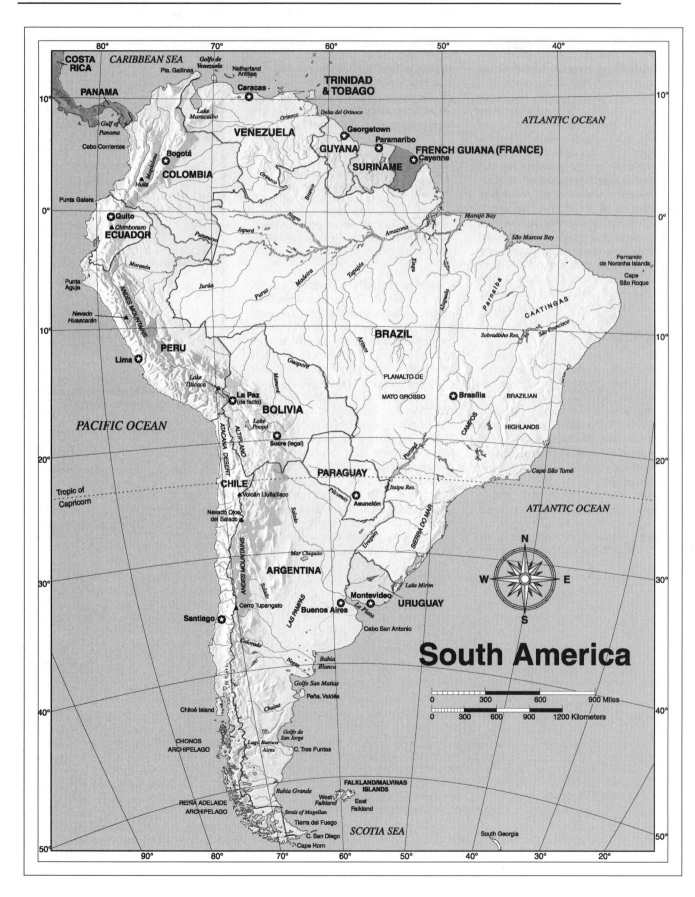

South America

South America: An Imperfect Prism

Any overview of South America must first address the incredible geographic and climatic diversity of the region. Equatorial rain forests are found in Brazil, Ecuador, Colombia, Venezuela, and other countries; and the coastal deserts in Peru and northern Chile are among the driest and most forbidding in the world (naturalist Charles Darwin described the area as "a complete and utter desert"). More hospitable are the undulating pampas and plains of Argentina, Uruguay, central Venezuela, eastern Colombia, and southeastern Brazil. The spine of the continent is formed by the Andes Mountains, majestic and snowcapped. Because of its topography and the many degrees of latitude in which it lies, South America has extremes of temperature, ranging from desert heat to the steaming humidity of the tropics to the cold gales of Tierra del Fuego, which lies close to the Antarctic Circle. To add further to the perils of generalization, wide-ranging differences often occur within a country. Geography has played a critical role in the evolution of each of the nations of South America; it has been one of several major influences in their histories and their cultures.

NATURE'S CHALLENGE

Nature has presented the inhabitants of South America with an unrelenting challenge. On the west coast, most of the major cities are located in geologically active zones. All too frequently, earthquakes, tidal waves, volcanic activity, and landslides have taken a staggering toll of human life. And throughout the region, floods and droughts make agriculture a risky business. Periodically, for example, the weather phenomenon known as *El Niño* brings devastating floods to Peru and Ecuador, with heavy loss of life and extensive damage to the infrastructure of the region.

REGIONALISM

South America's diverse topography has also helped to foster a deep-seated regionalism that has spawned innumerable civil wars and made national integration an extremely difficult task. In Colombia, for instance, the Andes fan out into three distinct ranges, separated by deep valleys. Each of the nation's three major cities—Bogotá, Medellín, and Cali—dominates a valley and is effectively isolated from the others by the mountains. The broad plains to the east have remained largely undeveloped because of the difficulty of access from the centers of population. Troubling to Colombian governments is the fact that, in terms of topography, the eastern plains are tied to Venezuela and not to the Colombian cities to the west.

(Photo Lisa Clyde)

The northern Andes Mountains meet the Caribbean Sea in Venezuela.

(United Nations photo)

The cultures of the countries of the southern Cone—Argentina, Uruguay, Paraguay, and Chile—have been profoundly influenced by the geography of their vast, fertile plains. These latter-day gauchos herd their animals to the auction pens.

Similarly, mountains divide Ecuador, Peru, Bolivia, and Venezuela. In all of these nations, there is a permanent tension between the capital cities and the hinterlands. As is the case in those republics that have large Indian populations, the tension often is as much cultural as it is a matter of geography. But in the entire region, physical geography interacts with culture, society, politics, and economics. Regionalism has been a persistent theme in the history of Ecuador, where there has been an often bitter rivalry between the capital city of Quito, located high in the central mountains, and the port city of Guayaquil. Commonly, port cities, with their window on the world outside, tend to be more cosmopolitan, liberal, and dynamic than cities that are more isolated. Such is the case with freewheeling Guayaquil, which stands in marked contrast to conservative, traditional, deeply Catholic Quito.

Venezuela boasts six distinct geographical regions, which include mountains and valleys, plains and deserts, rivers and jungles, and a coastline. Historian John Lombardi observes that each of these regions has had an important role in identifying and defining the character of Venezuela's past and present: "Over the centuries the geographical focus has shifted from one region to another in response to internal arrangements and external demands."

THE SOUTHERN CONE

The cultures of the countries of the so-called Southern Cone—Argentina, Uruguay, Paraguay, and Chile—have also been shaped by the geographical environment. Argentina, Uruguay,

and Brazil's southern state of Rio Grande do Sul developed subcultures that reflected life on the vast, fertile plains, where cattle grazed by the millions. The *gaucho* ("cowboy") became symbolic of the "civilization of leather." Fierce, independent, a law unto himself, the gaucho was mythologized by the end of the nineteenth century. At a time when millions of European immigrants were flooding into the region, the gaucho emerged as a nationalist symbol of Argentina and Uruguay, standing firm in the face of whatever natives viewed as "foreign."

Landlocked Paraguay, surrounded by powerful neighbors, has for most of its history been an introspective nation, little known to the outside world. Because of its geography, most of Paraguay's population is concentrated near the capital city of Asunción. A third of the nation is tropical and swampy—not suitable for settlement. To the west, the desolate Chaco region, with its lack of adequate sources of drinkable water, is virtually uninhabitable.

Chile, with a coastline 2,600 miles long, is a country of topographic and climatic extremes. If superimposed on a map of North America, Chile would stretch from Mexico's Baja California to the Yukon in Alaska. It is on Chile's border with Argentina that the Andes soar to their greatest heights. Several peaks, including Aconcagua, reach to nearly 23,000 feet in elevation. That mountain barrier has historically isolated Chile from eastern South America and from Europe. The central valley of Chile is the political, industrial, social, and cultural heart of the nation. With the capital city of Santiago and the large port of Valparaíso, the valley holds about 70 percent of Chile's population. The valley's Mediterranean climate and fertile soil have long acted as a magnet for Chileans from other, less hospitable, parts of the country.

BRAZIL

Historian Rollie Poppino has noted that the "major miracle of Brazil is its existence as a single nation." What he implies is that Brazil embraces regions that are so distinct that they could well be separate countries. "There are actually many Brazils within the broad expanse of the national territory, and the implication of uniformity conveyed by their common flag and language is often deceptive." In Brazil, there exists a tremendous range of geographical, racial, cultural, and economic contrasts. But part of the Brazilian "miracle" lies in the ability of its people to accept the diversity as normal. Many Brazilians were unaware of the great differences within their country for years, until the improvement of internal transportation and communications as well as the impact of the mass media informed them not only of their common heritage but also of their profound regional differences.

DIVERSE PEOPLES

In many respects, the peoples of South America are as diverse as its geography. While the populations of Argentina and Uruguay are essentially European, with virtually no Indian intermixture, Chilean society is descended from Spanish conquerors and the Indians they dominated. The Indian presence is strongest in the Andean republics of Bolivia, Peru, and Ecuador—the heart of the ancient Inca civilization. Bolivia is the most Indian, with well over half its population classified as such. Mestizos (mixed white and Indian) constitute about a quarter of the population, and whites make up only about one tenth.

Three ethnic groups are found among the populations of Colombia and Venezuela: Spanish and Indian predominate, and there are small black minorities. About 60 percent of the populations of both countries are of Mestizo or pardo (mixed blood) origin. One of Brazil's distinctive features is the rich racial mixture of its population. Peoples of Indian, European, African, and Japanese heritage live in an atmosphere largely free of racial enmity, if not degrees of prejudice.

Taken as a whole, the predominant culture is Iberian (that is, Spanish or Portuguese), although many mountain areas are overwhelmingly Indian in terms of ethnic makeup. With the conquest and colonization of South America in the sixteenth century, Spain and Portugal attempted to fasten their cultures, languages, and institutions on the land and its peoples. Spanish cities in South America—laid out in the familiar grid pattern consisting of a large central plaza bordered by a Catholic church, government buildings, and the dwellings of the ruling elite—represented the conscious intention of the conquerors to impose their will, not only on the defeated Indian civilizations but also on nature itself.

By way of contrast, the Brazilian cities that were laid out by early Portuguese settlers tended to be less formally structured, suggesting that their planners and builders were more flexible and adaptable to the new world around them. Roman Catholicism, however, was imposed on all citizens by the central authority. Government, conforming to Hispanic political culture, was authoritarian in the colonial period and continues to be so today. The conquerors created a stratified society of essentially two sectors: a ruling white elite and a ruled majority. But Spain and Portugal also introduced institutions that knit society together. Paternalistic patron–client relationships that bound the weak to the strong were common; they continue to be so today.

INDIAN CULTURE

Among the isolated Indian groups of Ecuador, Peru, and Bolivia, Spanish cultural forms were strongly and, for the most part, successfully resisted. Suspicious and occasionally hostile, the Indians refused integration into the white world outside their highland villages. By avoiding integration, in the words of historian Frederick Pike, "they maintain the freedom to live almost exclusively in the domain of their own language, social habits, dress and eating styles, beliefs, prejudices, and myths."

Only the Catholic religion was able to make some inroads, and that was (and still is) imperfect. The Catholicism practiced by Quechua- and Aymara-speaking Indians is a blend of Catholic teachings and ancient folk religion. For example, in an isolated region in Peru where eight journalists were massacred by Indians, a writer who investigated the incident reported in *The New York Times* that while Catholicism was "deeply rooted" among the Indians, "it has not displaced old beliefs like the worship of the *Apus*, or god mountains." When threatened, the Indians are "zealous defenders of their customs and mores." The societies' two cultures have had a profound impact on the literature of Ecuador, Peru, and Bolivia. The plight of the Indian, social injustice, and economic exploitation are favorite themes of these nations' authors.

(United Nations photo/Bruno J. Zehnder)

South America's Indian cultures and modern development have never really mixed. The native cultures persist in many areas, as exemplified by this Indian woman at a market in Ecuador.

Other Indian groups more vulnerable to the steady encroachment of "progress" did not survive. In the late nineteenth century, pampas Indians were virtually destroyed by Argentine cavalry armed with repeating rifles. Across the Andes, in Chile, the Araucanian Indians met a similar fate in the 1880s. Unfortunately, relations between the "civilized" world and the "primitive" peoples clinging to existence in the rain forests of Brazil, Peru, Bolivia, and Venezuela have generally improved little. Events in Brazil, Ecuador, and Venezuela in the early 1990s, however, signaled a shift toward greater Indian rights. Indigenous peoples throughout the Amazon Basin, however, are still under almost daily assault from settlers hungry for land, road builders, developers, and speculators—most of whom care little about the cultures they are annihilating.

AFRICAN-AMERICAN CULTURE

In those South American countries where slavery was widespread, the presence of a large black population has contributed yet another dimension to Hispanic culture (or, in the case of Guyana and Suriname, English and Dutch culture). Slaves, brutally uprooted from their cultures in Africa, developed new cultural forms that were often a combination of Christian and other beliefs. To insulate themselves against the rigors of forced labor and to forge some kind of common identity, slaves embraced folk religions that were heavily oriented toward magic. Magic helped blacks to face an uncertain destiny, and folk religions built bridges between peoples facing a similar, horrible fate. Folk religions not only survived the emancipation of slaves but have remained a common point of focus for millions of Brazilian blacks.

This phenomenon had become so widespread that in the 1970s, the Roman Catholic Church made a concerted effort to win Afro-Brazilians to a religion that was more Christian and less pagan. This effort was partly negated by the development of close relations between Brazil and Africa, which occurred at the same time as the Church's campaign. Brazilian blacks became more acutely aware of their African origins and began a movement of "re-Africanization." So pervasive had the folk religions become that one authority stated that Umbada (one of the folk religions) was now the religion of Brazil. The festival of *Carnaval* ("Carnival") in Rio de Janeiro, Brazil, is perhaps the best-known example of the blending of Christianity with spiritism. Even the samba, a dance form that is central to the Carnaval celebration, had its origins in black folk religions.

IMMIGRATION AND CULTURE

Italians, Eastern and Northern Europeans, Chinese, and Japanese have also contributed to the cultural, social, and economic development of several South American nations. The great outpouring of Europe's peoples that brought millions of immigrants to the shores of the United States also brought millions to South America. From the mid-1800s to the outbreak of World War I in 1914, great numbers of Italians and Spaniards, and much smaller numbers of Germans, Russians, Welsh, Scots, Irish, and English, boarded ships that would carry them to South America.

Many were successful in the "New World." Indeed, immigrants were largely responsible for the social restructuring of Argentina, Uruguay, and southern Brazil, as they created a large and dynamic middle class where none had existed before.

Italians

Many of the new arrivals came from urban areas, were literate, and possessed a broad range of skills. Argentina received the greatest proportion of immigrants. So great was the influx that an Argentine political scientist labeled the years 1890–1914 the "alluvial era" (flood). His analogy was apt, for by 1914, half the population of the capital city of Buenos Aires were foreign-born. Indeed, 30 percent of the total Argentine population were of foreign extraction. Hundreds of thousands of immigrants also flocked into Uruguay.

In both countries, they were able to move quickly into middle-class occupations in business and commerce. Others found work on the docks or on the railroads that carried the produce of the countryside to the ports for export to foreign markets. Some settled in the interior of Argentina, where they usually became sharecroppers or tenant farmers, although a sizable number were able to purchase land in the northern province of Santa Fe or became truck farmers in the immediate vicinity of Buenos Aires. Argentina's wine industry underwent a rapid transformation and expansion with the arrival of Italians in the western provinces of Mendoza and San Juan. In the major cities of Argentina, Uruguay, Chile, Peru, and Brazil, Italians built hospitals and established newspapers; they formed mutual aid societies and helped to found the first labor unions. Their presence is still strong today, and Italian words have entered into everyday discourse in Argentina and Uruguay.

(Photo Lisa Clyde)

The Spanish colonial influence is apparent in South America, as seen in this sixteenth-century building in Andean Venezuela. It was originally a monastery. Later, it was used as a hunting lodge; most recently, it was turned into a hotel.

Other Groups

Other immigrant groups also made their contributions to the formation of South America's societies and cultures. Germans colonized much of southern Chile and were instrumental in creating the nation's dairy industry. In the wilds of Patagonia, Welsh settlers established sheep ranches and planted apple, pear, and cherry trees in the Río Negro Valley.

In Buenos Aires, despite the 1982 conflict over the Falkland Islands, there remains a distinct British imprint. Harrod's is the largest department store in the city, and one can board a train on a railroad built with English capital and journey to suburbs with names such as Hurlingham, Temperley, and Thames. In both Brazil and Argentina, soccer was introduced by the English, and two Argentine teams still bear the names "Newell's Old Boys" and "River Plate." Collectively, the immigrants who flooded into South America in the late nineteenth and early twentieth centuries introduced a host of new ideas, methods, and skills. They were especially important in stimulating and shaping the modernization of Argentina, Uruguay, Chile, and southern Brazil.

In other countries that were bypassed earlier in the century, immigration has become a new phenomenon. Venezuela—torn by political warfare, its best lands long appropriated by the elite, and its economy developing only slowly—was far less attractive than the lands of opportunity to its north (the United States) and south (Argentina, Uruguay, and Brazil). In the early 1950s, however, Venezuela embarked on a broadscale development program that included an attempt to attract European immigrants. Thousands of Spaniards, Portuguese, and Italians responded to the economic opportunity. Most of the immigrants settled in the capital city of Caracas, where some eventually became important in the construction business, retail trade, and the transportation industry.

INTERNAL MIGRATION

Paralleling the movement of peoples from across the oceans to parts of South America has been the movement of populations from rural areas to urban centers. In every nation, cities have been gaining in population for years. What prompts people to leave their homes and strike out for the unknown? In the cases of Bolivia and Peru, the very real prospect of famine has driven people out of the highlands and into the larger cities. Frequently, families will plan the move carefully. Vacant lands around the larger cities will be scouted in advance, and suddenly, in the middle of the night, the new "settlers" will move in and erect a shantytown. With time, the seizure of the land is usually recognized by city officials and the new neighborhood is provided with urban services. Where the land seizure is resisted, however, violence and loss of life are common.

Factors other than famine also force people to leave their ancestral homes. Population pressure and division of the land into parcels too small to sustain families compel people to migrate. Others move to the cities in search of economic opportunities or chances for social advancement that do not exist in rural regions. Tens of thousands of Colombians illegally crossed into Venezuela in the 1970s and 1980s in search of employment. As is the case with Mexicans who enter the United States, Colombians experienced discrimination and remained on the margins of urban society, mired in low-paying, unskilled jobs. Those who succeeded in finding work in industry were a source of anger and frustration to Venezuelan labor-union members, who resented Colombians who accepted low rates of pay. Other migrants sought employment in the agricultural sector on coffee plantations or the hundreds of cattle ranches that dot the *llanos*, or plains. In summary, a combination of push and pull factors are involved in a person's decision to begin a new life.

Since World War II, indigenous migration in South America has rapidly increased urban populations and has forced cities to reorganize. Rural people have been exposed to a broad range of push–pull pressures to move to the cities. Land hunger, extreme poverty, and rural violence might be included among the push factors; while hope for a better job, upward social mobility, and a more satisfying life help to explain the attraction of a city. The phenomenon can be infinitely complex.

In Lima, Peru, there has been a twofold movement of people. While the unskilled and illiterate, the desperately poor and unemployed, the newly arrived migrant, and the delinquent have moved to or remained in inner-city slums, former slum dwellers have in turn moved to the city's perimeter. Although less centrally located, they have settled in more spacious and socially desirable shantytowns. In this way, some 16,000 families created a squatter settlement practically overnight in the south of Lima. Author Hernando DeSoto, in his groundbreaking and controversial book *The Other Path*, captures the essence of the shantytowns: "Modest homes cramped together on city perimeters, a myriad of workshops in their midst, armies of vendors hawking their wares on the street, and countless minibus lines crisscrossing them—all seem to have sprung from nowhere, pushing the city's boundaries ever outward."

Significantly, DeSoto notes, collective effort has increasingly been replaced by individual effort, upward mobility exists even for the inner-city slum dwellers, and urban culture and patterns of consumption have been transformed. Opera, theater, and *zarzuela* (comic opera) have gradually been replaced by movies, soccer, folk festivals, and television. Beer, rice, and table salt are now within the reach of much of the population; consumption of more expensive items, however, such as wine and meat, has declined.

On the outskirts of Buenos Aires there exists a *villa miseria* (slum) built on the bottom and sides of an old clay pit. Appropriately, the *barrio*, or neighborhood, is called La Cava (literally "The Digging"). The people of La Cava are very poor; most have moved there from rural Argentina or from Paraguay. Shacks seem to be thrown together from whatever is available—scraps of wood, packing crates, sheets of tin, and cardboard. There is no source of potable water, garbage litters the narrow alleyways, and there are no sewers. Because of the concave character of the barrio, the heat is unbearable in the summer. Rats and flies are legion. At times, the smells are repulsive. The visitor to La Cava experiences an assault on the senses; this is Latin America at its worst.

But there is another side to the slums of Buenos Aires, Lima, Santiago, and Rio de Janeiro. A closer look at La Cava, for example, reveals a community in transition. Some of the housing is more substantial, with adobe replacing the scraps of wood and tin; other homes double as places of business and sell general merchandise, food, and bottled drinks. One advertises itself as a food store, bar, and butcher shop. Another sells watches and repairs radios. Several promote their merchandise or services in a weekly newspaper that circulates in La Cava and two other *barrios de emergencia* ("emergency"—that is, temporary—neighborhoods). The newspaper addresses items of concern to the inhabitants. There are articles on hygiene and infant diarrhea; letters and editorials plead with people not to throw their garbage in the streets; births and deaths are recorded. The newspaper is a chronicle of progress as well as frustration: people are working together to create a viable neighborhood; drainage ditches are constructed with donated time and equipment; collections and raffles are held to provide materials to build sewers and, in some cases, to provide minimal street lighting; and men and women who have contributed their labor are singled out for special praise.

The newspaper also reproduces municipal decrees that affect the lives of the residents. The land on which the barrio sits was illegally occupied, the stores that service the neighborhood were opened without the necessary authorization, and the housing was built without regard to municipal codes, so city ordinances such as the following aimed at the barrios de emergencia are usually restrictive: "The sale, renting or transfer of *casillas* [homes] within the boundaries of the barrio de emergencia is prohibited; casillas can not be inhabited by single men, women or children; the opening of businesses within the barrio is strictly prohibited, unless authorized by the Munici-

(United Nations photo/M. Grant)

Colombia, as is the case with many other Latin American nations, has experienced rapid urbanization. Large numbers of migrants from rural areas have spread into slums on the outskirts of cities, as exemplified by this picture of a section of Colombia's capital, Bogotá. Most of the migrants are poorly paid, and the struggle to meet basic needs precludes political activism.

pality; dances and festivals may not be held without the express authorization of the Municipality." But there are also signs of accommodation: "The Municipality is studying the problem of refuse removal." For migrants, authority and the legal system typically are not helpful; instead, they are hindrances.

Hernando DeSoto found this situation to be true also of Peru, where "the greatest hostility the migrants encountered was from the legal system." Until the end of World War II, the system had either absorbed or ignored the migrants "because the small groups who came were hardly likely to upset the status quo." But when the rural-to-urban flow became a flood, the system could no longer remain disinterested. Housing and education were barred to them, businesses would not hire them. The migrants discovered over time that they would have to fight for every right and every service from an unwilling establishment. Thus, to survive, they became part of the informal sector, otherwise known as the underground or parallel economy.

On occasion, however, municipal laws can work to the advantage of newly arrived migrants. In the sprawling new communities that sprang up between Lima and its port city of Callao, there are thousands of what appear to be unfinished homes. In almost every instance, a second floor was begun but, curiously, construction ceased. The reason for the incomplete projects relates to taxes—they are not assessed until a building is finished.

These circumstances are true not only of the squatter settlements on the fringes of South America's great cities but also of the inner-city slums. Slum dwellers have been able to improve their market opportunities and have been able to acquire better housing and some urban services, because they have organized on their own, outside formal political channels. In the words of

sociologist Susan Eckstein, "They refused to allow dominant class and state interests to determine and restrict their fate. Defiance and resistance won them concessions which quiescence would not."

DeSoto found this to be the case with Lima: Migrants, "if they were to live, trade, manufacture, or even consume . . . had to do so illegally. Such illegality was not antisocial in intent, like trafficking in drugs, theft, or abduction, but was designed to achieve such essentially legal objectives as building a house, providing a service, or developing a business."

This is also the story of Buenos Aires's La Cava. To open a shop in the barrio with municipal approval, an aspiring businessperson must be a paragon of patience. Various levels of bureaucracy, with their plethora of paperwork and fees, insensitive municipal officials, inefficiency, and interminable waiting, drive people outside the system where the laws do not seem to conform to social need.

AN ECCLESIASTICAL REVOLUTION

During the past few decades, there have been important changes in the religious habits of many South Americans. Virtually everywhere, Roman Catholicism, long identified with the traditional order, has been challenged by newer movements such as Evangelical Protestantism and the Charismatics. Within the Catholic Church, the theology of liberation once gained ground. The creation of Christian communities in the barrios, people who bond together to discuss their beliefs and act as agents of change, has become a common phenomenon throughout the region. Base communities from the Catholic perspective instill Christian values in the lives of ordinary people. But it is an active form of religion that pushes for change and social justice. Hundreds of these communities exist in Peru, thousands in Brazil.

NATIONAL MYTHOLOGIES

In the midst of geographical and cultural diversity, the nations of South America have created national mythologies designed to unite people behind their rulers. Part of that mythology is rooted in the wars of independence that tore through much of the region between 1810 and 1830. Liberation from European colonialism imparted to South Americans a sense of their own national histories, replete with military heroes such as José de San Martín, Simón Bolívar, Bernardo O'Higgins, and Antonio José de Sucre, as well as a host of revolutionary myths. This coming to nationhood paralleled what the United States experienced when it won its independence from Britain. South Americans, at least those with a stake in the new society, began to think of themselves as Venezuelans, Chileans, Peruvians, or Brazilians. The architects of Chilean national mythology proclaimed the emergence of a new and superior being who was the result of the symbolic and physical union of Spaniards and the tough, heroic Araucanian Indians. The legacy of Simón Bolívar lives on in particular in Venezuela, his homeland; even today, the nation's foreign policymakers speak in Bolivarian terms about Venezuela's rightful role as a leader in Latin American affairs. In some instances, the mythology generated by the wars for independence became a shield against foreign ideas and customs and was used to force immigrants to become "Argentines" or "Chileans." It was an attempt to bring national unity out of diversity.

Argentines have never solved the question of their identity. Many consider themselves European and hold much of the rest of Latin America in contempt. Following Argentina's loss in the Falklands War with Britain, one scholar suggested that perhaps Argentines should no longer consider themselves as "a forlorn corner of Europe" but should wake up to the reality that they are Latin Americans. Much of Argentine literature reflects this uncertain identity and may help to explain author Jorge Luis Borges's affinity for English gardens and Icelandic sagas. It was also an Argentine military government that invoked Western Catholic civilization in its fight against a "foreign" and "godless" communism in the 1970s.

THE ARTIST AND SOCIETY

There is a strongly cultured and humane side of South America. Jeane Franco, an authority on Latin American cultural movements, observes that to "declare oneself an artist in Latin America has frequently involved conflict with society." The art and literature of South America in particular and Latin America in general represent a distinct tradition within the panorama of Western civilization.

The art of South America has as its focus social questions and ideals. It expresses love for one's fellow human beings and "has kept alive the vision of a more just and humane form of society." It rises above purely personal relationships and addresses humanity.

Much change is also evident at the level of popular culture. Andean folk music, for example, is being replaced by the more urban and upbeat chicha music in Peru; and in Argentina, the traditional tango has lost much of its early appeal. Radio and television programs are more and more in the form of soap operas, adventure programs, or popular entertainment, once considered vulgar by cosmopolitan city dwellers. South America is rather like a prism. It can be treated as a single object or region. Yet when exposed to a shaft of sunlight of understanding, it throws off a brilliant spectrum of colors that exposes the diversity of its lands and peoples.

Argentina (Argentine Republic)

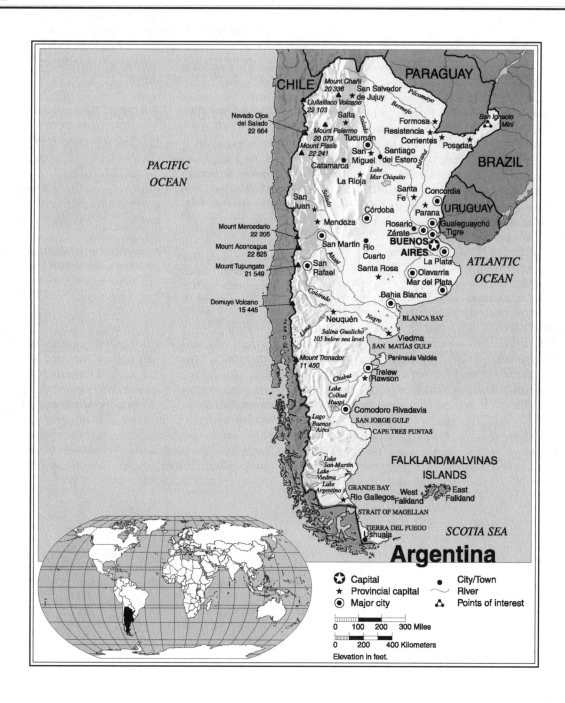

Argentina Statistics

GEOGRAPHY

Area in Square Miles (Kilometers): 1,100,000 (2,771,300) (about 4 times the size of Texas)

Capital (Population): Buenos Aires (11,802,000)

Environmental Concerns: soil erosion and degradation; air and water pollution; desertification.

Geographical Features: rich plains of the Pampas in the north; the flat to rolling plateau of Patagonia in the south; the rugged Andes along western border

Climate: varied; mostly temperate; subantarctic in southwest

PEOPLE

Population

Total: 39,144,753

Annual Growth Rate: 1.02%

Rural/Urban Population Ratio: 12/88

Major Languages: Spanish; Italian Ethnic Makeup: 97% white; 3% Mestizo, Indian, and others

Religions: 90% Roman Catholic (fewer than 20% practicing); 2% Protestant; 2% Jewish; 6% others

Health

Life Expectancy at Birth: 72 years (male); 76 years (female)

Infant Mortality Rate (Ratio): 15.66/1,000

Physicians Available (Ratio): 1/376

Education

Adult Literacy Rate: 97.1%

Compulsory (Ages): 6–14; free

COMMUNICATION

Telephones: 8,009,400 main lines

Daily Newspaper Circulation: 138 per 1,000 people

Televisions: 347 per 1,000 people

Internet Users: 41 million

TRANSPORTATION

Highways in Miles (Kilometers): 135,549 (218,276)

Railroads in Miles (Kilometers): 23,542 (37,910)

Usable Airfields: 1,359

Motor Vehicles in Use: 6,100,000

GOVERNMENT

Type: republic

Independence Date: July 9, 1816 (from Spain)

Head of State/Government: President Néstor Kirchner is both head of state and head of government

Political Parties: Radical Civic Union; Justicialist Party (Peronist); Union of the Democratic Center; others

Suffrage: universal at 18

MILITARY

Military Expenditures (% of GDP): 1.3%

Current Disputes: indefinite boundary with Chile; claims UK-administered South Georgia and South Sandwich Islands, and Falkland Islands (Islas Malvinas); territorial claim in Antarctica

ECONOMY

Currency ($ U.S. Equivalent): 2.9 pesos = $1

Per Capita Income/GDP: $11,200/$432 billion

GDP Growth Rate: 8%

Inflation Rate: 3.7%

Unemployment Rate: 16.3%; substantial underemployment

Labor Force: 15,000

Natural Resources: fertile plains; lead; zinc; tin; copper; iron ore; manganese; petroleum; uranium

Agriculture: wheat; corn; sorghum; fruits; soybeans; tobacco; tea; livestock

Industry: food processing; motor vehicles; consumer durables; textiles; chemicals and petrochemicals; printing; metallurgy; steel

Exports: $29.5 billion (primary partners Brazil, European Union, United States)

Imports: $13.2 billion (primary partners European Union, United States, Brazil)

SUGGESTED WEBSITES

http://www.cia.gov/cia/
publications/factbook/index.html

Argentina Country Report

ARGENTINA: THE DIVIDED LAND

Writers as far back as the mid-1800s have perceived two Argentinas. Domingo F. Sarmiento, the president of Argentina in the 1860s, entitled his classic work about the country *Civilization and Barbarism*. More contemporary writers speak of Argentina as a divided land or as a city and a nation. All address the relationship of the capital city, Buenos Aires, to the rest of the country. Buenos Aires is cultured, cosmopolitan, modern, and dynamic. The rural interior is in striking contrast in terms of living standards, the pace of life, and, perhaps, expectations as well. For many years, Buenos Aires and other urban centers have drawn population away from the countryside: Today, Argentina is 88 percent urban.

There are other contrasts. The land is extremely rich and produces a large share of the world's grains and beef. Few Argentines are malnourished, and the annual per capita consumption of beef is comparable to that of the United States. Yet this land of promise, which seemed in the 1890s to have a limitless future, has slowly decayed. Its greatness is now more mythical than real. Since the Great Depression of the

1930s, the Argentine economy has, save for brief spurts, never been able to return to the sustained growth of the late nineteenth and early twentieth centuries.

In the 1900s, the Argentine economy enjoyed a brief period of stability and growth. Inefficient and costly state enterprises were privatized, with the exception of the petroleum industry, traditionally a strategic sector reserved to the state. A peso tied to the dollar brought inflation under control, and the pace of business activity, employment, and foreign investment quickened.

The nation's economy is vulnerable to events in other parts of the world, however. The collapse of the Mexican peso in the early 1990s and the economic crises in Russia and, especially, Asia in the late 1990s had profound negative effects in Argentina. The global slowdown in the new millenium further complicated the economic situation.

By the first quarter of 2002, the economy was in crisis. A foreign debt of $142 billion (which works out to $3,000 for every man, woman, and child in the country), declining export revenues, high unemployment, and the inability of the government to win International Monetary Fund sup-

port for additional loans forced a devaluation of the currency.

Argentine economic history has been typified by unrealized potential and unfulfilled promises. Much depends on the confidence of the Argentine people in the leadership and policies of their elected representatives. Five changes of government between December 2001 and March 2002 suggest a wholesale *lack* of confidence.

DEVELOPMENT

Argentina convinced the IMF to help its economic recovery without following the strict full fiscal discipline measures usually required. Economic recovery was well underway in 2004.

AUTHORITARIAN GOVERNMENT

In political terms, Argentina has revealed a curious inability to bring about the kind of stable democratic institutions that seemed assured in the 1920s. Since 1930, the military has seized power at least half a dozen times. It must be noted, however, that it has been civilians who have encouraged the generals to play an active role in politics.

Historian Robert Potash writes: "The notion that Argentine political parties or other important civilian groups have consistently opposed military takeovers bears little relation to reality."

Argentina has enjoyed civilian rule since 1983, but the military is still a presence. Indeed, one right-wing faction, the *carapintadas* ("painted faces"), responsible for mutinies against President Raúl Alfonsín in 1987 and 1988, have organized a nationwide party and have attracted enough votes to rank as an important political force. An authoritarian tradition is very much alive in Argentina, as is the bitter legacy of the so-called Dirty War.

THE DIRTY WAR

What made the latest era of military rule different is the climate of political violence that gripped Argentina starting in the late 1960s. The most recent period of violence began with the murder of former president Pedro Aramburu by left-wing guerrillas (Montoneros) who claimed to be fighting on behalf of the popular but exiled leader Juan Perón (president from 1946 to 1955 and from 1973 to 1974). The military responded to what it saw as an armed challenge from the left with tough antisubversion laws and official violence against suspects. Guerrillas increased their activities and intensified their campaign to win popular support.

Worried by the possibility of a major popular uprising and divided over policy, the military called for national elections in 1973, hoping that a civilian government

would calm passions. The generals could then concentrate their efforts on destroying the armed left. The violence continued, however, and even the brief restoration of Juan Perón to power failed to bring peace.

FREEDOM

Some 40 retired military officers accused of human rights violations during the Dirty War have been stripped of their immunity from extradition to Spain where they face criminal charges. This is seen as a major advance in international law.

In March 1976, with the nation on the verge of economic collapse and guerrilla warfare spreading, the military seized power once again and declared a state of internal war, popularly called the Dirty War. Between 1976 and 1982, approximately 6,000 Argentine citizens "disappeared." Torture, the denial of basic human rights, harsh press censorship, officially directed death squads, and widespread fear came to characterize Argentina.

The labor movement—the largest, most effective, and most politically active on the continent—was, in effect, crippled by the military. Identified as a source of leftist subversion, the union movement was destroyed as an independent entity. Collective-bargaining agreements were dismantled, pension plans were cut back, and social-security and public-health programs were eliminated. The military's intent was to destroy a labor movement capable of operating on a national level.

The press was one of the immediate victims of the 1976 coup. A law was decreed warning that anyone spreading information derived from organizations "dedicated to subversive activities or terrorism" would be subject to an indefinite sentence. To speak out against the military was punishable by a 10-year jail term. The state also directed its terrorism tactics against the media, and approximately 100 journalists disappeared. Hundreds more received death threats, were tortured and jailed, or fled into exile. Numerous newspapers and magazines were shut down, and one, *La Opinión*, passed to government control.

The ruling junta justified these excesses by portraying the conflict as the opening battle of "World War III," in which Argentina was valiantly defending Western Christian values and cultures against hordes of Communist, "godless" subversives. It was a "holy war," with all of the unavoidable horrors of such strife.

By 1981, leftist guerrilla groups had been annihilated. Argentines slowly began to recover from the shock of internal war and talked of a return to civilian government. The military had completed its task; the nation needed to rebuild. Organized labor attempted to re-create its structure and threw the first tentative challenges at the regime's handling of the economy. The press carefully criticized both the economic policies of the government and the official silence over the fate of *los desaparecidos* ("the disappeared ones"). Human-rights groups pressured the generals with redoubled efforts.

(United Nations photo/P. Teuscher)

Few people are malnourished in Argentina. Well known for its abundant grains and beef, Argentina also has a large fishing industry. These fishing boats are in the bay of the Plata River in Buenos Aires.

OPPOSITION TO THE MILITARY

Against this backdrop of growing popular dissatisfaction with the regime's record, together with the approaching 150th anniversary of Great Britain's occupation of Las Islas Malvinas (the Falkland Islands), President Leopoldo Galtieri decided in 1982 to regain Argentine sovereignty and attack the Falklands. A successful assault, the military reasoned, would capture the popular imagination with its appeal to Argentine nationalism. The military's tarnished image would regain its luster. Forgiven would be the excesses of the Dirty War. But the attack ultimately failed.

In the wake of the fiasco, which cost thousands of Argentine and British lives, the military lost its grip on labor, the press, and the general population. Military and national humiliation, the continuing economic crisis made even worse by war costs, and the swelling chorus of discontent lessened the military's control over the flow of information and ideas. Previously forbidden subjects—such as the responsibility for the disappearances during the Dirty War—were raised in the newspapers.

The labor movement made a rapid and striking recovery and is now in the forefront of renewed political activity. Even though the movement is bitterly divided into moderate and militant wings, it is a force that cannot be ignored by political parties on the rebound.

The Falklands War may well prove to be a watershed in recent Argentine history. A respected Argentine observer, Torcuato DiTella, argues that the Falklands crisis was a "godsend," for it allowed Argentines to break with "foreign" economic models that had failed in Argentina. Disappointed with the United States and Europe over their support of Great Britain, he concludes: "We belong in Latin America and it is better to be a part of this strife-torn continent than a forlorn province of Europe."

HEALTH/WELFARE

 In recent years, inflation has had an adverse impact on the amount of state spending on social services. Moreover, the official minimum wage falls significantly lower than the amount considered necessary to support a family.

Popularly elected in 1983, President Raúl Alfonsín's economic policies initially struck in bold new directions. He forced the International Monetary Fund to renegotiate Argentina's huge multi-billion-dollar debt in a context more favorable to Argentina, and he was determined to bring order out of chaos.

One of his most difficult problems centered on the trials for human-rights abuses against the nation's former military rulers. According to *Latin American Regional Reports*, Alfonsín chose to "distinguish degrees of responsibility" in taking court action against those who conducted the Dirty War. Impressively, Alfonsín put on trial the highest authorities, to be followed by action against those identified as responsible for major excesses.

Almost immediately, however, extreme right-wing nationalist officers in the armed forces opposed the trials and engineered a series of mutinies that undermined the stability of the administration. In 1987, during the Easter holiday, a rebellion of dissident soldiers made its point, and the Argentine Congress passed legislation that limited the prosecution of officers who killed civilians during the Dirty War to those only at the highest levels. Mini-mutinies in 1988 resulted in further concessions to the mutineers by the Alfonsín government, including reorganization of the army high command and higher wages.

Political scientist Gary Wynia aptly observed: "The army's leadership is divided between right-wing officers willing to challenge civilian authorities with force and more romantic officers who derive gratification from doing so. Many of the latter refuse to accept the contention that they are 'equal' to civilians, claiming that they have a special role that prevents their subordination to civilian authorities." To this day, the Argentine military has come to terms neither with itself nor with democratic government.

Carlos Menem was supported by the military in the presidential election of May 1989, with perhaps 80 percent of the officer corps casting their votes for the Peronist Party. Menem adopted a policy of rapprochement with the military, which included the 1990 pardon of former junta members convicted of human-rights abuses. Historian Peter Calvert argues that Menem chose the path of amnesty because elements in the armed forces "would not be content until they got it." Rebellious middle-rank officers were well disposed toward Peronists, and Menem's pardon was "a positive gain in terms of the acceptance of the Peronists among the military themselves." In essence, then, Menem's military policy was consistent with other policies in terms of its pragmatic core. And the military seems to have been contained; military spending has been halved, the army has been reduced from 100,000 to 20,000 soldiers, military enterprises have been divested, and mandatory service has been abandoned in favor of a professional force.

Significant progress has been made with regard to "disappeared" people. In 1992, President Menem agreed to create a commission to deal with the problem of children of the disappeared who were adopted by other families. Many have had their true identities established as a result of the patient work of "The Grandmothers of the Plaza de Mayo" and by the technique of cross-generational genetic analysis. (In 1998, former junta chief Admiral Emilio Massera was arrested on charges of kidnapping—that is, the distribution to families of babies born to victims of the regime.) In 1995, the names of an additional 1,000 people were added to the official list of the missing. Also, a retired military officer revealed his part in pushing drugged prisoners out of planes over the South Atlantic Ocean.

ECONOMIC TRAVAIL

The Argentine economy under President Alfonsín was virtually out of control. Inflation soared. The sorry state of the economy and spreading dissatisfaction among the electorate forced the president to hand over power to Carlos Menem six months early.

ACHIEVEMENTS

 Argentine citizens have won four Nobel Prizes—two for peace and one each for chemistry and medicine. The nation's authors—Jorge Luis Borges, Julio Cortazar, Manuel Puig, and Ricardo Guiraldes, to name only a few—are world-famous.

Menem's new government worked a bit of an economic miracle, despite an administration nagged by corruption and early policy indecision, which witnessed the appointment of 21 ministers to nine cabinet positions during his first 18 months in office. In Menem's favor, he was not an ideologue but, rather, an adept politician whose acceptance by the average voter was equaled by his ability to do business with almost anyone. He quickly identified the source of much of Argentina's chronic inflation: the state-owned enterprises. From the time of Perón, these industries were regarded as wellsprings of employment and cronyism rather than as instruments for the production of goods or the delivery of services such as electric power and telephone service. "Ironically," says Luigi Manzetti, writing in *North-South FOCUS*, "it took a Peronist like Menem to dismantle Perón's legacy." While Menem's presidential campaign stressed "traditional Peronist themes like social justice and government investments" to revive the depressed economy, once he was in power, "having inherited a bankrupt state and under pressure from foreign banks and domestic business circles to enact a stiff adjustment program, Menem reversed his stand." He embraced the market-oriented policies of his political adversaries, "only in a much harsher fashion." State-owned enterprises were sold off in rapid-fire order. Argentina thus underwent a rapid transformation, from one of the world's most closed economies to one of the most open.

Economic growth began again in 1991, but the social costs were high. Thousands of public-sector workers lost their jobs; a third of Argentina's population lived below the poverty line, and the gap between the rich and poor tended to increase. But both inflation and the debt were eventually contained, foreign investment increased, and confidence began to return to Argentina.

In November 1993, former president Alfonsín supported a constitutional reform that allowed Menem to serve another term. Menem accepted some checks on executive power, including reshuffling the Supreme Court, placing members of the political opposition in charge of certain state offices, creating a post similar to that of prime minister, awarding a third senator to each province, and shortening the presidential term from six to four years. With these reforms in place, Menem easily won another term in 1995.

Convinced that his mandate should not end with the conclusion of his second term, Menem lobbied hard in 1998 for yet another constitutional reform to allow him to run again. This was not supported by the Supreme Court.

The Radical Party won the elections in 1999. Almost immediately President Fernando de la Rua confronted an economy mired in a deepening recession. Rising unemployment, a foreign debt that stood at 50 percent of gross domestic product, and fears of a debt default prompted the government to announce tax increases and spending cuts to meet IMF debt targets, because the peso was tied to the strong U.S. dollar. At the end of 2001, the economic crisis triggered rioting in the streets and brought down the de la Rua ad-

ministration and three others that followed in rapid succession.

By the end of 2002 the economy was in such shambles that some provinces began to issue their own currencies, farmers resorted to barter—exchanging soy beans for agricultural equipment—and many Argentines seriously considered emigration. Crime rates rose and people lost faith in governments that seemed incapable of positive policies and all-to-susceptible to corruption.

This dismal picture began to change with the election of Néstor Kirchner in May, 2003. During his first year in office he called on Congress to begin impeachment proceedings against the widely hated Supreme Court. The justices were accused to producing verdicts that reflected payoffs and political favors. Kirchner also laid siege to Argentina's security forces: he ordered more than 50 admirals and generals into early retirement and dismissed 80 percent of the high command of the notoriously corrupt Federal Police.

Finally, after years of severe malaise, the economy began to turn around in 2003. Kirchner noted that the IMF had abandoned Argentina in 2001 as its economy spiraled downward. Consequently the Argentine president, in the words of *New York Times* reporter Tony Smith, "felt justified in resolutely refusing to make a series of concessions that negotiators for the monetary fund wanted in exchange for refinancing $21.6 billion in debt" that Argentina owes to multilateral institutions …" In effect Argentina worked out a deal in accord with Argentine economic realities.

FOREIGN POLICY

The Argentine government's foreign policy has usually been determined by realistic appraisals of the nation's best interests. From 1946, the country moved between the two poles of pro-West and nonaligned. President Menem firmly supported the foreign-policy initiatives of the United States and the UN. Argentine participation in the Persian Gulf War and the presence of Argentine troops under United Nations command in Croatia, Somalia, and other trouble spots paid dividends: Washington

agreed to supply Argentina with military supplies for the first time since the Falklands War in 1982. President Kirchner has assumed an independent posture. The U.S. invasion of Iraq was cast as a violation of international law and Argentina has moved closer to Latin American regimes not in the good graces Washington, i.e., Brazil, Venezuela, and Cuba.

ARGENTINA'S FUTURE

Renewed confidence in government and economic recovery auger well for Argentina. Experts predict a 4 percent annual growth from 2004 to 2006 while inflation should decline from double digits to about 7 percent.

Timeline: PAST

1536
Pedro de Mendoza establishes the first settlement at Buenos Aires

1816
Independence of Spain

1865–1870
War with Paraguay

1912
Electoral reform: Compulsory male suffrage

1946–1955 and **1973–1974**
Juan Perón is in power

1976–1982
The Dirty War

1980s
War with Great Britain over the Falkland Islands; military mutinies and economic chaos

1990s
Economic crises in Mexico, Russia, and Asia slow the economy

PRESENT

2000s
Argentina Struggles to climb out of recession

Néstor Kirchner elected president in May 2003

Bolivia (Republic of Bolivia)

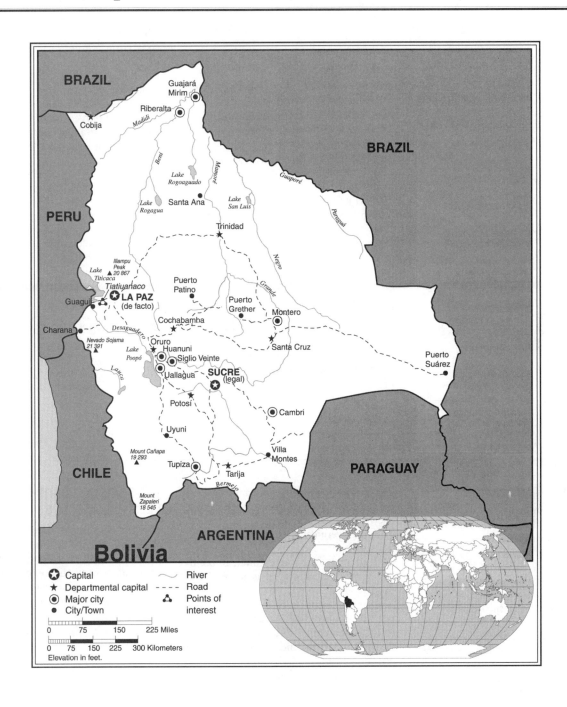

Map legend:
- ✪ Capital
- ★ Departmental capital
- ◉ Major city
- ● City/Town
- ～ River
- - - - Road
- ⚠ Points of interest

0   75   150   225 Miles
0   75   150   225   300 Kilometers
Elevation in feet.

Bolivia Statistics

GEOGRAPHY

Area in Square Miles (Kilometers): 424,162 (1,098,160) (about 3 times the size of Montana)

Capital (Population): La Paz (de facto) (740,000); Sucre (legal)

Environmental Concerns: deforestation; soil erosion; desertification; loss of biodiversity; water pollution

Geographical Features: rugged Andes Mountains with a highland plateau (Altiplano), hills, lowland plains of the Amazon Basin

Climate: varies with altitude; from humid and tropical to semiarid and cold

PEOPLE

Population

Total: 8,724,156

67

Annual Growth Rate: 1.56%

Rural/Urban Population Ratio: 39/61

Major Languages: Spanish; Quechua; Aymara

Ethnic Makeup: 30% Quechua; 25% Aymara; 30% Mestizo; 15% white

Religions: 95% Roman Catholic; Protestant

Health

Life Expectancy at Birth: 65 years (male); 67 years (female)

Infant Mortality Rate (Ratio): 60.4/1,000

Physicians Available (Ratio): 1/3,663

Education

Adult Literacy Rate: 87.2%

Compulsory (Ages): 6–14; free

COMMUNICATION

Telephones: 563,900 main lines

Daily Newspaper Circulation: 69 per 1,000 people

Televisions: 202 per 1,000

Internet Users: 270,000

TRANSPORTATION

Highways in Miles (Kilometers): 32,426 (52,216)

Railroads in Miles (Kilometers): 2,292 (3,691)

Usable Airfields: 1,109

Motor Vehicles in Use: 432,000

GOVERNMENT

Type: republic

Independence Date: August 6, 1825 (from Spain)

Head of State/Government: President Carlos Diego Mesa Gisbert is both head of state and head of government

Political Parties: Free Bolivia Movement; Revolutionary Front of the Left; Nationalist Revolutionary Movement; Christian Democratic Party; Nationalist Democratic Action; popular Patriotic Movement; Unity and Progress Movement; many others

Suffrage: universal and compulsory at 18 if married, at 21 if single

MILITARY

Military Expenditures (% of GDP): 1.6%

Current Disputes: dispute with Chile over water rights; seeks sovereign corridor to the South Pacific Ocean

ECONOMY

Currency ($ U.S. Equivalent): 7.66 bolivianos = $1

Per Capita Income/GDP: $2,400/$20.88 billion

GDP Growth Rate: 2.1%

Inflation Rate: 2.1%

Unemployment Rate: 7.6%; substantial underemployment

Labor Force: 2,500,000

Natural Resources: tin; natural gas; petroleum; zinc; tungsten; antimony; silver; iron; lead; gold; timber; hydropower

Agriculture: soybeans; coffee; coca; cotton; corn; sugarcane; rice; potatoes; timber

Industry: mining; smelting, petroleum; food and beverages; tobacco; handicrafts; clothing

Exports: $1.5 billion (primary partners United Kingdom, United States, Peru)

Imports: $1.6 billion (primary partners United States, Japan, Brazil)

SUGGESTED WEBSITE

http://www.cia.gov/cia/
publications/factbook/geos/
bl.html

Bolivia Country Report

BOLIVIA: AN INDIAN NATION

Until recently, the images of Bolivia captured by the world's press were uniformly negative. Human-rights abuses were rampant, a corrupt and brutal military government was deeply involved in cocaine trafficking, and the nation was approaching bankruptcy.

Other images might include Bolivia's complex society. So intermixed has this multiethnic culture become that one's race is defined by one's social status. So-called whites, who look very much like the Indians with whom their ancestors intermarried, form the upper classes only because of their economic, social, and cultural positions—that is, the degree to which they have embraced European culture.

DEVELOPMENT

Bolivia's economic development stalled as a result of indigenous demands with regard to basic rights.

Another enduring image fixed in the literature is Bolivia's political instability.

The actual number of governments over the past 200 years is about 80, however, and not the 200 commonly noted. Indeed, elected governments have been in power for the past two decades. What outsiders perceive as typical Latin American political behavior clouds what is unusual and positive about Bolivia.

One nineteenth-century leader, Manuel Belzu, played an extremely complex role that combined the forces of populism, nationalism, and revolution. Belzu encouraged the organization of the first trade unions, abolished slavery, promoted land reform, and praised Bolivia's Indian past.

In 1952, a middle-class–led and popularly supported revolution swept the country. The ensuing social, economic, and political reforms, while not erasing an essentially dual society of "whites" and Indians, did significantly ease the level of exploitation. Most of the export industries, including those involved with natural resources, were nationalized. Bolivia's evolution—at times progressive, at times regressive—continues to reflect the impulse for change.

THE SOCIETY: POSITIVE AND NEGATIVE ASPECTS

Bolivia, despite the rapid and startling changes that have occurred in the recent past, remains an extremely poor society. In terms of poverty, life expectancy, death rates, and per capita income, the country ranks among the worst in the Western Hemisphere.

Rights for women have made slow progress, even in urban areas. In 1975, a woman was appointed to the Bolivian Supreme Court; and in 1979, the Bolivian Congress elected Lidia Gueiler Tejada, leader of the lower house, as president. Long a supporter of women's rights, Tejada had drafted and pushed through Congress a bill that created a government ministry to provide social benefits for women and children. That remarkably advanced legislation has not guaranteed that women enjoy a social status equal to that of men, however. Furthermore, many women are likely unaware of their rights under the law.

Bolivia's press is reasonably free, although many journalists are reportedly paid by politicians, drug traffickers, and

(United Nations photo)

Bolivia has a complex society, tremendously affected by the continued interplay of multiethnic cultures. The influence of indigenous peoples on Bolivia remains strong.

officials to increase their exposure or suppress negative stories. A few journalists who experienced repression under previous governments still practice self-censorship.

URBANIZATION

Santa Cruz has been transformed in the last 50 years from an isolated backwater into a modern city with links to the other parts of the country and to the rest of South America. From a population of 42,000 in 1950, the number of inhabitants quickly rose to half a million in the mid-1980s and is now growing at the rate of about 8 percent a year. Bolivia's second largest city, its population now exceeds that of the de facto capital, La Paz.

Most of the city's population growth has been the result of rural-to-urban migration, a phenomenon closely studied by geographer Gil Green. On paper, Santa Cruz is a planned city, but, since the 1950s, there has been a running battle between city planners and new settlers wanting land. "Due to the very high demand for cheap land and the large amount of flat, empty, nonvaluable land surrounding it, the city has tended to expand by a process of land invasion and

squatting. Such invasions are generally overtly or covertly organized by political parties seeking electoral support of the low-income population." In the wake of a successful "invasion," the land is divided into plots that are allocated to the squatters, who then build houses from whatever materials are at hand. Then begins the lengthy process of settlement consolidation and regularization of land tenure. Once again the new land is subdivided and sold cheaply to the low-income population.

FREEDOM

A corrupt judicial system, overcrowded prisons, and violence and discrimination against women and indigenous peoples are perennial problems in Bolivia, despite protective legislation.

Perhaps the pace of urbanization as a result of internal migration is most pronounced in El Alto, which hardly existed on maps 30 years ago. It is now a "city" of 700,000 and overlooks La Paz. The rapid growth actually reflects a profound crisis in Bolivia. Tens of thousands of Aymara and Quechua-speaking miners and peasant

farmers have been driven to El Alto by their inability to make a living in rural areas. Over the past two years it has become, in the works of a local newspaper editor, "the capital of social protest in Bolivia." In fact, a rebellion centered in El Alto succeeded in driving President Gonzalo Sánchez from power in 2003 and threatens to do the same with his successor, Carlos Mesa.

The character of what has been termed the "Ideology of Fury" is complex and springs from a broad range of contexts. Perhaps most important, Bolivia's indigenous majority has suffered centuries of neglect and abuse. President Mesa noted as much when he explained the uprising as an "eruption of deeply held positions, over many centuries, that have been accumulating." There appear to have been two more immediate catalysts: U.S. insistence on the eradication of coca and the governments proposal to export natural gas to the U.S. through the construction of a pipeline to Chile.

With respect to coca cultivation, Bolivian politicians for years have promised to put an end to the trade and substitute other crops such as pineapples, coffee, black pepper, oregano, and passion fruit. Unfortunately most of the government's efforts were put on eradication and not alternative development. The United States, according to economist Jeffrey Sachs, "has constantly made demands on an impoverished country without any sense of reality or an economic framework and strategy to help them in development." The net result was the impoverishment of thousands of peasant farmers, who have since migrated to El Alto and become the taproot of the "Ideology of Fury". Indigenous uprisings first backed program of coca eradication. What is not appreciated by Washington and Bolivian politicians who are fearful of losing U.S. aid money, is that coca is central to indigenous culture. Certainly much is exported in the form of coca past or cocaine. In the 1990s it was calculated that illegal exports contributed the equivalent of 13 to 15 percent of Bolivia's gross domestic product and that coca by-products accounted for as much as 40 percent of total exports, both legal and illicit. Today, about 400,000 Bolivians are estimated to live off coca and cocaine production. U.S. wishes run afoul of the multifaceted heritage of coca, the sacred plant of the Incas. There is virtually no activity in domestic, social, or religious life in which coca does not play a role; thus, attempts to limit its cultivation have had profound repercussions among the peasantry.

HEALTH/WELFARE

Provisions against child labor in Bolivia are frequently ignored; many children may be found shining shoes, selling lottery tickets, and as street vendors.

Indigenous resistance to coca eradication now centers on Evo Morales, the head of the coca growers' federation, who finished second in the presidential elections of 2002. Morales' new part, the Movement Toward Socialism, now demands that Mesa's government modify the laws against coca cultivation, regardless of the wishes of the United States. "There has to be a change, to a policy that is truly Boliv-

ian, not one that is imposed by foreigners with the pretext that eradication will put an end to narcotics trafficking," said a member of congress and an ally of Morales.

ACHIEVEMENTS

The Bolivian author Armando Chirveches, in his political novel *La Candidatura de Rojas* (1909), produced one of the best examples of this genre in all of Latin America. The book captures the politics of the late nineteenth century extraordinarily well.

The second catalyst involved the proposed gas pipeline. While there are good historical reasons for Bolivian antipathy towards Chile (Chile deprived Bolivia its access to the Pacific as a result of territorial adjustments following the War of the Pacific in 1879–1880), resistance to the pipeline also ahs a social dimension. As reported in *The New York Times*, a Chilean pollster noted: "Part of the democratic process is assuring that people are going to get a piece of the cake, and that has been lacking in Bolivia. Bolivians are suspicious of whoever is making the deal because they think, "The elite always puts money in its own pockets, and we are left on the streets with nothing to eat." Regionalism also plays a role in the controversy. Gas-producing regions, those with large deposits of lithium, and even farmers in Santa Cruz who have experienced a boom in soybean exports, now demand a significant voice in the distribution of the wealth they produce as well as a degree of autonomy from La Paz.

The success of the indigenous majority in Bolivia in toppling one government and threatening another has both emboldened their leaders, who relish their new found power, and awakened a sense of racial pride among the Aymara and Quechua. As lone unemployed carpenter told reporter Larry Rohter: "They may still say that we are only Indians, but now we can see what is happening and what the Aymara nation can do when it is united."

Any action now has to take into account the demands of an awakened and angry indigenous population. It will be impossible for Bolivia to develop its extensive natural resources in the face of an indigenous and regional resistance that does not take into account their needs. If the government can link development to the creation of jobs, if it finally begins to deliver on years of unfulfilled promises with regard to health care and education, and if it can pursue a coca policy that respects the culture of the majority of the population and stems from Bolivian reality and not Washington's wishes, then perhaps a modus vivendi can be reached. If not the future will bring further economic malaise and political upheaval.

Timeline: PAST

1538
Spanish settle the altiplano (high plain)

1825
Bolivian declaration of independence of Spain

1879–1880
The War of the Pacific with Chile; Bolivia loses access to the sea

1932–1935
The Chaco War with Paraguay

1952–1964
Reforms: nationalization of mines, land reform, universal suffrage, creation of labor federation

1990s
Privatization of the economy accelerates; labor unrest grips the mining sector Bolivia's indigenous people achieve a new political voice

PRESENT

2000s
President Sánchez de Lozada forced to resign

Bolivia's indigenous majority demands economic, political and social reform

Brazil (Federative Republic of Brazil)

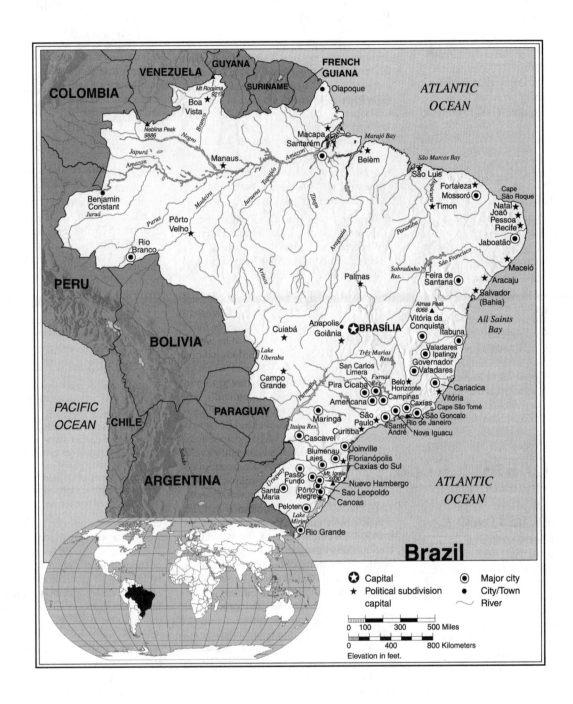

Brazil

- ✪ Capital
- ★ Political subdivision capital
- ◉ Major city
- ● City/Town
- ∿ River

```
0   100    300    500 Miles
0      400      800 Kilometers
Elevation in feet.
```

Brazil Statistics

GEOGRAPHY

Area in Square Miles (Kilometers):
3,285,670 (8,512,100) (slightly smaller than the United States)
Capital (Population): Brasília (1,738,000)

Environmental Concerns: deforestation; water and air pollution; land degradation

Geographical Features: mostly flat to rolling lowlands in the north; some plains, hills, mountains, and a narrow coastal belt

Climate: mostly tropical or semitropical; temperate zone in the south

71

PEOPLE*

Population

Total: 184,101,109

Annual Growth Rate: 1.11%

Rural/Urban Population Ratio: 21/79

Ethnic Makeup: 55% white; 38% mixed; 6% black; 1% others

Major Languages: Portuguese; Spanish; English; French

Religions: 70% nominal Roman Catholic; 30% others

Health

Life Expectancy at Birth: 67 years (male); 75 years (female)

Infant Mortality Rate (Ratio): 30.66/1,000

Physicians Available (Ratio): 1/681

Education

Adult Literacy Rate: 86.4%

Compulsory (Ages): 7–14; free

COMMUNICATION

Telephones: 38.81 million main lines

Daily Newspaper Circulation: 47 per 1,000 people

Televisions: 193 per 1,000 people

Internet Users: 14.3 million

TRANSPORTATION

Highways in Miles (Kilometers): 1,229,580 (1,980,000)

Railroads in Miles (Kilometers): 16,702 (26,895)

Usable Airfields: 3,277

Motor Vehicles in Use: 18,030,000

GOVERNMENT

Type: federal republic

Independence Date: September 7, 1822 (from Portugal)

Head of State/Government: President Luiz Inacio "Lula" Da Silva is both head of state and head of government

Political Parties: Brazilian Democratic Movement Party; Liberal Front Party; Workers' Party; Brazilian Workers' Party; Democratic Labor Party; Popular Socialist Party; Liberal Party; others

Suffrage: voluntary at 16; compulsory between 19 and 70; voluntary over 70

MILITARY

Military Expenditures (% of GDP): 2.1%

Current Disputes: boundary disputes with Uruguay

ECONOMY

Currency ($ U.S. Equivalent): 3.08 reals = $1

Per Capita Income/GDP: $7,600/$1.38 trillion

GDP Growth Rate: 0.1%

Inflation Rate: 9.3%

Unemployment Rate: 12.2%

Labor Force: 74,000,000

Natural Resources: bauxite; gold; iron ore; manganese; nickel; phosphates; platinum; tin; uranium; petroleum; hydropower; timber

Agriculture: coffee; rice; corn; sugarcane; soybeans; cotton; manioc; oranges

Industry: textiles; shoes; chemicals; cement; lumber; iron ore; tin; steel; aircraft; motor vehicles and parts; other machinery and equipment

Exports: $73.3 billion (primary partners United States, Argentina, Germany)

Imports: $48.7 billion (primary partners United States, Argentina, Germany)

SUGGESTED WEBSITE

http://www.cia.gov/cia/
publications/factbook/geos/
br.html

---

*Note: Estimates explicitly take into account the effects of excess mortality due to AIDS.

Brazil Country Report

BRAZIL: A TROUBLED GIANT

In 1977, Brazilian president Ernesto Geisel stated that progress was based on "an integrated process of political, social, and economic development." Democracy, he argued, was the first necessity in the political arena. But democracy could only be achieved "if we also further social development …, if we raise the standard of living of Brazilians." The standard of living, he continued, "can only be raised through economic development."

It was clear from his remarks that the three broad objectives of democratization, social progress, and economic development were interconnected. He could not conceive of democracy in a poor country or in a country where there were "gaps, defects, and inadequacies in the social realm."

CONCEPTS OF PROGRESS

Geisel's comments offer a framework within which to consider not only the cur-

rent situation in Brazil but also historical trends that reach back to the late nineteenth century—and, in some instances, to Portugal. Historically, most Brazilians have believed that progress would take place within the context of a strong, authoritarian state. In the nineteenth century, for example, a reform-minded elite adapted European theories of modernization that called for government-sponsored changes. The masses would receive benefits from the state; in this way, the elite reasoned, pressure for change from the poorer sectors of society would be eliminated. There would be progress with order. *Ordem e Progresso* ("Order and Progress") is the motto that graces the Brazilian flag; the motto is as appropriate today as it was in 1889, when the flag first flew over the new republic.

The tension among modernization, social equity, and order and liberty was first obvious in the early 1920s, when politically isolated middle-class groups united with junior military officers (*tenentes*) to

challenge an entrenched ruling class of coffee-plantation owners. By the mid-1920s, the tenentes, bent on far-reaching reforms, conceived a new role for themselves. With a faith that bordered at times on the mystical and a philosophy that embraced change in the vaguest of terms, they felt that only the military could shake Brazil from its lethargy and force it to modernize. Their program demanded the ouster of conservative, tradition-minded politicians; an economic transformation of the nation; and, eventually, a return to strong, centralized constitutional rule. The tenentes also proposed labor reforms that included official recognition of trade unions, a minimum wage and maximum work week, restraints on child labor, land reform, nationalization of natural resources, and a radical expansion of educational facilities. Although the tenentes were frustrated in their attempts to mold policy, many of their reforms were taken up by Getulio Vargas, who seized power in 1930 and imposed a strong, authoritarian state on Brazil.

(United Nations photo/Jef Foxx)

Certain areas of Brazil attract enormous numbers of visitors from all over the world. This beach in Rio de Janeiro has one of the most famous skylines in South America.

THE 1964 REVOLUTION

In some respects, the goals of the tenentes were echoed in 1964, when a broad coalition of civilians—frustrated by an economy that seemed to be disintegrating, concerned with the "leftist" slant of the government of João Goulart, and worried about a social revolution that might well challenge the status and prestige of the wealthy and the middle classes—called on the military to impose order on the country.

The military leaders did not see their intervention as just another coup but, rather, as a revolution. They foresaw change but believed that it would be dictated from above. Government was highly centralized, the traditional parties were virtually frozen out of the political process, and the military and police ruthlessly purged Brazil of elements considered "leftist" or "subversive." (The terms were used interchangeably.) Order and authority triumphed over liberty and freedom. The press was muzzled, and human-rights abuses were rampant.

Brazil's economic recovery eventually began to receive attention. The military gave economic growth and national security priority over social programs and political liberalization. Until the effects of the oil crisis generated by the Organization of Petroleum Exporting Countries (OPEC) in 1973 began to be felt, the recovery of the Brazilian economy was dubbed a "miracle," with growth rates averaging 10 percent a year.

The benefits of that growth went primarily to the upper and middle classes, who enjoyed the development of industries based largely on consumer goods. Moreover, Brazil's industrialization was flawed. It was heavily dependent on foreign investment, foreign technology, and foreign markets. It required large investments in machinery and equipment but needed little labor, and it damaged the environment through pollution of the rivers and air around industrial centers. Agriculture was neglected to the point that even basic foodstuffs had to be imported.

THE IMPACT OF RURAL–URBAN MIGRATION

The stress on industrialization tremendously increased rural-to-urban migration and complicated the government's ability to keep up with the expanded need for public health and social services. In 1970, nearly 56 percent of the population were concentrated in urban areas; by the late 1990s, 79 percent of the population were so classified. These figures also illustrate the inadequacies of an agrarian program based essentially on a "moving frontier."

73

(United Nations Photo)

By the late 1980s, agrarian reforms that were designed to establish peasants in plots of workable land had caused the depletion of Brazilian jungle and, as space and opportunities diminished, there was a large movement of these people to the cities. The profound urban crowding in Brazil is illustrated by this photo of a section of Rio de Janeiro.

Peasants evicted from their plots have run out of new lands to exploit, unless they move to the inhospitable Amazon region. As a result, many have been attracted by the cities.

The pressure of the poor on the cities, severe shortages of staple foods, and growing tension in rural areas over access to the land forced the government to act. In 1985, the civilian government of José Sarney announced an agrarian-reform plan to distrib-

ute millions of acres of unused private land to peasants. Implementation of the reform was not easy, and confrontations between peasants and landowners occurred.

MILITARY RULE IS CHALLENGED

Nineteen seventy-four was a crucial year for the military government of Brazil. The virtual elimination of the urban-guerrilla threat challenged the argument that demo-

cratic institutions could not be restored because of national security concerns.

Pressure grew from other quarters as well. Many middle- and upper-class Brazilians were frightened by the huge state-controlled sector in the economy that had been carved out by the generals. The military's determination to promote the rapid development of the nation's resources, to control all industries deemed vital to the nation's security, and to compete with multinational corporations concerned Brazilian businesspeople, who saw their role in the economy decreasing.

Challenges to the military regime also came from the Roman Catholic Church, which attacked the government for its brutal violations of human rights and constantly called for economic and social justice. One Brazilian bishop publicly called the government "sinful" and in "opposition to the plans of God" and noted that it was the Church's moral and religious obligation to fight it. After 1974, as Brazil's economic difficulties mounted, the chorus of complaints grew insistent.

DEVELOPMENT

Since 2000 trade with China has tripled. China is now Brazil's fourth largest market. The bulk of the trade consists of soya beans, iron ore, and steel.

THE RETURN OF DEMOCRACY

The relaxation of political repression was heralded by two laws passed in 1979. The Amnesty Bill allowed for the return of hundreds of political exiles; the Party Reform Bill in essence reconstructed Brazilian politics. Under the provisions of the Party Reform Bill, new political parties could be established—provided they were represented in nine states and in 20 percent of the counties of those states. The new parties were granted the freedom to formulate political platforms, as long as they were not ideological and did not favor any single economic class. The Communist Party was outlawed, and the creation of a workers' party was expressly forbidden. (Communist parties were legalized again in 1985.)

The law against the establishment of a workers' party reflected the regime's concern that labor, increasingly anxious about the state of the economy, might withdraw its traditional support for the state. Organized labor had willingly cooperated with the state since the populist regime of Getulio Vargas (1937–1945). For Brazilian workers in the 1930s, the state was their "patron," the source of benefits. This dependence on the government, deeply rooted in Portuguese political culture, replaced the formation of a more independent labor movement and minimized industrial conflict. The state played the role of mediator between workers and management. President Vargas led the workers to believe that the state was the best protector of their interests. (Polls have indicated that Brazilian workers still cling to that belief.)

If workers expect benefits from the state, however, the state must then honor those expectations and allocate sufficient resources to assure labor's loyalty. A deep economic crisis, such as the one that occurred in the early 1960s and again in the early 1990s, endangers the state's control of labor. In 1964, organized labor supported the coup, because workers felt that the civilian regime had failed to perform its protective function. This phenomenon also reveals the extremely shallow soil in which Brazilian democracy has taken root.

Organized labor tends not to measure Brazilian governments in political terms, but within the context of the state's ability to address labor's needs. For the rank-and-file worker, it is a question not of democracy or military authoritarianism, but of bread and butter. President Sarney, in an effort to keep labor loyal to the government, sought the support of union leaders for a proposal to create a national pact with businesspeople, workers, and his government. But pervasive corruption, inefficient government, and a continuing economic crisis eventually eroded the legitimacy of the elites and favored nontraditional parties in the 1989 election. The candidacy of Luís Inácio da Silva, popularly known as Lula and leader of the Workers' Party, "was stunning evidence of the Brazilian electorate's dissatisfaction with the conduct of the country's transition to democracy and with the political class in general." He lost the election by a very narrow margin. In 2002, he won the election and promised to "end hunger."

Workers continue to regard the state as the source of benefits, as do other Brazilians. Many social reformers, upset with the generals for their neglect of social welfare, believe that social reform should be dispensed from above by a strong and paternalistic state. Change is possible, even welcome—but it must be the result of compromise and conciliation, not confrontation or non-negotiable demands.

THE NEW CONSTITUTION

The *abertura* (political liberalization) of Brazil climaxed in January 1985 with the election of President Sarney, a civilian, following 21 years of military rule. Importantly, the Brazilian military promised to respect the Constitution and promised a policy of nonintervention in the political process. In 1987, however, with the draft of a new constitution under discussion, the military strongly protested language that removed its responsibility for internal law and order and restricted the military's role to that of defense of the nation against external threats. According to *Latin American Regional Reports: Brazil*, the military characterized the draft constitution as "confused, inappropriate, at best a parody

of a constitution, just as Frankenstein was a gross and deformed imitation of a human being."

Military posturing aside, the new Constitution went into effect in October 1988. It reflects the input of a wide range of interests: The Constituent Assembly—which also served as Brazil's Congress—heard testimony and suggestions from Amazonian Indians, peasants, and urban poor as well as from rich landowners and the military. The 1988 Constitution is a document that captures the byzantine character of Brazilian politics and influence peddling and reveals compromises made by conservative and liberal vested interests.

The military's fears about its role in internal security were removed when the Constituent Assembly voted constitutional provisions to grant the right of the military independently to ensure law and order, a responsibility it historically has claimed. But Congress also arrogated to itself the responsibility for appropriating federal monies. This is important, because it gives Congress a powerful check on both the military and the executive office.

FREEDOM

Violence against street children, indigenous peoples, homosexuals, and common criminals at the hands of the police, landowners, vigilante groups, gangs, and hired thugs is commonplace. Homicide committed by police is the third-leading cause of death among children and adolescents. Investigation of such crimes is lax and prosecution of the perpetrators sporadic. Indians continue to clash with miners and landowners.

Nationalists won several key victories. The Constituent Assembly created the concept of "a Brazilian company of national capital" that can prevent foreigners from engaging in mining, oil-exploration risk contracts, and biotechnology. Brazilian-controlled companies were also given preference in the supply of goods and services to local, state, and national governments. Legislation reaffirmed and strengthened the principle of government intervention in the economy should national security or the collective interest be at issue.

Conservative congressional representatives were able to prevail in matters of land reform. They defeated a proposal that would have allowed the compulsory appropriation of property for land reform. Although a clause that addressed the "social function" of land was included in the Constitution, it was clear that powerful landowners and agricultural interests had triumphed over Brazil's landless peasantry.

In other areas, however, the Constitution is remarkably progressive on social and economic issues. The work week was reduced to a maximum of 44 hours, profit sharing was established for all employees, time-and-a-half was promised for overtime work, and paid vacations were to include a bonus of 30 percent of one's monthly salary. Day-care facilities were to be established for all children under age six, maternity leave of four months and paternity leave of five days were envisaged, and workers were protected against "arbitrary dismissal." The Constitution also introduced a series of innovations that would increase significantly the ability of Brazilians to claim their guaranteed rights before the nation's courts and ensure the protection of human rights, particularly the rights of Indians and peasants involved in land disputes.

Despite the ratification of the 1988 Constitution, a functioning Congress, and an independent judiciary, the focus of power in Brazil is still the president. A legislative majority in the hands of the opposition in no way erodes the executive's ability to govern as he or she chooses. Any measure introduced by the president automatically becomes law after 40 days, even without congressional action. Foreign observers perceive "weaknesses" in the new parties, which in actuality are but further examples of well-established political practices. The parties are based on personalities rather than issues, platforms are vague, goals are so broad that they are almost illusions, and party organization conforms to traditional alliances and the "rules" of patronage. Democratic *forms* are in place in Brazil; the *substance* remains to be realized.

The election of President Fernando Collor de Mello, who assumed office in March 1990, proves the point. As political scientist Margaret Keck explains, Collor fit well into a "traditional conception of elite politics, characterized by fluid party identifications, the predominance of personal relations, a distrust of political institutions, and reliance on charismatic and populist appeals to *o povo*, the people." Unfortunately, such a system is open to abuse; revelations of widespread corruption that reached all the way to the presidency brought down Collor's government in 1992 and gave Brazilian democracy its most difficult challenge to date. The scandal brought to light a range of strengths and weaknesses that presents insights into the Brazilian political system.

THE PRESS AND THE PRESIDENCY

Brazil's press was severely censored and harassed from the time of the military coup of 1964 until 1982. Not until passage of the Constitution of 1988 was the right of free speech and a free press guaranteed. It was the press, and in particular the news magazine *Veja*, that opened the door to President Collor's impeachment. In the words of *World Press Review*, "Despite government pressure to ease off, the magazine continued to uncover the president's malfeasance, tugging hard at the threads of Collor's unraveling administration. As others in the media followed suit, Congress was forced to begin an investigation and, in the end, indict Collor." The importance of the event to Brazil's press, according to *Veja* editor Mario Sergio Conti, is that "It will emerge with fewer illusions about power and be more rigorous. Reporting has been elevated to a higher plane...."

While the failure of Brazil's first directly elected president in 29 years was tragic, it should not be interpreted as the demise of Brazilian democracy. Importantly, according to Brazilian journalist Carlos Eduardo Lins da Silva, writing in *Current History*, many "Brazilians and outside observers saw the workings of the impeachment process as a sign of the renewed strength of democratic values in Brazilian society. They were also seen as a healthy indicator of growing intolerance to corruption in public officials."

The military, despite persistent rumors of a possible coup, has to date allowed the constitutional process to dictate events. For the first time, most civilians do not see the generals as part of the solution to political shortcomings. And Congress, to its credit, has chosen to act responsibly and not be "bought off" by the executive office.

THE RIGHTS OF WOMEN AND CHILDREN

Major changes in Brazilian households have occurred over the last decade as the number of women in the workforce has dramatically increased. In 1990, just over 35 percent of women were in the workforce, and the number was expected to grow. As a result, many women are limiting the size of their families. More than 20 percent use birth-control pills, and Brazil is second only to China in the percentage of women who have been sterilized. The traditional family of 5.0 or more children has shrunk to an average of 3.4. With two wage earners, the standard of living has risen slightly for some families. Many homes now have electricity and running water. Television sales increased by more than 1,000 percent in the last decade.

In relatively affluent, economically and politically dynamic urban areas, women are more evident in the professions, educa-

HEALTH/WELFARE

The quality of education in Brazil varies greatly from state to state, in part because there is no system of national priorities. The uneven character of education has been a major factor in the maintenance of a society that is profoundly unequal. The provision of basic health needs remains poor, and land reform is a perennial issue.

tion, industry, the arts, media, and political life. In rural areas, however, especially in the northeast, traditional cultural attitudes, which call upon women to be submissive, are still well entrenched.

Women are routinely subjected to physical abuse in Brazil. Americas Watch, an international human-rights group, reports that more than 70 percent of assault, rape, and murder cases take place in the home and that many incidents are unreported. Even though Brazil's Supreme Court struck down the outmoded concept of a man's "defense of honor," local courts routinely acquit men who kill unfaithful wives. Brazil, for all intents and purposes, is still a patriarchy.

Children are also in many cases denied basic rights. According to official statistics, almost 18 percent of children between the ages of 10 and 14 are in the labor force, and they often work in unhealthy or dangerous environments. Violence against urban street children has reached frightening proportions. Between January and June 1992, 167 minors were killed in Rio de Janeiro; 306 were murdered in São Paulo over the first seven months of the year. In July 1993, the massacre in a single night of seven street children in Rio de Janeiro resulted, for a time, in cries for an investigation of the matter. In February 1997, however, five children were murdered on the streets of Rio.

THE STATUS OF BLACKS

Scholars continue to debate the actual status of blacks in Brazil. Not long ago, an elected black member of Brazil's federal Congress blasted Brazilians for their racism. However, argues historian Bradford Burns, Brazil probably has less racial tension and prejudice than other multiracial societies.

A more formidable barrier, Burns says, may well be class. "Class membership depends on a wide variety of factors and their combination: income, family history and/or connections, education, social behavior, tastes in housing, food and dress, as well as appearance, personality and talent." But, he notes, "The upper class traditionally has been and still remains mainly white, the

The status of blacks in Brazil is considered better than most other multiracial societies. The class structure is determined by a number of factors: income, family history, education, social behavior, cultural tastes, and talent. Still, the upper class remains mainly white, and the lower class principally of color.

lower class principally colored." Upward mobility exists and barriers can be breached. But if such advancement depends upon a symbolic "whitening out," does not racism still exist?

This point is underscored by the 1988 celebration of the centennial of the abolition of slavery in Brazil. In sharp contrast to the government and Church emphasis on racial harmony and equality were the public protests by militant black groups claiming that Brazil's much-heralded "racial democracy" was a myth. In 1990, blacks earned 40 percent less than whites in the same professions.

THE INDIAN QUESTION

Brazil's estimated 200,000 Indians have suffered greatly in recent decades from the gradual encroachment of migrants from the heavily populated coastal regions and from government efforts to open the Amazon region to economic development. Highways have penetrated Indian lands, diseases for which the Indians have little or no immunity have killed thousands, and additional thousands have experienced a profound culture shock. Government efforts to protect the Indians have been largely ineffectual.

The two poles in the debate over the Indians are captured in the following excerpts from *Latin American Regional Reports: Brazil.* A Brazilian Army officer observed that the "United States solved the problem with its army. They killed a lot of Indians. Today everything is quiet there, and the country is respected throughout the world." And in the words of a Kaingang Indian woman: "Today my people see their lands invaded, their for-

ests destroyed, their animals exterminated and their hearts lacerated by this brutal weapon that is civilization."

Sadly, the assault against Brazil's Indian peoples has accelerated, and disputes over land have become more violent. One case speaks for itself. In the aftermath of a shooting incident in which several Yanomamö Indians were killed by prospectors, the Brazilian federal government declared that all outsiders would be removed from Yanomamö lands, ostensibly to protect the Indians. Those expelled by the government included anthropologists, missionaries, doctors, and nurses. A large number of prospectors remained behind. By the end of 1988, while medical personnel had not been allowed back in, the number of prospectors had swelled to 50,000 in an area peopled by 9,000 Yanomamö. The Indians have been devastated by diseases, particularly malaria, and by mercury poisoning as a result of prospecting activities upriver from Yanomamö settlements. In 1991, cholera began to spread among indigenous Amazon peoples, due to medical waste dumped into rivers in cholera-ridden Peru and Ecuador.

The new Constitution devotes an entire chapter to the rights of Indians. For the first time in the country's history, Indians have the authority to bring suits in court to defend their rights and interests. In all such cases, they will be assisted by a public prosecutor. Even though the government established a large protected zone for Brazil's Yanomamö Indians in 1991, reports of confrontations between Indians and prospectors have persisted. There are also Brazilian nationalists who insist that a 150-mile-wide strip along the border with Venezuela be excluded from the reserve as a matter of national security. The Yanomamö cultural area extends well into Venezuela; such a security zone would bisect Yanomamö lands.

THE BURNING OF BRAZIL

Closely related to the destruction of Brazil's Indians is the destruction of the tropical rain forests. The burning of the forests by peasants clearing land in the traditional slash-and-burn method, or by developers and landowners constructing dams or converting forest to pasture, has become a source of worldwide concern and controversy.

Ecologists are horrified by the mass extinction of species of plants, animals, and insects, most of which have not even been catalogued. The massive annual burning (equivalent in one recent year to the size of Kansas) also fuels the debate on the greenhouse effect and global warming. The problem of the burning of Brazil is indeed global, because we are all linked to the tropics by climate and the migratory patterns of birds and animals.

ACHIEVEMENTS

Brazil's cultural contributions to the world are many. Authors such as Joaquim Maria Machado de Assis, Jorge Amado, and Graciliano Ramos are evidence of Brazil's high rank in terms of important literary works. Brazilian music has won millions of devotees throughout the world, and Brazil's *Cinema Novo* (New Cinema) has won many awards.

World condemnation of the destruction of the Amazon basin has produced a strong xenophobic reaction in Brazil. Foreign Ministry Secretary-General Paulo Tarso Flecha de Lima informed a 24-nation conference on the protection of the environment that the "international community cannot try to strangle the development of Brazil in the name of false ecological theories." He further noted that foreign criticism of his government in this regard was "arrogant, presumptuous and aggressive." The Brazilian military, according to Latin American Regional Reports: Brazil, has adopted a high-profile posture on the issue. The military sees the Amazon as "a kind of strategic reserve vital to national security interests." Any talk of transforming the rain forests into an international nature reserve is rejected out of hand.

Over the next decade, however, Brazilian and foreign investors will create a 2.5 million-acre "green belt" in an already devastated area of the Amazon rain forest. Fifty million seedlings have been planted in a combination of natural and commercial zones. It is hoped that responsible forestry will generate jobs to maintain and study the native forest and to log the commercial zones. Steady employment would help to stem the flow of migrants to cities and to untouched portions of the rain forest. On the other hand, to compound the problem, landless peasants in 16 of Brazil's states launched violent protests in May 2000 to pressure the government to provide land for 100,000 families, as well as to grant millions of dollars in credits for poor rural workers.

FOREIGN POLICY

If Brazil's Indian and environmental policies leave much to be desired, its foreign policy has won it respect throughout much of Latin America and the developing world. Cuba, Central America, Angola, and Mozambique seemed far less threatening during the cold war to the Brazilian government than they did to Washington. Brazil is more concerned about its energy needs, capital requirements, and trade opportunities. Its foreign policy, in short, is one of pragmatism.

ECONOMIC POLICY

In mid-1993, Finance Minister Fernando Henrique Cardoso announced a plan to restore life to an economy in shambles. The so-called Real Plan, which pegged the new Brazilian currency (the real) to the dollar, brought an end to hyperinflation and won Cardoso enough popularity to carry him to the presidency. Inflation, which had raged at a rate of 45 percent per month in July 1994, was only 2 percent per month in February 1995. His two-to-one victory in elections in October 1994 was the most one-sided win since 1945.

President Cardoso transformed the economy through carefully conceived and brilliantly executed constitutional reforms. A renovated tax system, an overhauled social-security program, and extensive privatization of state-owned enterprises were supported by a new generation of legislators pledged to support broad-based reform.

Timeline: PAST

1500
Pedro Alvares Cabral discovers and claims Brazil for Portugal

1822
Declaration of Brazil's independence

1888
The Golden Law abolishes slavery

1889
The republic is proclaimed

1944
The Brazilian Expeditionary Force participates in the Italian campaign

1964
The military seizes power

1980s–1990s
Economic, social, and ecological crises

1990s
President Fernando Collor de Mello is convicted; the Asian financial crisis plunges Brazil into deep recession

PRESENT

2000s
Brazil wins praise for its handling of its HIV/AIDS problem

Luis Inacio da Silva, "Lula", elected president in 2002. Brazil pursues independent foreign and economic policies.

But, as was the case in much of Latin America in 1995, Mexico's financial crisis spread quickly to affect Brazil's economy, in large measure because foreign investors were unable to distinguish between Mexico and other Latin American nations. A similar problem occurred in 1998 with the collapse of Asian financial markets. Again, foreign investors shied away from Brazil's economy, and President Cardoso was forced to back away from a promise not to devalue the real. With devaluation in 1999 and signs of recovery in Asianmarkets, Brazil's economic prospects brightened considerably. Exports rose, and Brazil was able to finance its foreign debtthrough bond issues. In 2000 and 2001, however, the economy slowed, and concerns were expressed about energy supplies and costs, and the default of Brazil's major trading partner, Argentina, on its foreign debt. Economic uncertainty emboldened Congress to initiate a probe against corruption in government. Life for average Brazilians remained difficult. Cardoso's loss of popularity opened the door to the political opposition who were able to capitalize on presidential elections in 2002, when Luis Inacio da Silva, or "Lula" as he is popularly known, won a resounding triumph at the polls.

Lula, who worried many foreign observers because of his "leftist" ideology has begun to tackle Brazil's myriad problems in a pragmatic fashion. Labor unions, who supported his presidency and expected all of the benefits of political patronage, have been somewhat disillusioned. Lula, in attempt to bring the nation's spending under control, significantly culled the public work force. With regard to the economy, his policies have not been "leftist" but have more closely adhered to classical economic approaches. This has calmed the fears of foreign investors.

Cardoso's laudable economic reforms did not succeed in transforming the quality of Brazilian democracy.

The lament of Brazilian journalist Lins da Silva is still accurate: "Brazilian elites have once again shown how capable they are of solving political crises in a creative and peaceful manner but also how unwilling to promote change in inequitable social structures." The wealth of the nation still remains in the hands of a few, and the educational system has failed to absorb and train as many citizens as it should. Police continue routinely to abuse their power. Lula, who's own family roots lie in the favelas, is deeply sensitive to the needs of Brazil's poor and disadvantaged. He has made a point of visiting the slums, of listening to the complaints and needs of people, of behaving, in short, like the classic "patron".

On a positive note, Brazil's progress in the struggle against AIDS, a disease that contributed to the deaths of 9,600 people in Brazil in 1996, is among the best in the world. In simple terms, the government uses language in the Paris Convention of 1883 to produce low-cost generic drugs similar to costlier medications manufactured abroad. Everyone in Brazil infected with the HIV virus is provided with a "cocktail" of drugs, and with training in how to take them effectively. More than 100,000 Brazilians are on the drug regimen, at an annual cost of $163 million. In 2000, AIDS–related deaths declined to 1,200, and the rate of transmission was sharply reduced.

At a broader level, Brazil has prospered from its membership in Mercosur, a regional trade organization that consists of Argentina, Brazil, Paraguay, and Uruguay.

The success of Mercosur has expanded relations with other countries, especially Chile, which became an "associate" member in 1997. Lula, like Cardoso before him, is opposed to Washington's efforts to forge a Free Trade Area of the Americas (FTAA), in part because Mercosur and Brazil consider Europe a more important market and do not send a high percentage of their exports to the United States. Brazil has kept the pressure on other South American governments to convince them to join with Mercosur, not only in a "South American Free Trade Agreement", but in closer ties with the European Union. This independent policy has provided Brazil with leverage in the era of globalization.

With regard to foreign policy, Brazil has made common cause with other populist or independent-countries in South America, much to the consternation of the U.S. Lula has a warm relationship with Venezuela's Hugo Chávez and Cuba's Fidel Castro. He has strongly attacked the United States' invasion of Iraq. His foreign policy, both in terms of its economic and political contexts, has another dimension. Standing up to the United States plays well at home and may be used to balance domestic policies that fall short of the radical solutions many of his supporters expected.

Chile (Republic of Chile)

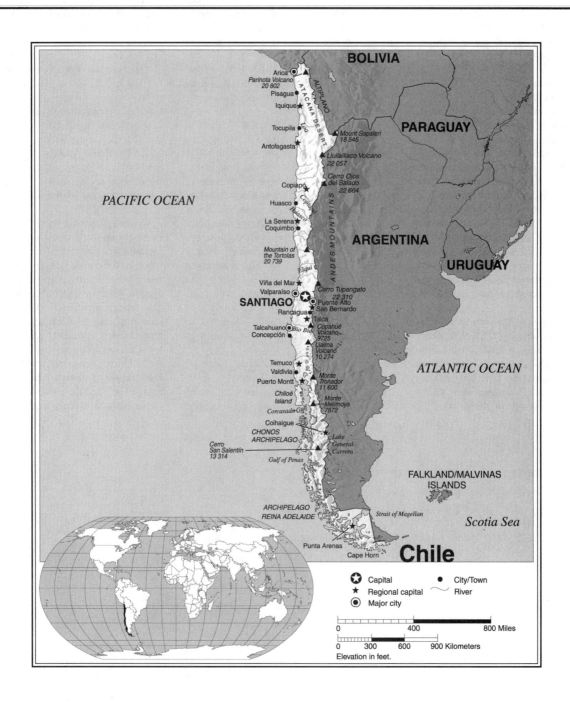

Chile Statistics

GEOGRAPHY

Area in Square Miles (Kilometers):
292,280 (756,945) (about twice the size of Montana)

Capital (Population): Santiago (4,642,000)

Environmental Concerns: air and water pollution; deforestation; loss of biodiversity; soil erosion; desertification

Geographical Features: low coastal mountains; a fertile central valley; rugged Andes Mountains in the east

Climate: temperate; desert in the north; Mediterranean in the center; cool and damp in the south

PEOPLE

Population

Total: 15,823,957

Annual Growth Rate: 1.01%

Rural/Urban Population Ratio: 16/84

Major Language: Spanish

80

Ethnic Makeup: 95% European and Mestizo; 3% Indian; 2% others
Religions: 89% Roman Catholic; 11% Protestant

Health

Life Expectancy at Birth: 72 years (male); 79 years (female)
Infant Mortality Rate (Ratio): 9.6/1,000
Physicians Available (Ratio): 1/875

Education

Adult Literacy Rate: 96.2%
Compulsory (Ages): for 8 years; free

COMMUNICATION

Telephones: 3.5 million main lines
Daily Newspaper Circulation: 101 per 1,000 people
Televisions: 280 per 1,000 people
Internet Users: 3.6 million

TRANSPORTATION

Highways in Miles (Kilometers): 49,556 (79,800)
Railroads in Miles (Kilometers): 4,212 (6,782)

Usable Airfields: 370
Motor Vehicles in Use: 1,375,000

GOVERNMENT

Type: republic
Independence Date: September 18, 1810 (from Spain)
Head of State/Government: President Ricardo Lagos Escobar is both head of state and head of government
Political Parties: Christian Democratic Party; Party for Democracy; Socialist Party; National Renewal; Independent Democratic Union; others
Suffrage: universal and compulsory at 18

MILITARY

Military Expenditures (% of GDP): 4.1%
Current Disputes: boundary or territorial disputes with Argentina, and Bolivia; territorial claim in Antarctica

ECONOMY

Currency ($ U.S. Equivalent): 691.43 pesos = $1

Per Capita Income/GDP: $9,900/$154.6 billion
GDP Growth Rate: 3.2%
Inflation Rate: 1.1%
Unemployment Rate: 8.5%
Labor Force: 5,800,000
Natural Resources: copper; timber; iron ore; nitrates; precious metals; molybdenum; fish; hydropower
Agriculture: wheat; corn; grapes; beans; sugar beets; potatoes; fruit; beef; poultry; wool; timber; fish
Industry: copper and other minerals; foodstuffs; fish processing; iron and steel; wood and wood products; transport equipment; cement; textiles
Exports: $20.44 billion (primary partners, European Union, United States, Japan)
Imports: $17 billion (primary partners United States, European Union, Argentina)

SUGGESTED WEBSITE

http://www.cia.gov/cia/
  publications/factbook/
  geos.ci.html

Chile Country Report

CHILE: A NATION ON THE REBOUND

In September 1973, the Chilean military, with the secret support of the U.S. Central Intelligence Agency (CIA), seized power from the constitutionally elected government of President Salvador Allende. Chile, with its long-standing traditions of free and honest elections, respect for human rights, and freedom of the press, was quickly transformed into a brutal dictatorship that arrested, tortured, and killed thousands of its own citizens. In the larger sweep of Chilean history, however, the coup seemed to be the most recent and severe manifestation of a lengthy conflict between social justice, on the one hand, and the requirements of order dictated by the nation's ruling elite, on the other. This was true in the colonial period, when there was conflict between the Roman Catholic Church and landowners over Indian rights. It was also apparent in later confrontations among Marxists, reformers, and conservatives.

FORM, NOT SUBSTANCE

Form, as opposed to substance, had characterized the rule of the Christian Democrats in the 1960s, when they created many separate rural unions, supposedly to address

the needs of *campesinos* ("peasants"). A divided union movement in effect became a form of government control that prevented the emergence of a single powerful rural organization.

In the early 1970s, President Allende—despite his talk of socialism and his genuine attempt to destroy the institutions and values of an old social order—used as his weapon of transformation, a centralized bureaucracy that would have been recognized by sixteenth-century viceroys and nineteenth-century presidents. Allende's attempts to institute far-reaching social change led to a strong reaction from powerful sectors of Chilean society who felt threatened.

THE 1973 COUP D'ETAT

When the military ousted Allende, it had the support of many Chileans, including the majority of the middle class, who had been hurt by the government's economic policies, troubled by continuous political turmoil, and infuriated by official mismanagement. The military, led by General Augusto Pinochet, began a new experiment with another form of centrist rule: military authoritarianism. The generals made it clear that they had not restored order merely to return it to the "discredited" constitutional practices of the past. They spoke of regeneration, of a new Chile, and of an end to the immorality, corruption, and incompetence of all civilian politics. The military announced in 1974 that, "guided by the inspiration of [Diego] Portales"—one of nineteenth-century Chile's greatest civilian leaders—"the government of Chile will energetically apply the principle of authority and drastically punish any outburst of disorder and anarchy."

The political, economic, and social reforms proposed by the military aimed at restructuring Chile to such an extent that there would no longer be a need for traditional political parties. Economic policy favored free and open competition as the main regulator

(Reuters/Bettmann)

On October 5, 1988, the voters of Chile denied the brutal dictator General Augusto Pinochet an additional eight-year term as president. To his credit, his military regime accepted defeat peacefully.

of economic and social life. The Chilean state rid itself of hundreds of state-owned corporations, struck down tariffs designed to protect Chilean industry from foreign competition, and opened the economy to widespread foreign investment. The changes struck deeply at the structure of the Chilean economy and produced a temporary but sharp recession, high unemployment, and hundreds of bankruptcies. A steep decline in the standard of living for most Chileans was the result of the government's anti-inflation policy.

FREEDOM

With the restoration of democracy in 1990 successive Chilean presidents have worked hard to resolve human rights issues stemming from the Pinochet dictatorship.

Social-welfare programs were reduced to a minimum. The private sector was encouraged to assume many functions and services once provided by the state. Pensions were moved entirely to the private sector as all state programs were phased out. In this instance, the state calculated that workers tied through pensions and other benefits to the success of private enterprise would be less likely to be attracted to "non-Chilean" ideologies such as Marxism, socialism, and even Christian democracy. State-sponsored health programs were also cut to the bone, and many of the poor now paid for services once provided by the government.

THE DEFEAT OF A DICTATOR

To attain a measure of legitimacy, Chileans expected the military government to produce economic achievement. By 1987, and continuing into 1989, the regime's economic policies seemed successful; the economic growth rate for 1988 was an impressive 7.4 percent. However, it masked critical weaknesses in the Chilean economy. For example, much of the growth was overdependent on exports of raw materials—notably, copper, pulp, timber, and fishmeal.

Modest economic success and an inflation rate of less than 20 percent convinced General Pinochet that he could take his political scenario for Chile's future to the voters for their ratification. But in the October 5, 1988, plebiscite, Chile's voters upset the general's plans and decisively denied him an additional eight-year term. (He did, however, continue in office until the next presidential election determined his suc-

cessor.) The military regime (albeit reluctantly) accepted defeat at the polls, which signified the reemergence of a deep-rooted civic culture and long democratic tradition.

Where had Pinochet miscalculated? Public-opinion surveys on the eve of the election showed a sharply divided electorate. Some political scientists even spoke of the existence of "two Chiles." In the words of government professor Arturo Valenzuela and *Boston Globe* correspondent Pamela Constable, one Chile "embraced those who had benefited from the competitive economic policies and welfare subsidies instituted by the regime and who had been persuaded that power was best entrusted to the armed forces." The second Chile "consisted of those who had been victimized by the regime, who did not identify with Pinochet's anti-Communist cause, and who had quietly nurtured a belief in democracy." Polling data from the respected Center for Public Policy Studies showed that 72 percent of those who voted against the regime were motivated by economic factors. These were people who had lost skilled jobs or who had suffered a decrease in real wages. While Pinochet's economic reforms had helped some, it had also created a disgruntled mass of downwardly mobile wage earners.

The rural areas of Chile have presented challenges for community development. Here, volunteers work on a road that will link the village of Tincnamar to a main road.

Valenzuela and Constable explain how a dictator allowed himself to be voted out of power. "To a large extent Pinochet had been trapped by his own mythology. He was convinced that he would be able to win and was anxious to prove that his regime was not a pariah but a legitimate government. He and other officials came to believe their own propaganda about the dynamic new Chile they had created." The closed character of the regime, with all lines of authority flowing to the hands of one man, made it "impossible for them to accept the possibility that they could lose." And when the impossible occurred and the dictator lost an election played by his own rules, neither civilians on the right nor the military were willing to override the constitutional contract they had forged with the Chilean people.

In March 1990, Chile returned to civilian rule for the first time in almost 17 years, with the assumption of the presidency by Patricio Aylwin. His years in power revealed that tensions still existed between civilian politicians and the military. In 1993, for example, General Pinochet mobilized elements of the army in Santiago—a move that, in the words of the independent newspaper *La Época*, "marked the crystallization of long-standing hostility" between the Aylwin government and the army. The military had reacted both to investigations into human-rights abuses during the Pinochet dictatorship and proposed legislation that would have subordinated the military to civilian control. On the other

hand, the commanders of the navy and air force as well as the two right-wing political parties refused to sanction the actions of the army.

President Aylwin regained the initiative when he publicly chastised General Pinochet. Congress, in a separate action, affirmed its supremacy over the judiciary in 1993, when it successfully impeached a Supreme Court justice for "notable dereliction of duty." The court system had been notorious for transferring human-rights cases from civil to military courts, where they were quickly dismissed. The impeachment augured well for further reform of the judicial branch.

HEALTH/WELFARE

Since 1981, all new members of Chile's labor force have been required to contribute 10% of their monthly gross earnings to private-pension-fund accounts, which they own. By 1995, more than 93% of the labor force were enrolled in 20 separate and competing private pension funds. The reforms increased the domestic savings rate to 26% of GDP.

Further resistance to the legacy of General Pinochet was expressed by the people when, on December 11, 1993, the center-left coalition candidate Eduardo Frei Ruiz-Tagle won the Chilean presidential election, with 58 percent of the vote. As part of his platform, Frei had promised to bring the military under civilian rule. The parlia-

mentary vote, however, did not give him the two-thirds majority needed to push through such a reform. The trend toward civilian government, though, seemed to be continuing.

Perhaps the final chapter in Pinochet's career began in November 1998, while the former dictator was in London for medical treatment. At that time, the British government received formal extradition requests from the governments of Spain, Switzerland, and France. The charges against Pinochet included attempted murder, conspiracy to murder, torture, conspiracy to torture, hostage taking, conspiracy to take hostages, and genocide, based on Pinochet's alleged actions while in power.

British courts ruled that the general was too ill to stand trial, and Pinochet returned to Chile. In May 2004 a Chilean appeals court revoked Pinochet's immunity from prosecution, a decision that renders a trial possible, if not probable. In the words of one lawyer, "It is as likely that he will stand trial as it is that he will get into Heaven." The government of Ricardo Lagos has made progress against human-rights offenders but Lagos himself admits that "this chapter can never be closed."

THE ECONOMY

By 1998, the Chilean economy had experienced 13 consecutive years of strong growth. But the Asian financial crisis of that year hit Chile hard, in part because 33 percent of the nation's exports in 1997 went to Asian markets. Copper prices tumbled; and because the largest copper mine is government-owned, state revenues contracted sharply. Following a sharp recession in 1999, the economy once again began to grow. However, domestic recovery has been slow. Unemployment remains high at 9 percent of the workforce, and a growth rate of 5.5 percent does not produce sufficient revenue to finance President Lagos's planned social programs and education initiatives. The sluggish global economy in 2001 was partly to blame, as prices fell for copper, Chile's number-one export.

Although there is still a large gap between the rich and poor in Chile, those living in poverty has been reduced from 40 percent to 20 percent over the course of the last decade. The irony is that Chile's economic success story is built on the economic model imposed by the Pinochet regime. "Underlying the current prosperity", writes *New York Times* reporter Larry Rohter, "is a long trail of blood and suffering that makes the thought of reversing course too difficult to contemplate." Many Chileans want to bury the past and move on—but the persistence of memory will not allow closure at this time. Chile has chosen to follow its own course with respect to economic policy. While many of its neighbors in the Southern Cone—notably Argentina, Brazil, Bolivia, Peru, Ecuador, and Venezuela—have moved away from free trade and open markets, Chile remains firmly wed to both.

ACHIEVEMENTS

Chile's great literary figures, such as Gabriela Mistrál and Pablo Neruda, have a great sympathy for the poor and oppressed. Other major Chilean writers, such as Isabel Allende and Ariel Dorfman have won worldwide acclaim.

Peruvian novelist and politician Mario Vargas Llosa observes that while Chile "is not paradise," it does have a "stability and economic dynamism unparalled in Latin America." Indeed, "Chile is moving closer to Spain and Australia and farther from Peru or Haiti." He suggests that there has been a shift in Chile's political culture. "The ideas of economic liberty, a free market open to the world, and private initiative as the motor of progress have become embedded in the people of Chile."

Chilean novelist Ariel Dorfman has a different perspective: "Obviously it is better to be dull and virtuous than bloody and Pinochetista, but Chile has been a very gray country for many years now. Modernization doesn't always have to come with a lack of soul, but I think there is a degree of that happening."

SIGNS OF CHANGE

Although the Chilean Constitution was essentially imposed on the nation by the military in 1980, there are signs of change. The term for president was reduced from eight to six years in 1993; and in 1997, the Chamber of Deputies, the lower house of the Legislature, approved legislation to further reduce the term of a president to four years, with a prohibition on reelection. Military courts, which have broader peacetime jurisdiction than most other countries in the Western Hemisphere, have also come under scrutiny by politicians. According to the *Revista Hoy*, as summarized by *CHIP News*, military justice reaches far beyond the ranks. If, for example, several people are involved in the commission of a crime and one of the perpetrators happens to be a member of the military, all are tried in a military court. Another abuse noted by politicians is that the military routinely uses the charge of sedition against civilians who criticize it. A group of Christian Democrats wants to limit the jurisdiction of the military to military crimes committed by military personnel; eliminate the participation of the army prosecutor in the Supreme Court, where he sits on the bench in cases related to the military; grant civilian courts the authority to investigate military premises; and accord civilian courts jurisdiction over military personnel accused of civilian-related crimes. The military itself, in 2004, in an effort to improve its tarnished image has worked in the background to hold accountable those officers involved in human rights abuses in the past.

Another healthy sign of change is a concerted effort by the Chilean and Argentine governments to discuss issues that have been a historical source of friction between the two nations. Arms escalation, mining exploration and exploitation in border areas, and trade and investment concerns were on the agenda. The Chilean foreign relations minister and the defense minister sat down with their Argentine counterparts in the first meeting of its kind in the history of Argentine–-Chilean relations.

Timeline: PAST

1541
The founding of Santiago de Chile

1818
Independence of Spain is proclaimed

1964–1970
Revolution in Liberty dramatically alters Chilean society

1973
A military coup ousts President Salvador Allende; General Augusto Pinochet becomes president

1988
Pinochet is voted out—and goes

1990s
Asian financial woes cut into Chilean economic growth

PRESENT

2000s
Ricardo Lagos, a moderate Socialist, wins the presidency in December 1999–January 2000 elections

Lagos government accelerates prosecution of human rights abusers

Colombia (Republic of Colombia)

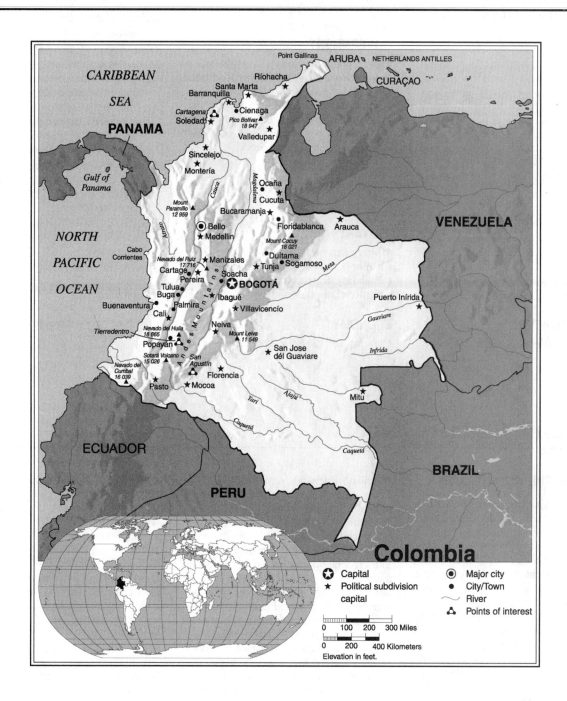

Colombia Statistics

GEOGRAPHY

Area in Square Miles (Kilometers):

440,000 (1,139,600) (about 3 times the size of Montana)

Capital (Population): Bogotá (6,005,000)

Environmental Concerns: deforestation; soil damage; air pollution

Geographical Features: flat coastal lowlands; central highlands; high Andes Mountains; eastern lowland plains

Climate: tropical on coast and eastern plains; cooler in highlands

PEOPLE

Population

Total: 42,310,775

Annual Growth Rate: 1.53%

Rural/Urban Population Ratio: 27/73

Major Language: Spanish

Ethnic Makeup: 58% Mestizo; 20% white; 14% mulatto; 4% African; 3% African-Indian; 1% Indian

Religions: 95% Roman Catholic; 5% others

Health

Life Expectancy at Birth: 66 years (male); 74 years (female)

Infant Mortality Rate (Ratio): 24.7/1,000; Indians 233/1,000

Physicians Available (Ratio): 1/1,078

Education

Adult Literacy Rate: 91.3%

Compulsory (Ages): for 5 years between 6 and 12; free

COMMUNICATION

Telephones: 7.7 million main lines

Daily Newspaper Circulation: 55 per 1,000 people

Televisions: 188 per 1,000 people

Internet Users: 2 million

TRANSPORTATION

Highways in Miles (Kilometers): 69, 338 (115,564)

Railroads in Miles (Kilometers): 2,103 (3,386)

Usable Airfields: 1,101

Motor Vehicles in Use: 1,700,000

GOVERNMENT

Type: republic

Independence Date: July 10, 1810 (from Spain)

Head of State/Government: President Alvaro Uribe Velez is both head of state and head of government

Political Parties: Liberal Party; Conservative Party; New Democratic Force; Democratic Alliance M-19; Patriotic Union

Suffrage: universal at 18

MILITARY

Military Expenditures (% of GDP): 3.4%

Current Disputes: civil war; maritime boundary dispute with Venezuela; territorial disputes with Nicaragua

ECONOMY

Currency ($ U.S. Equivalent): 2,877 pesos = $1

Per Capita Income/GDP: $6,300/$262.5 billion

GDP Growth Rate: 3.4%

Inflation Rate: 7.2%

Unemployment Rate: 13.6%

Labor Force: 16,800,000

Natural Resources: petroleum; natural gas; coal; iron ore; nickel; gold; copper; emeralds; hydropower

Agriculture: coffee; cut flowers; bananas; rice; tobacco; corn; sugarcane; cocoa beans; oilseed; vegetables; forest products; shrimp farming

Industry: textiles; food processing; petroleum; clothing and footwear; beverages; chemicals; cement; gold; coal; emeralds

Exports: $12.96 billion (primary partners United States, European Union, Andean Community)

Imports: $13.06 billion (primary partners United States, European Union, Andean Community)

SUGGESTED WEBSITE

http://www.cia.gov/cia/
publications/factbook/index.html

Colombia Country Report

COLOMBIA: THE VIOLENT LAND

Colombia has long been noted for its violent political history. The division of political beliefs in the mid-nineteenth century into conservative and liberal factions produced not only debate but also civil war. To the winner went the presidency and the spoils of office. That competition for office came to a head during the savage War of the Thousand Days (1899–1902). Nearly half a century later, Colombia was again plagued by political violence, which took perhaps 200,000 lives. Although on the surface it is distinct from the nineteenth-century civil wars, *La Violencia* ("The Violence," 1946–1958) offers striking parallels to the violence of the 1800s. Competing factions were again led by conservatives and liberals, and the presidency was the prize. Explanations for this phenomenon have tended to be at once simple and powerful. Colombian writers blame a Spanish heritage and its legacy of lust for political power.

Gabriel García Márquez, in his classic novel *One Hundred Years of Solitude*, spoofed the differences between liberals and conservatives. "The Liberals," said Aureliano Buendia's father-in-law, "were Freemasons, bad people, wanting to hang priests, to institute civil marriage and divorce, to recognize the rights of illegitimate children as equal to those of legitimate ones, and to cut the country up into a federal system that would take power away from the supreme authority." On the other hand, "the Conservatives, who had received their power directly from God, proposed the establishment of public order and family morality. They were the defenders of the faith of Christ, of the principle of authority, and were not prepared to permit the country to be broken down into autonomous entities." Aureliano, when later asked if he was a Liberal or a Conservative, quickly replied: "If I have to be something I'll be a Liberal, because the Conservatives are tricky."

THE ROOTS OF VIOLENCE

The roots of the violence are far more complex than a simple quest for spoils caused by a flaw in national character. Historian Charles Bergquist has shown that "divisions within the upper class and the systematic philosophical and programmatic positions that define them are not merely political manifestations of cultural traits; they reflect diverging economic interests within the upper class." These opposing interests developed in both the nineteenth and twentieth centuries. Moreover, to see Colombian politics solely as a violent quest for office ignores long periods of relative peace (1902–1935). But whatever the underlying causes of the violence, it has profoundly influenced contemporary Colombians.

La Violencia was the largest armed conflict in the Western Hemisphere since the Mexican Revolution (1910–1917). It was a civil war of ferocious intensity that cut through class lines and mobilized people from all levels of society behind the banner of either liberalism or conservatism. That elite-led parties were able to win popular support was evidence of their strong organization rather than their opponents' political weakness.

These multiclass parties still dominate Colombian political life, although the fierce interparty rivalry that characterized the civil wars of the nineteenth century as well as La Violencia has been stilled. In 1957, Colombia's social elite decided to bury partisan differences and devised a plan to end the widespread strife. Under this National Front agreement, the two parties agreed to divide legislative and bureaucratic positions equally and to alternate the presidency every four years from 1958 to 1974. This form of coalition government proved a highly successful means of elite compromise.

THE IMPACT OF LA VIOLENCIA

The violence has left its imprint on the people of Colombia in other ways. Some scholars have suggested that peasants now shun political action because of fear of renewed violence. Refugees from La Violencia generally experienced confusion and a loss of values. Usually, rising literacy rates, improved transportation and communications, and integration into the nation's life produce an upsurge of activism as people clamor for more rapid change. This has not been the case in rural Colombia. Despite guerrilla activity in the countryside—some of which is a spin-off from La Violencia, some of which until recently had a Marxist orientation, and some of which is banditry—the guerrillas have not been able to win significant rural support.

La Violencia also led to the professionalization and enlargement of the Colombian armed forces in the late 1950s and early 1960s. Never a serious participant in the nation's civil wars, the military acquired a new prestige and status unusual for Colombia. It must be considered an important factor in any discussion of Colombian politics today.

DEVELOPMENT

In 1996, the government continued its economic liberalization program and the privatization of selected public industries. Exports of crude petroleum have increased to the point that they almost equaled coffee exports. Two new oil fields will likely increase the importance of petroleum to the Colombian economy.

A standoff between guerrillas and the military prompted the government of Virgilio Barco to engage reluctantly in a dialogue with the insurgents, with the ultimate goal of peace. In 1988, he announced a three-phase peace plan to end the violence, to talk about needed reforms, and ultimately to reincorporate guerrillas into society. This effort came to fruition in 1991, when the guerrilla movement M-19 laid down its arms after 16 years of fighting and engaged in political dialogue. Other guerrilla groups, notably the long-lived (since 1961) Colombian Revolutionary Armed

Forces (FARC) and the National Liberation Army (ELN), led by a Spanish priest, chose to remain in the field.

Numbering perhaps 10,000, the guerrillas claim that their armed insurgency is about social change; but as *The Economist* has observed, lines between revolution and crime are increasingly blurred. Guerrillas ambush army units, attack oil pipelines, engage in blackmail, and kidnap rich ranchers and foreign oil executives for ransom. Some guerrillas are also apparently in the pay of the drug traffickers and collect a bounty for each helicopter they shoot down in the government's campaign to eradicate coca-leaf and poppy fields.

DRUGS AND DEATH

The guerrillas have a different perspective. One FARC leader asserted in an interview with the Colombian news weekly *Semana* that the guerrillas had both political and social objectives. Peace would come only if the government demilitarized large portions of the country and took action against the paramilitary organizations, some of them private and some of them supported by elements within the government. President Andrés Pastrana, who feared losing control of the country as well the credibility of his government, began to press for peace talks in January 1999 and, as a precondition to peace, agreed to demilitarize—that is, to withdraw government soldiers from a number of municipalities in southern Colombia. The United States objected that any policy of demilitarization would result in looser counter-narcotics efforts and urged a broader program to eradicate coca crops through aerial spraying. Critics of the policy claim that crop eradication plays into the hands of the guerrillas, who come to the support of the peasants who grow the coca. There is substance in the criticism, for by late 1999 FARC guerrillas controlled about 40 percent of the countryside.

FARC leaders, contrary to reports of foreign news media, disingenuously claim not to be involved in drug trafficking and have offered their own plan to counter the drug problem. It would begin with a government development plan for the peasants. In the words of a FARC leader: "Thousands of peasants need to produce and grow drugs to live, because they are not protected by the state." Eradication can succeed only if alternative crops can take the place of coca. Rice, corn, cacao, or cotton might be substituted. "Shooting the people, dropping bombs on them, dusting their sown land, killing birds and leaving their land sterile" is not the solution.

The peace talks scheduled between the government and the guerrillas in 1999 stalled and then failed, in large measure because of distrust on the part of FARC. Although a large portion of southern Colombia was demilitarized, the activities of paramilitary organizations were not curbed, and the United States sought to intensify its eradication policy. In the meantime, the Colombian Civil War entered its fourth decade.

FREEDOM

Colombia continues to have the highest rate of violent deaths in Latin America. Guerrillas, the armed forces, right-wing vigilante groups and drug traffickers are responsible for many deaths. On the positive side, a 1993 law that accorded equal rights to black Colombians resulted in the 1995 election of Piedad de Castro, the first black woman to hold the position, to the Senate.

In addition to the deaths attributed to guerrilla warfare, literally hundreds of politicians, judges, and police officers have been murdered in Colombia. It has been estimated that 10 percent of the nation's homicides are politically motivated. Murder is the major cause of death for men between ages 15 and 45. The violence resulted in 250,000 deaths in the 1990s; 300,000 people have left the country; and, since the late 1980s, 1½ million have been internally displaced or become refugees. While paramilitary violence accounts for many deaths, drug trafficking and the unraveling of Colombia's fabric of law are responsible for most. As political scientist John D. Martz writes: "Whatever the responsibility of the military or the rhetoric of government, the penetration of Colombia's social and economic life by the drug industry [is] proving progressively destructive of law, security and the integrity of the political system." Colombian political scientist Juan Gabriel Tokatlian echoed these sentiments in 2001 when he wrote: "The state is losing sovereignty and legitimacy. The left-wing guerrillas and the right-wing paramilitaries control more territory than the government."

Drug traffickers, according to *Latin American Update*, "represent a new economic class in Colombia; since 1981 'narcodollars' have been invested in real estate and large cattle ranches." The newsmagazine Semana noted that drug cartels had purchased 2.5 million acres of land since 1984 and now own one twelfth of the nation's productive farmland in the Magdalena River Basin. More than 100,000 acres of forest have been cut down to grow marijuana, coca, and opium poppies. Of particular concern to environmentalists is the fact that opium poppies are usually planted in the forests of the Andes at elevations above 6,000 feet. "These forests," according to Semana, "do not have great commercial value, but their tree cover is vital to the conservation of the sources of the water supply." The cartels also bought up factories, newspapers, radio stations, shopping centers, soccer teams, and bullfighters. The emergence of Medellín as a modern city of gleaming skyscrapers and expensive cars also reflects the enormous profits of the drug business.

Political scientist Francisco Leal Buitrago argues that while trafficking in narcotics in the 1970s was economically motivated, it had evolved into a social phenomenon by the 1980s. "The traffickers represent a new social force that wants to participate like other groups—new urban groups, guerrillas and peasant movements. Like the guerrillas, they have not been able to participate politically…."

HEALTH/WELFARE

Rape and other acts of violence against women are pervasive but seldom prosecuted. Spousal abuse was not considered a crime until 1996. Law 294 on family violence identifies as crimes violent acts committed within families, including spousal rape. Although the Constitution of 1991 prohibits it, discrimination against women persists in terms of access to employment and equal pay for equal work.

Domestic drug consumption has also emerged as a serious problem in Colombia's cities. *Latin American Regional Reports* notes that the increase in consumption of the Colombian form of crack, known as *bazuko*, "has prompted the growth of gangs of youths in slum areas running the bazuko business for small distributors." In Bogotá, police reported that more than 1,500 gangs operated from the city's slums.

URBANIZATION

As is the case in other Andean nations, urbanization has been rapid in Colombia. But the constantly spreading slums on the outskirts of the larger cities have not produced significant urban unrest or activism. Most of the migrants to the cities are first generation and are less frustrated and demanding than the general urban population. The new migrants perceive an improvement in their status and opportunities simply because they have moved into a more hopeful urban environment. Also, since most of the mi-

grants are poorly paid, their focus tends to be on daily survival, not political activism.

Migrants make a significant contribution to the parallel Colombian economy. As is the case in Peru and other South American countries, the informal sector amounts to approximately 30 percent of gross domestic product.

The Roman Catholic Church in Colombia has also tended to take advantage of rapid urbanization. Depending on the individual beliefs of local bishops, the Church has to a greater or lesser extent embraced the migrants, brought them into the Church, and created or instilled a sense of community where none existed before. The Church has generally identified with the expansion and change taking place and has played an active social role.

Marginalized city dwellers are often the targets of violence. Hired killers, called *sicarios*, have murdered hundreds of petty thieves, beggars, prostitutes, indigents, and street children. Such "clean-up" campaigns are reminiscent of the activities of the Brazilian death squads since the 1960s. An overloaded judicial system and interminable delays have contributed to Colombia's high homicide rate. According to government reports, lawbreakers have not been brought to justice in 97 to 99 percent of *reported* crimes. (Perhaps three quarters of all crimes remain unreported to the authorities.) Increasingly, violence and murder have replaced the law as a way to settle disputes; private "justice" is now commonly resorted to for a variety of disputes.

SOCIAL CHANGE

Government has responded to calls for social change and reform. President Virgilio Barco sincerely believed that the eradication of poverty would help to eliminate guerrilla warfare and reduce the scale of violence in the countryside. Unfortunately, his policies lacked substance, and he was widely criticized for his indecisiveness.

President César Gaviria felt that political reform must precede social and economic change and was confident that Colombia's new Constitution would set the process of national reconciliation in motion. The constitutional debate generated some optimism about the future of liberal democracy in Colombia. As Christopher Abel writes, it afforded a forum for groups ordinarily denied a voice in policy formulation—"to civic and community movements in the 40 and more intermediate cities angry at the poor quality of basic public services; to indigenous movements…; and to cooperatives, blacks, women, pensioners, small businesses, consumer and sports groups."

ACHIEVEMENTS

Colombia has a long tradition in the arts and humanities and has produced international figures such as the Nobel Prize–winning author Gabriel García Márquez; the painters and sculptors Alejandro Obregón, Fernando Botero, and Edgar Negret; the poet León de Greiff; and many others well known in music, art, and literature.

Violence and unrest have thwarted all of these efforts. Since the mid-1980s, according to a former Minister of Defence writing in 2000, 200 car bombs had exploded in Colombian cities, an entire democratic left of center party (the Unión Patriotica) had been eliminated by right-wing paramilitaries, and 4 presidential candidates, 200 judges and investigators, half of the supreme court justices, 1,200 police, 151 journalists, and 300,000 ordinary Colombians had been murdered.

While some scholars have described Colombia as a "failed state" others perceptively note that the focus should be on what holds the nation together in the face of unprecedented assaults. In the words of political scientist Malcolm Deas, Colombia is more united than fragmented, ethnically and religiously homogeneous, and its regional differences, while real, are not especially divisive. President Alvaro Uribe, a tough-minded pragmatist, has worked hard to restore the rule of law to Colombia. His first year in office resulted in a significant reduction in murder and kidnapping and attacks by guerrillas, as well acreage devoted to coca cultivation. Economic recovery was underway, as is indicated by the increased amount of highway traffic. Colombians, for the first time in years, felt more secure and, in 2004, 80 percent of the voting population supported Uribe.

ECONOMIC POLICIES

Colombia has a mixed economy. While state enterprises control domestic participation in the coal and oil industries and play a commanding role in the provision of electricity and communications, most of the economy is dominated by private business. At this point, Colombia is a moderate oil producer. A third of the nation's legal exports comes from the coffee industry, while exports of coal, cut flowers, seafood, and other nontraditional exports have experienced significant growth. In that Colombia is not saddled with an onerous foreign debt, its economy is relatively prosperous.

Contributing to economic success is the large informal sector. Also of tremendous importance are the profits from the illegal-drug industry. *The Economist* estimated that Colombia grossed perhaps $1.5 billion in drug sales in 1987, as compared to official export earnings of $5.5 billion. Indeed, over the past 15 years, profits from drug trafficking have grown to encompass between 25 and 35 percent of Colombia's legal exports. Perhaps half the profits are repatriated—that is, converted from dollars into local currency. An unfortunate side effect of the inflow of cash is an increase in the inflation rate.

FOREIGN POLICY

In the foreign-policy arena, President Barco's policies were attacked as low-profile, shallow, and too closely aligned to the policies of the United States. While Presidents Gaviria and Samper tried to adopt more independent foreign policy lines, especially in terms of the drug trade, Presidents Pastrana and Uribe have welcomed United States aid against drug trafficking and its attendant evils.

With an uneasy peace reigning in Central America, Colombia's focus has turned increasingly toward its neighbors and a festering territorial dispute with Venezuela over waters adjacent to the Guajira Peninsula. Colombia has proposed a multilateral solution to the problem, perhaps under the auspices of the International Court of Justice. Venezuela continues to reject a multilateral approach and seeks to limit any talks to the two countries concerned. It is likely that a sustained deterioration of internal conditions in either Venezuela or Colombia will keep the territorial dispute in the forefront. A further detriment to better relations with Venezuela is the justified Venezuelan fears that Colombian violence

as a result of guerrilla activity, military sweeps, and drugs will cross the border. As it is, thousands of Colombians have fled to Venezuela to escape their violent homeland. Venezuela's president recently infuriated Colombia's government when he independently opened negotiations with guerrillas and implied that they had more power than did President Pastrana.

THE CLOUDED FUTURE

Francisco Leal Buitrago, a respected Colombian academic, argues forcefully that his nation's crisis is, above all, "political": "It is the lack of public confidence in the political regime. It is not a crisis of the state itself…, but in the way in which the state sets the norms—the rules for participation—for the representation of public opinion…."

Constitutional reforms have taken place in Colombia, but changes in theory must reflect the country's tumultuous realities. Many of those in opposition have looked for a political opening but in the meantime continue to wage an armed insurgency against the government. Other problems, besides drugs, that dog the government include corruption, violence, slow growth, high unemployment, a weak currency, inflation, and the need for major reforms in banking. To get the economy on track, the International Monetary Fund has recommended that Colombia broaden its tax base, enhance municipal tax collections, get tough on tax evasion, and reduce spending.

Endemic violence and lawlessness, the continued operation of guerrilla groups, the emergence of mini-cartels in the wake of the eclipse of drug kingpins, and the attitude of the military toward conditions in Colombia all threaten any kind of progress.

The hard-line antidrug trafficking policy of the United States adds another complicated, and possibly counterproductive, dimension to the difficult task of governing Colombia.

Timeline: PAST

1525
The first Spanish settlement at Santa Marta

1810
Independence from Spain

1822
The creation of Gran Colombia (including Venezuela, Panama, and Ecuador)

1830
Independence as a separate country

1899–1902
War of the Thousand Days

1946–1958
La Violencia; nearly 200,000 lose their lives

1957
Women's suffrage

1980s
The drug trade becomes big business

1990s
Violence hampers progress; an earthquake kills or injures thousands in central Colombia

PRESENT

2000s
Colombia's violence threatens to involve its neighbors

Alvaro Uribe elected president in 2002

Ecuador (Republic of Ecuador)

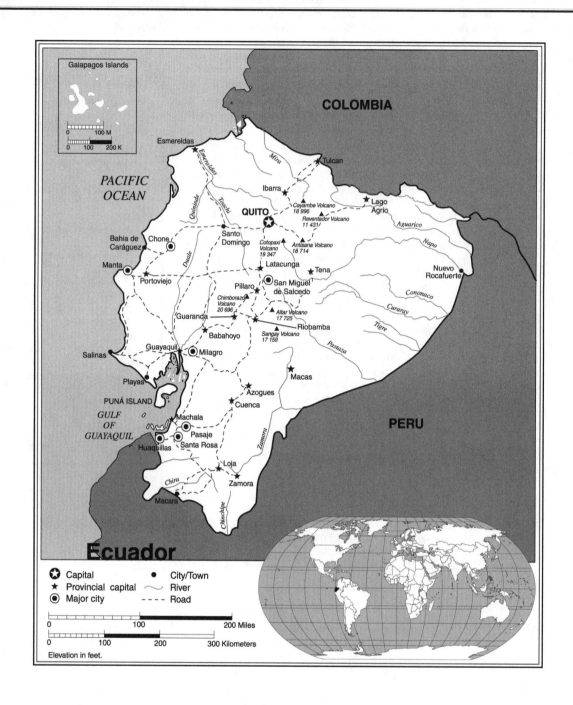

Ecuador

★ Capital	● City/Town
★ Provincial capital	～ River
◉ Major city	--- Road

0   100   200 Miles
0   100   200   300 Kilometers
Elevation in feet.

Ecuador Statistics

GEOGRAPHY

Area in Square Miles (Kilometers):
109,454 (283,560) (about the size of Nevada)

Capital (Population): Quito (1,445,000)

Environmental Concerns: deforestation; soil erosion; desertification; water pollution; pollution from petroleum wastes

Geographical Features: coastal plain; inter-Andean central highlands; flat to rolling eastern jungle

Climate: varied; tropical on the coast and in the inland jungle; cooler inland at higher elevations

PEOPLE

Population

Total: 13,212,742
Annual Growth Rate: 1.03%
Rural/Urban Population Ratio: 40/60
Major Languages: Spanish; Quechua and other Amerindian languages

Ethnic Makeup: 65% Mestizo; 25% Indian; 10% Spanish, black, and others

Religions: 95% Roman Catholic; 5% indigenous and others

Health

Life Expectancy at Birth: 73 years (male); 79 years (female)

Infant Mortality Rate (Ratio): 24.49/1,000

Physicians Available (Ratio): 1/904

Education

Adult Literacy Rate: 90%

Compulsory (Ages): for 6 years between 6 and 14; free

COMMUNICATION

Telephones: 1,426,200 main lines

Daily Newspaper Circulation: 72 per 1,000 people

Televisions: 79 per 1,000

Internet Users: 537,900

TRANSPORTATION

Highways in Miles (Kilometers): 26,858 (43,249)

Railroads in Miles (Kilometers): (812)

Usable Airfields: 182

Motor Vehicles in Use: 480,000

GOVERNMENT

Type: republic

Independence Date: May 24, 1822 (from Spain)

Head of State/Government: President Lucio Gutierrez is both head of state and head of government

Political Parties: Democratic Left; Social Christian Party; Pachakutik; Popular Democracy; Popular Democratic Movement; others

Suffrage: universal and compulsory for literate people ages 18–65; optional for other eligible voters

MILITARY

Military Expenditures (% of GDP): 2.4%

Current Disputes: none

ECONOMY

Currency ($ U.S. Equivalent): 1.00 dollar = $1

Per Capita Income/GDP: $3,300/$45.46 billion

GDP Growth Rate: 2.6%

Inflation Rate: 6.1%

Unemployment Rate: 9.8%; plus widespread underemployment

Labor Force: 4,200,000

Natural Resources: petroleum; fish; timber; hydropower

Agriculture: bananas; coffee; cocoa; rice; potatoes; manioc; plantains; sugarcane; livestock; balsa wood; fish; shrimp

Industry: petroleum; food processing; textiles; metalwork; paper products; wood products; chemicals; plastics; fishing; lumber

Exports: $6 billion (primary partners United States, Colombia, Italy)

Imports: $6.2 billion (primary partners United States, Colombia, Japan)

SUGGESTED WEBSITE

http://www.cia.gov/cia/
publications/factbook/
geos.ec.html

Ecuador Country Report

ECUADOR: A LAND OF CONTRASTS

Several of Ecuador's great novelists have had as the focus of their works the exploitation of the Indians. Jorge Icaza's classic *Huasipungo* (1934) describes the actions of a brutal landowner who first forces Indians to work on a road so that the region might be "developed" and then forces them, violently, from their plots of land so that a foreign company's operations will not be impeded by a troublesome Indian population.

That scenario, while possible in some isolated regions, is for the most part unlikely in today's Ecuador. In recent years, despite some political and economic dislocation, Ecuador has made progress in health care, literacy, human rights, freedom of the press, and representative government. Indigenous peoples have been particularly active and over the past decade have demanded cultural rights. An indigenous political party, Pachakutik, has identified with Ecuador's nonindigenous poor and won several seats in Congress. In protest against an economic program of austerity and reflecting ethnic and social conflict, several of these groups in league with midlevel army officers moved to top-

ple President Jamil Mahuad from power in January 2000. It was South America's first successful coup in a quarter of a century. Current President Lucio Gutierrez, who was behind the coup, has since lost support from indigenous leaders who are angry over his inability to deliver on promises made. Austerity policies have hurt the indigenous poor and Ecuador's large public debt has hamstrung social programs.

DEVELOPMENT

Ecuador's development problems have been exacerbated by problems with debt negotiations with and interest payments to foreign creditors. The privatization policy continues despite protests.

Although Ecuador is still a conservative, traditional society, it has shown an increasing concern for the plight of its rural inhabitants, including the various endangered Indian groups inhabiting the Amazonian region. The new attention showered on rural Ecuador—traditionally neglected by policymakers in Quito, the capital city—reflects in part the government's concern with patterns

of internal migration. Even though rural regions have won more attention from the state, social programs continue to be implemented only sporadically.

Two types of migration are currently taking place: the move from the highlands to the coastal lowlands and the move from the countryside to the cities. In the early 1960s, most of Ecuador's population was concentrated in the mountainous central highlands. Today, the population is about equally divided between that area and the coast, with more than half the nation's people crowded into the cities. So striking and rapid has the population shift been that the director of the National Institute of Statistics commented that it had assumed "alarming proportions" and that the government had to develop appropriate policies if spreading urban slums were not to develop into "potential focal points for insurgency." What has emerged is a rough political parity between regions that has led to parliamentary paralysis and political crisis.

The large-scale movement of people has not rendered the population more homogenous but, because of political parity, has instead fractured the nation. Political rivalry has always characterized relations between Quito, in the sierras, and cosmopolitan

(United Nations photo)

The migration of the poor to the urban areas of Ecuador has been very rapid and of great concern to the government. The increase in inner-city population can easily lead to political unrest. The graffiti in the photo says, "We don't want a dictatorship."

Guayaquíl, on the coast. The presidential election of 1988 illustrated the distinctive styles of the country. Rodrigo Borja's victory was regionally based, in that he won wide support in Ecuador's interior provinces. Usually conservative in its politics, the interior voted for the candidate of the Democratic Left, in part because of the extreme populist campaign waged by a former mayor of Guayaquíl, Abdalá Bucaram. Bucaram claimed to be a man of the people who was persecuted by the oligarchy. He spoke of his lower-class followers as the "humble ones," or, borrowing a phrase from former Argentine president Juan Perón, *los descamisados* ("the shirtless ones"). Bucaram, in the words of political scientist Catherine M. Conaghan, "honed a political style in the classic tradition of coastal populism. He combined promises of concrete benefits to the urban poor with a colorful anti-oligarchic style." Bucaram's style finally triumphed in 1996, when he won election to the presidency.

EDUCATION AND HEALTH

Central to the government's policy of development is education. Twenty-nine percent of the national budget was set aside for education in the early 1980s, with increases proposed for the following years. Adult literacy improved from 74 percent in 1974 to 87 percent by 1995. In the central highlands, however, illiteracy rates of more than 35 percent are still common,

largely because Quechua is the preferred language among the Indian peasants.

The government has approached this problem with an unusual sensitivity to indigenous culture. Local Quechua speakers have been enlisted to teach reading and writing in both Quechua and Spanish. This approach has won the support of Indian leaders who are closely involved in planning local literacy programs built around indigenous values.

FREEDOM

Ecuador's media, with the exception of two government-owned radio stations, are in private hands and represent a broad range of opinion. They are often critical of government policies, but they practice a degree of self-censorship in coverage involving corruption among high-level military personnel.

Health care has also shown steady improvement, but the total statistics hide sharp regional variations. Infant mortality and malnutrition are still severe problems in rural areas. In this sense, Ecuador suffers from a duality found in other Latin American nations with large Indian populations: Social and racial differences persist between the elite-dominated capitals and the Indian hinterlands. Income, services, and resources tend to be concentrated in the capital cities. Ecuador, at least, is attempting to correct the imbalance.

The profound differences between Ecuador's highland Indian and its European cultures is illustrated by the story of an Indian peasant who, when brought to a clinic, claimed that he was dying as the result of a spell. He told the doctor, trained in Western medicine, that, while traveling a path from his highland village down to a valley, he passed by a sacred place, where a witch cast a spell on him. The man began to deteriorate, convinced that this had happened. The doctor, upon examination of the patient, could find no physical reason for the man's condition. Medicine produced no improvements. The doctor finally managed to save his patient, but only after a good deal of compromise with Indian culture. "Yes," he told the peasant, "a witch has apparently cast a spell on you and you are indeed dying." And then the doctor announced: "Here is a potion that will remove the spell." The patient's recovery was rapid and complete. Thus, though modern medicine can work miracles, health-care workers must also be sensitive to cultural differences.

THE ECONOMY

Between 1998 and 2000, the Ecuadoran economy was hit hard by two crises. Falling petroleum prices in combination with the ravages of the El Niño weather phenomenon transformed a $598 million surplus in 1997 into a troubling $830 million deficit in 1998. Petroleum revenues fell to third place, behind exports of bananas and shrimp, which themselves were devastated by bad weather (in the case of shrimp, due to the dramatic warming of waters in the eastern Pacific as a result of El Niño).

HEALTH/WELFARE

Educational and economic opportunities in Ecuador are often not made available to women, blacks, and indigenous peoples. Most of the nation's peasantry, overwhelmingly Indian or Mestizo, are poor. Infant mortality, malnutrition, and epidemic disease are common among these people.

Newly elected president Jamil Mahuad was confronted from the outset of his administration with some daunting policy decisions. A projected growth rate for 1998 of only 1 percent and an inflation rate that soared to 40 percent resulted in budget austerity and an emergency request to Congress to cut spending and prepare legislation for the privatization of Ecuador's telecommunications and electrical industries. The privatization plans raised the ire of nationalists. In the mid-1990s, the government privatized more than 160 state-owned enterprises and, in an effort to modernize and streamline the economy, cut the number of public employees from 400,000 to 260,000.

The sharp economic downturn resulted in severe belt-tightening by the Mahuad government, threw people out of work, produced social and political upheaval, and led to a coup. The military quickly handed over power to the civilian vice president, Gustavo Noboa, to finish out Mahuad's term. Noboa took steps to restore Ecuador's economic viability and adopted some of Mahuad's unpopular policies, including "dollarization" of the economy and continued privatization of state enterprises. Rising oil prices have helped ease the shock of his policies, but Ecuador's economic, political, and social problems remain unresolved and will likely produce further political and social unrest.

BITTER NEIGHBORS

A long legacy of boundary disputes that reached back to the wars for independence created a strained relationship between Ecuador and Peru which erupted in violence in July 1941. Ecuador initiated an undeclared war against Peru in an attempt to win territory along its southeastern border, in the Marañón River region, and, in the southwest, around the town of Zaramilla. In the 1942 Pact of Peace, Amity, and Limits, which followed a stunning Peruvian victory, Ecuador lost about 120,000 square miles of territory. The peace accord was guaranteed by Argentina, Brazil, Chile, and the United States. In January 1995, the usual tensions that grew each year as the anniversary of the conflict approached were given foundation when fighting again broke out between Peru and Ecuador; Peruvian soldiers patrolling the region had stumbled upon well-prepared and waiting Ecuadoran soldiers. Three weeks later, with the intervention of the guarantors of the original pact, the conflict ended. The Peruvian armed forces were shaken from their smug sense of superiority over the Ecuadorans, and the Ecuadoran defense minister used the fight to support his political pretensions.

ACHIEVEMENTS

Ecuadoran poets have often made their poetry an expression of social criticism. The so-called Tzántzicos group has combined avant-garde techniques with social commitment and has won a measure of attention from literary circles.

The border war sent waves of alarm through the rest of Latin America, in that it reminded more than a dozen nations of boundary problems with their neighbors. Of particular concern were revelations made in 1998 and 1999 that individuals within the Argentine government and the military had sold arms to the Ecuadoran military during the conflict. Argentina was embarrassed because it was one of the

original guarantors of the 1942 Pact of Peace.

On October 16, 1998, the Legislatures of Ecuador and Peru supported an agreement worked out by other governments in the region to end the border dispute. Under the terms of the agreement, Peru's sovereignty of the vast majority of the contested territory was affirmed. Ecuador won a major concession when it was granted navigation rights on the Amazon River and its tributaries within Peru and the right to establish trading centers on the river. In that both parties benefited from the negotiation, it is hoped that a lasting peace will have been effected.

Timeline: PAST

1528
First Spanish contact

1822
Ecuador is part of Gran Colombia (with Panama, Venezuela, and Colombia); independence as a separate state

1929
Women's suffrage

1941
A border war with Peru

1990s
Modernization laws aim to speed the privatization of the economy

Popular dissatisfaction with the government's handling of the economy rises

PRESENT

2000s
El Niño devastates the coastal economy
Refugees and drug activity spill into Ecuador from Colombia

Lucio Gutierrez elected president in 2003

Guyana (Co-operative Republic of Guyana)

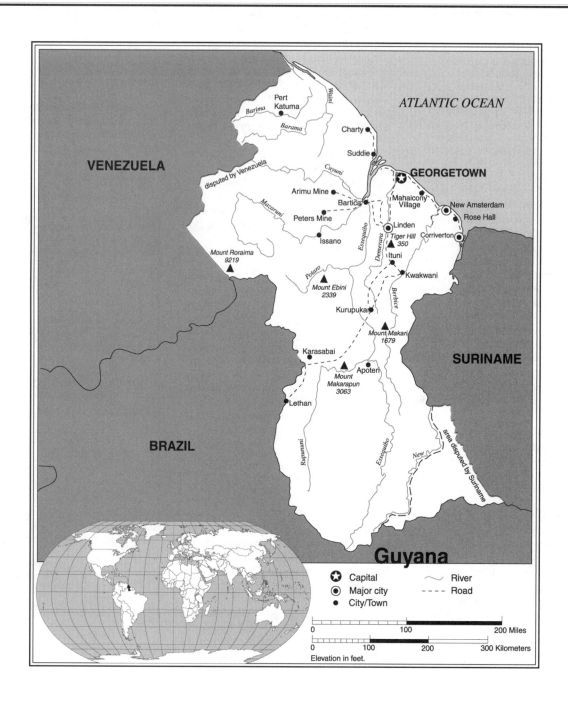

Guyana Statistics

GEOGRAPHY

Area in Square Miles (Kilometers): 82,990 (215,000) (about the size of Idaho)

Capital (Population): Georgetown (248,500)

Environmental Concerns: water pollution; deforestation

Geographical Features: mostly rolling highlands; low coastal plain; savannah in the south

Climate: tropical

PEOPLE*

Population

Total: 705,813

*Note: Estimates explicitly take into account the effects of excess mortality due to AIDS.

Annual Growth Rate: 0.61%

Rural/Urban Population Ratio: 74/36

Major Languages: English; indigenous dialects; Creole; Hindi; Urdu

Ethnic Makeup: 51% East Indian, 30% black; 14% mixed; 4% Amerindian; 2% white and Chinese

Religions: 50% Christian; 33% Hindu; 9% Muslim; 8% others

Health

Life Expectancy at Birth: 60 years (male); 64 (female)

Infant Mortality Rate (Ratio): 37.22/1,000

Physicians Available (Ratio): 1/3,000

Education

Adult Literacy Rate: 98%

Compulsory (Ages): 6–14; free

COMMUNICATION

Telephones: 80,400 main lines

Daily Newspaper Circulation: 97 per 1,000 people

Televisions: 1 per 26 people

Internet Users: 125,000

TRANSPORTATION

Highways in Miles (Kilometers): 4,949 (7,970)

Railroads in Miles (Kilometers): (187)

Usable Airfields: 1

Motor Vehicles in Use: 33,000

GOVERNMENT

Type: republic

Independence Date: May 26, 1966 (from the United Kingdom)

Head of State/Government: President Bharrat Jagdeo; Prime Minister Samuel Hinds

Political Parties: People's National Congress; Alliance for Guyana People's Progressive Party; United Force; Democratic Labour Movement; People's Democratic Movement; National Democratic Front; others

Suffrage: universal at 18

MILITARY

Military Expenditures (% of GDP): 0.8%

Current Disputes: territorial disputes with Venezuela and Suriname

ECONOMY

Currency ($ U.S. Equivalent): 190.672002) Guyanese dollars = $1

Per Capita Income/GDP: $4,000/$2.8 billion

GDP Growth Rate: 0.3%

Inflation Rate: 4.7%

Unemployment Rate: 9.1%

Natural Resources: bauxite; gold; diamonds; hardwood timber; shrimp; fish

Agriculture: sugar; rice; wheat; vegetable oils; livestock; potential for fishing and forestry

Industry: bauxite; sugar; rice milling; timber; fishing; textiles; gold mining

Exports: $5.12 million (primary partners United States, Canada, United Kingdom)

Imports: $612 million (primary partners United States, Trinidad and Tobago, Netherland Antilles)

SUGGESTED WEBSITE

http://www.cia.gov/cia/ publications/factbook/geos/ gy.html

Guyana Country Report

GUYANA: RACIAL AND ETHNIC TENSIONS

Christopher Columbus, who cruised along what are now Guyana's shores in 1498, named the region *Guiana*. The first European settlers were the Dutch, who settled in Guyana late in the sixteenth century, after they had been ousted from Brazil by a resurgent Portuguese Crown. Dutch control ended in 1796, when the British gained control of the area. In 1815, as part of the treaty arrangements that brought the Napoleonic Wars to a close, the Dutch colonies of Essequibo, Demerera, and Berbice were officially ceded to the British. In 1831, the former Dutch colonies were consolidated as the Crown Colony of British Guiana.

DEVELOPMENT

Moderate economic growth was achieved in 2001–2003 by the expansion of the agricultural and mining sectors, a favorable climate for business, a more realistic exchange rate, and modest inflation.

Guyana is a society deeply divided along racial and ethnic lines. East Indians make up the majority of the population. They predominate in rural areas, constituting the bulk of the labor force on the sugar plantations, and they comprise nearly all of the rice-growing peasantry. They also dominate local businesses and are prominent in the professions. Blacks are concentrated in urban areas, where they are employed in clerical and secretarial positions in the public bureaucracy, in teaching, and in semiprofessional jobs. A black elite dominates the state bureaucratic structure.

Before Guyana's independence in 1966, plantation owners, large merchants, and British colonial administrators consciously favored some ethnic groups over others, providing them with a variety of economic and political advantages. The regime of President Forbes Burnham revived old patterns of discrimination for political gain.

Burnham, after ousting the old elite when he nationalized the sugar plantations and the bauxite mines, built a new regime that simultaneously catered to lower-class blacks and discriminated against East Indians. In an attempt to address the blacks' basic human needs, the Burnham government greatly expanded the number of blacks holding positions in public administration. To demonstrate his largely contrived black-power ideology, Burnham spoke out strongly in support of African liberation movements. The government played to the fear of communal strife in order to justify its increasingly authoritarian rule.

FREEDOM

One of the priorities of the Jagan governments was the elimination of all forms of ethnic and racial discrimination, a difficult task in a country where political parties are organized along racial lines. It was hoped that Guyana's indigenous peoples would be offered accelerated development programs to enhance their health and welfare.

In the mid-1970s, a faltering economy and political mismanagement generated an increasing opposition to Burnham that cut across ethnic lines. The government increased the size of the military, packed Parliament through rigged elections, and amended the Constitution so that the president held virtually imperial power.

There has been some improvement since Burnham's death in 1985. The appearance of newspapers other than the government-controlled *Guyana Chronicle* and the public's dramatically increased access to television have served to curtail official control of the media. In politics, the election of Indo-Guyanese leader Cheddi Jagan to the presidency reflected deep-seated disfavor with the behavior and economic policies of the previous government of Desmond Hoyte. President Jagan identified the nation's foreign debt of $2 billion as a "colossally big problem, because the debt overhang impedes human development."

HEALTH/WELFARE

The government has initiated policies designed to lower the cost of living for Guyanese. Prices for essentials have been cut. Money has been allocated for school lunch programs and for a "food-for-work" plan. Pensions have been raised for the first time in years. The minimum wage, however, will not sustain an average family.

While president, Hoyte once pledged to continue the socialist policies of the late Forbes Burnham; but in the same breath, he talked about the need for privatization of the crucial sugar and bauxite industries. Jagan's economic policies, according to *Latin American Regional Reports*, outlined an uncertain course. During his campaign, Jagan stated that government should not be involved in sectors of the economy where private or cooperative ownership would be more efficient. In 1993, however, he backed away from the sale of the Guyana Electric Company and had some doubts about selling off the sugar industry. In Jagan's words: "Privatisation and divestment must be approached with due care. I was not elected president to preside over the liquidation of Guyana. I was mandated by the Guyanese people to rebuild the national economy and to restore a decent standard of living." Jagan's policies stimulated rapid socioeconomic progress as Guyana embarked on the road to economic recovery.

Following Jagan's death, new elections were held in December 1997, and Janet Jagan, the ex-president's 77-year-old widow, was named president. In August 1999, she stepped down due to health reasons. She named Finance Minister Bharrat Jagdeo to succeed her.

ACHIEVEMENTS

The American Historical Association selected Walter Rodney for the 1982 Beveridge Award for his study of the Guyanese working people. The award is for the best book in English on the history of the United States, Canada, or Latin America. Rodney, the leader of the Working People's Alliance, was assassinated in 1980.

Jagdeo's presidency has only exacerbated ethnic tensions. The Afro-Guyanese, who represent less than half of Guyana's population, tend to support the opposition People's National Congress party, which had held power from 1964 to 1992, and have responded to their lack of power by confronting the government on its policies, sometimes violently.

In the meantime, a divided Guyana may soon be confronted by an aggressive Venezuela, whose president seems intent on re-igniting its long-standing border dispute with Guyana.

Timeline: PAST

1616
The first permanent Dutch settlements on Essequibo River

1815
The Netherlands cedes the territory to Britain

1966
Independence

1985
President Forbes Burnham dies

1990s
The government promises to end racial and ethnic discrimination

PRESENT

2000s
Territorial disputes with Suriname and Venezuela persist

Politics remains bitterly divided along ethnic lines

Paraguay (Republic of Paraguay)

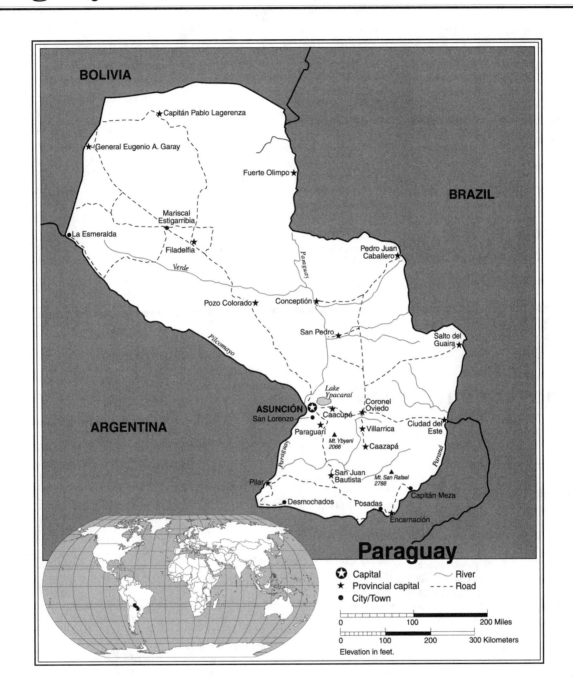

Paraguay Statistics

GEOGRAPHY

Area in Square Miles (Kilometers): 157,048 (406,752) (about the size of California)

Capital (Population): Asunción (548,000)

Environmental Concerns: deforestation; water pollution; problems with waste disposal

Geographical Features: grassy plains and wooded hills east of Rio Paraguay; Gran Chaco region west of the river; mostly marshy plain near the river; dry forest and thorny scrub elsewhere

Climate: subtropical to temperate

PEOPLE

Population

Total: 6,191,368

Annual Growth Rate: 2.6%

Rural/Urban Population Ratio: 47/53

Major Languages: Spanish; Guaraní; Portuguese

Ethnic Makeup: 95% Mestizo; 5% white and Indian

Religions: 90% Roman Catholic; 10% Protestant

Health

Life Expectancy at Birth: 72 years (male); 77 years (female)

Infant Mortality Rate (Ratio): 26.67/1,000

Physicians Available (Ratio): 1/1,406

Education

Adult Literacy Rate: 94%

Compulsory (Ages): 6–12

COMMUNICATION

Telephones: 273,200 main lines

Daily Newspaper Circulation: 40 per 1,000 people

Televisions: 144 per 1,000

Internet Users: 100,000

TRANSPORTATION

Highways in Miles (Kilometers): 18,320 (29,500)

Railroads in Miles (Kilometers): 602 (970)

Usable Airfields: 937

Motor Vehicles in Use: 125,000

GOVERNMENT

Type: republic

Independence Date: May 14, 1811 (from Spain)

Head of State/Government: President Nicanor Duarte Frutos is both head of state and head of government

Political Parties: Colorado Party; Authentic Radical Liberal Party; Christian Democratic Party; Febrerist Revolutionary Party; National Encounter

Suffrage: universal and compulsory from 18 to 75

MILITARY

Military Expenditures (% of GDP): 0.9%

Current Disputes: none

ECONOMY

Currency ($ U.S. Equivalent): 6,424 guaranis = $1

Per Capita Income/GDP: $4,600/$28.03 billion

GDP Growth Rate: 1.3%

Inflation Rate: 10.5%

Unemployment Rate: 16.4%

Labor Force: 1,700,000

Natural Resources: hydropower; timber; iron ore; manganese; limestone

Agriculture: cotton; sugarcane; soybeans; corn; wheat; tobacco; cassava (tapioca); fruits; vegetables; livestock; timber

Industry: sugar; cement; textiles; beverages; wood products

Exports: $2.7 billion (primary partners Brazil, Argentina, European Union)

Imports: $2.8 billion (primary partners Brazil, United States, Argentina)

SUGGESTED WEBSITE

http://www.cia.gov/cia/publications/factbook/index.html

Paraguay Country Report

PARAGUAY

Paraguay is a country of paradox. Although there is little threat of foreign invasion and guerrilla activity is insignificant, a state of siege was in effect for 35 years, ending only in 1989 with the ouster of President (General) Alfredo Stroessner, who had held the reins of power since 1954. Government expenditures on health care in Paraguay are among the lowest in the Western Hemisphere, yet life expectancy is impressive, and infant mortality reportedly has fallen to levels comparable to more advanced developing countries. On the other hand, nearly a third of all reported deaths are of children under five years of age. Educational achievement, especially in rural areas, is low.

DEVELOPMENT

The devaluation of the Brazilian real in 1999 sent some shock waves through the Paraguyan economy. Also problematic for development was the decision of the U.S. Congress to refuse Paraguay "certification" in the war against drug trafficking, as well as the nation's incredibly contentious and unpredictable political scene.

Paraguayan politics, economic development, society, and even its statistical base are comprehensible only within the context of its geography and Indo–Hispanic culture. Its geographic isolation in the midst of powerful neighbors has encouraged Paraguay's tradition of militarism and self-reliance—of being led by strongmen who tolerate little opposition. There is no tradition of constitutional government or liberal democratic procedures upon which to draw. Social values influence politics to the extent that politics is an all-or-nothing struggle for power and its accompanying prestige and access to wealth. These political values, in combination with a population that is largely poor and politically ignorant, contribute to the type of paternalistic, personal rule characteristic of a dictator such as Stroessner.

The paradoxical behavior of the Acuerdo Nacional—a block of opposition parties under Stroessner—was understandable within the context of a quest for power, or at least a share of power. Stroessner, always eager to divide and conquer, identified the Acuerdo Nacional as a fruitful field for new alliances. Leaping at the chance for patronage positions but anxious to demonstrate to Stroessner that they were

a credible political force worthy of becoming allies, Acuerdo members tried to win the support of unions and the peasantry. At the same time, the party purged its youth wing of leftist influences.

FREEDOM

Monolingual Guaraní speakers suffer from a marked disadvantage in the labor market. Where Guaraní speakers are employed, their wages are much lower than for monolingual Spanish speakers. This differential is accounted for by the "educational deficiencies" of the Guaraní speakers as opposed to those who speak Spanish.

Just when it seemed certain that Stroessner would rule until his death, Paraguayans were surprised in February 1989 when General Andrés Rodríguez—second-in-command of the armed forces, a member of the Traditionalist faction of the Colorado Party, which was in disfavor with the president, and a relative of Stroessner—seized power. Rodríguez's postcoup statements promised the democratization of Paraguay, respect for human rights, repudiation of drug trafficking, and the scheduling of presidential elections. Not surprisingly, General Rodríguez emerged as President Rodríguez. When asked about voting irregularities, Rodríguez indicated that "real" democracy would begin with elections in 1993 and that his rule was a necessary "transition."

HEALTH/WELFARE

The Paraguayan government spends very little on human services and welfare. As a result, its population is plagued by health problems—including poor levels of nutrition, lack of drinkable water, absence of sanitation, and a prevalence of fatal childhood diseases.

"Real" democracy, following the 1993 victory of President Juan Carlos Wasmosy, had a distinct Paraguayan flavor. Wasmosy won the election with 40 percent of the vote; and the Colorado Party, which won most of the seats in Congress, was badly divided. When an opposition victory seemed possible, the military persuaded the outgoing government to push through legislation to reorganize the armed forces. In effect, they were made autonomous.

Poitical turmoil has continued to characterize Paraguayan politics. Assassination, an attempted coup in 2000, endemic corruption and back room deals are stock in trade. The victory of Paraguay's new president, Nicanor Duarte Frutos, will continue the Colorado's half century lock on political power.

The problems he faces are serious. Corruption, counterfeiting, contraband, money laundering, and organized crime are entrenched. Despite campaign promises that "there will be no place for people who believe the party and state are there to be abused to the detriment of the country" few Paraguayans expect change. There are other issues that cloud the future. The commercialization of agriculture and high population growth have led to a dramatic increase in the number of landless families who have begun to migrate to urban areas where they resettle in shanty towns. Poverty effects nearly 60 percent of the population.

ACHIEVEMENTS

Paraguay has produced several notable authors, including Gabriel Casaccia and Augusto Roa Bastos. Roa Bastos makes extensive use of religious symbolism in his novels as a means of establishing true humanity and justice.

THE ECONOMY

It is difficult to acquire accurate statistics about the Paraguayan economy, in part because of the large informal sector and in part because of large-scale smuggling and drug trafficking. It is estimated that 20 percent of the nation's economy has been driven by illicit cross-border trafficking and that almost all of Paraguay's tobacco exports are illicit, counterfeit, or both. Officially, the country experienced negative growth in 1998. The new government's

privatization plans, needed to raise revenue, must confront the military, which controls the most important state-owned enterprises. There is also concern about the "Brazilianization" of the eastern part of Paraguay, which has developed to the point at which Portuguese is heard as frequently as Spanish or Guaraní, the most common Indian language.

Timeline: PAST

1537
The Spanish found Asunción

1811
Independence is declared

1865–1870
War against the "Triple Alliance": Argentina, Brazil, and Uruguay

1954
General Alfredo Stroessner begins his rule

1961
Women win the vote

1989
Stroessner is ousted in a coup

1990s
A new Constitution is promulgated

PRESENT

2000s
Attempted coup in 2000

Nicanor Durate Frutos elected president in 2003

Peru (Republic of Peru)

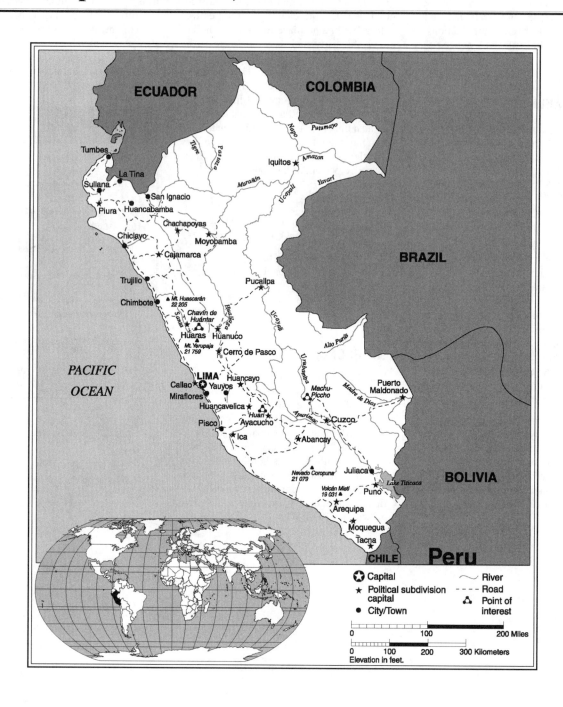

Peru Statistics

GEOGRAPHY

Area in Square Miles (Kilometers):

496,087 (1,285,200) (about the size of Alaska)

Capital (Population): Lima (6,743,000)

Environmental Concerns: deforestation; overgrazing; soil erosion; desertification; air and water pollution

Geographical Features: western coastal plain; high and rugged Andes Mountains in the center; eastern lowland jungle of Amazon Basin

Climate: temperate to tropical

PEOPLE

Population

Total: 27,544,305
Annual Growth Rate: 1.39%

Rural/Urban Population Ratio: 29/71
Major Languages: Spanish; Quechua; Aymara
Ethnic Makeup: 45% Indian; 37% Mestizo; 15% white, and others
Religions: more than 90% Roman Catholic; others

Health

Life Expectancy at Birth: 67 years (male); 71 years (female)
Infant Mortality Rate (Ratio): 33/1,000
Physicians Available (Ratio): 1/1,116

Education

Adult Literacy Rate: 90.9%
Compulsory (Ages): 6–11; free

COMMUNICATION

Telephones: 1,766,100 main lines
Daily Newspaper Circulation: 86 per 1,000 people
Televisions: 85 per 1,000
Internet Users: 2.5 million

TRANSPORTATION

Highways in Miles (Kilometers): 44,803 (72,146)

Railroads in Miles (Kilometers): 1,267 (2,041)
Usable Airfields: 234
Motor Vehicles in Use: 775,000

GOVERNMENT

Type: republic
Independence Date: July 28, 1821 (from Spain)
Head of State/Government: President Alejandro Toledo is both head of state and head of government
Political Parties: Change 90–New Majority; Union for Peru; Popular Action Party; Popular Christian Party; United Left; Civic Works Movement; Renovation Party; Alianza Popular Revolucionaria Americana (APRA); others
Suffrage: universal at 18

MILITARY

Military Expenditures (% of GDP): 1.3%
Current Disputes: a boundary dispute with Ecuador was resolved in 1999

ECONOMY

Currency ($ U.S. Equivalent): 3.48 nuevo sols = $1
Per Capita Income/GDP: $5,200/$146.9 billion
GDP Growth Rate: 4%
Inflation Rate: 2.2%
Unemployment Rate: 13.4% (official rate); extensive underemployment
Labor Force: 7.5 million
Natural Resources: copper; silver; gold; petroleum; timber; fish; iron ore; coal; phosphate; potash
Agriculture: coffee; sugarcane; rice; wheat; potatoes; plantains; coca; livestock; wool; fish
Industry: mining; petroleum; fishing; textiles and clothing; food processing; cement; auto assembly; steel; shipbuilding; metal fabrication
Exports: $8.9 billion (primary partners United States, Japan, China)
Imports: $8.2 billion (primary partners United States, Colombia, Venezuela)

SUGGESTED WEBSITE

http://www.cia.org/cia/
publications/factbook/index.html

Peru Country Report

PERU: HEIR TO THE INCAS

The culture of Peru, from pre-Hispanic days to the present, has in many ways reflected the nation's variegated geography and climate. While 55 percent of the nation is covered with jungle, coastal Peru boasts one of the world's driest deserts. Despite its forbidding character, irrigation of the desert is made possible by run-offs from the Andes. This allows for the growing of a variety of crops in fertile oases that comprise about 5.5 percent of the land area.

Similarly, in the highlands, or *sierra*, there is little land available for cultivation. Because of the difficulty of the terrain, only about 7 percent of the land can produce crops. Indeed, Peru contains the lowest per capita amount of arable land in South America. The lack of fertile land has had—and continues to have—profound social and political repercussions, especially in the southern highlands near the city of Ayacucho.

THE SUPREMACY OF LIMA

Historically, coastal Peru and its capital city of Lima have attempted to dominate the sierra—politically, economically, and, at

times, culturally. Long a bureaucratic and political center, in the twentieth century Lima presided over the economic expansion of the coast. Economic opportunity in combination with severe population pressure in the sierra caused Lima and its port of Callao to grow tremendously in population, if not in services.

DEVELOPMENT

President Toledo's attempts to stimulate the economy with an infusion of foreign investments, privatization, a renegotiation of outstanding agreements with the IMF have been overshadowed by political scandal.

Ironically, the capital city has one of the worst climates for dense human settlement. Thermal inversions are common; between May and September, they produce a cloud ceiling and a pervasive cool fog.

Middle- and upper-class city dwellers have always been ignorant of the people of the highlands. Very few know either Quechua or Aymara, the Indian languages spoken daily by millions of Peruvians.

Yet this ignorance of the languages—and, by extension, of the cultures—has not prevented government planners or well-meaning intellectuals from trying to impose a variety of developmental models on the inhabitants of the sierra. In the late nineteenth century, for example, modernizers known collectively in Latin America as Positivists sought in vain to transform indigenous cultures by Europeanizing them. Other reformers sought to identify with the indigenous peoples. In the 1920s, a young intellectual named Victor Raúl Haya de la Torre fashioned a political ideology called APRISMO, which embraced the idea of an alliance of Indoamerica to recover the American states for their original inhabitants. While his broader vision proved to be too idealistic, the specific reforms he recommended for Peru were put into effect by reform-minded governments in the 1960s and 1970s. Sadly, reform continued to be developed and imposed from Lima, without an understanding of the rationale behind existing agrarian systems or an appreciation of a peasant logic that was based not on production of a surplus but on attaining a satisfying level of well-being. Much of the turmoil in rural Peru today stems from the agrarian reform of 1968–1979.

AGRARIAN REFORM

From the mid-1950s, rural laborers in the central and southern highlands and on the coastal plantations demonstrated an increasingly insistent desire for agrarian reform. Peasant communities in the sierra staged a series of land invasions and challenged the domination of the large estate, or *hacienda*, from outside. Simultaneously, tenants living on the estates pressured the hacienda system from within. In both cases, peasants wanted land.

The Peruvian government responded with both the carrot and the stick. A military regime, on the one hand, tried to crush peasant insurgency in 1962 and, on the other, passed agrarian reform legislation. The laws had no practical effect, but they did give legal recognition to the problem of land reform. In the face of continued peasant unrest in the south, the military enacted more substantial land laws in 1963, confiscating some property and redistributing it to peasants. The trend toward reform continued with the election of Francisco Belaunde Terry as president of a civilian government.

FREEDOM

President Toled's anti-corruption campaign ground to a halt in 2004 as the archaic court system proved totally incapable of handling the case load.

In the face of continued peasant militancy, Belaunde promised far-ranging reforms, but a hostile Congress refused to provide sufficient funds to implement the proposed reforms. Peasant unrest increased, and the government feared the development of widespread rural guerrilla warfare.

Against this backdrop of rural violence, the Peruvian military again seized power in 1968. To the astonishment of most observers, the military chose not to crush popular unrest but, rather, to embrace reforms. Clearly, the military had become sensitive to the political, social, and economic inequalities in Peru that had bred unrest. The military was intent on revolutionizing Peru from the top down rather than waiting for revolution from below.

In addition to land reform, the military placed new emphasis on Peru's Indian heritage. Tupac Amaru, an Incan who had rebelled against Spanish rule in 1780–1781, became a national symbol. In 1975, Quechua, the ancient language of the Inca, became Peru's second official language (along with Spanish). School curricula were revised and approached Peru's Indian heritage in a new and positive light.

NATIONALIZATION AND INTEGRATION

Behind the reforms, which were extended to industry and commerce and included the nationalization of foreign enterprises, lay the military's desire to provide for Peru a stable social and political order. The military leaders felt that they could provide better leadership in the quest for national integration and economic development than could "inefficient" civilians. Their ultimate goal was to construct a new society based on citizen participation at all levels.

As is so often the case, however, the reform model was not based on the realities of the society. It was naively assumed by planners that the Indians of the sierra were primitive socialists and wanted collectivized ownership of the land. In reality, each family's interests tended to make it competitive, rather than cooperative, with every other peasant family. Collectivization in the highlands failed because peasant communities outside the old hacienda structure clamored for the return of traditional lands that had been taken from them over the years. The Peruvian government found itself, awkwardly, attempting to defend the integrity of the newly reformed units from peasants who wanted their own land.

THE PATRON

Further difficulties were caused by the disruption of the patron–client relationship in the more traditional parts of the sierra. Hacienda owners, although members of the ruling elite, often enjoyed a tight bond with their tenants. Rather than a boss–worker relationship, the patron–client tie came close to kinship. Hacienda owners, for example, were often godparents to the children of their workers. A certain reciprocity was expected and given. But with the departure of the hacienda owners, a host of government bureaucrats arrived on the scene, most of whom had been trained on the coast and were ignorant of the customs and languages of the sierra. The peasants who benefited from the agrarian reform looked upon the administrators with a good deal of suspicion. The agrarian laws and decrees, which were all written in Spanish, proved impossible for the peasants to understand. Not surprisingly, fewer than half of the sierra peasants chose to join the collectives; and in a few places, peasants actually asked for the return of the hacienda owner, someone to whom they could relate. On the coast, the cooperatives did not benefit all agricultural workers equally, since permanent workers won the largest share of the benefits. In sum, the reforms had little impact on existing trends in agricultural production, failed to reverse income inequalities within the peasant population, and did not ease poverty.

The shortcomings of the reforms—in combination with drought, subsequent crop failures, rising food prices, and population pressure—created very difficult and tense situations in the sierra. The infant mortality rate rose 35 percent between 1978 and 1980, and caloric intake dropped well below the recommended minimum. More than half of the children under age six suffered from some form of malnutrition. Rural unrest continued.

RETURN TO CIVILIAN RULE

Unable to solve Peru's problems and torn by divisions within its ranks, the military stepped aside in 1980, and Belaunde was again elected as Peru's constitutional president. Despite the transition to civilian government, unrest continued in the highlands, and the appearance of a left-wing guerrilla organization known as Sendero Luminoso ("Shining Path") led the government to declare repeated states of emergency and to lift civil guarantees.

In an attempt to control the situation, the Ministry of Agriculture won the power to restructure and, in some cases, to liquidate the cooperatives and collectives established by the agrarian reform. Land was divided into small individual plots and given to the peasants. Because the plots can be bought, sold, and mortgaged, some critics argue that the undoing of the reform may hasten the return of most of the land into the hands of a new landed elite.

HEALTH/WELFARE

Peru's poor and the unemployed expect President Toledo to adopt policies that will stimulate the economy in ways that will generate employment and provide the revenue necessary for health care, social programs, and education.

Civilian rule, however, has not necessarily meant democratic rule for Peru's citizens. This helps to explain the spread of Sendero Luminoso despite its radical strategy and tactics of violence. By 1992, according to Diego García-Sayán, the executive director of the Andean Commission of Jurists, the Sendero Luminoso controlled "many parts of Peruvian territory. Through its sabotage, political assassinations, and terrorist actions, Sendero Luminoso has helped to make political violence, which used to be rather infre-

(United Nations photo)

Machu Picchu, a famous Inca ruin, stands atop a 6,750-foot mountain in the Peruvian Andes.

quent, one of the main characteristics of Peruvian society."

Violence was not confined to the guerrillas of Sendero Luminoso or of the Tupac Amaru Revolutionary Movement (MRTA). Economist Javier Iguíñiz, of the Catholic University of Lima, argued that a solution to the violence required an understanding that it flowed from disparate, autonomous, and competing sources, including guerrillas, right-wing paramilitary groups, the Peruvian military and police forces, and cocaine traffickers, "particularly the well-armed Colombians active in the Huallaga Valley." Sendero Luminoso, until recently, was also active in the Huallaga Valley and profited from taxing drug traffickers. Raúl González, of Lima's Center for Development Studies, observed that as both the drug traffickers and the guerrillas "operate[d] outside the law, there has evolved a relationship of mutual convenience in certain parts of Huallaga to combat their common enemy, the state."

President Alan García vacillated on a policy toward the Sendero Luminoso insurgency. But ultimately, he authorized the launching of a major military offensive against Sendero Luminoso bases thought to be linked to drug trafficking. Later, determined to confront an insurgency that claimed 69,000 victims, President Alberto Fujimori armed rural farmers, known as *rondas campesinas,* to fight off guerrilla incursions. (The arming of peasants is not new to Peru; it is a practice that dates to the colonial period.) Critics correctly feared

that the accelerated war against insurgents and drug traffickers would only strengthen the Peruvian military's political power.

A BUREAUCRATIC REVOLUTION?

Peruvian author Hernando DeSoto's bestselling and controversial book *The Other Path* (as opposed to Sendero Luminoso, or Shining Path), argues convincingly that both left- and right-wing governments in Latin America in general and in Peru in particular are neo-mercantile—that is, both intervene in the economy and promote the expansion of state activities. "Both strengthened the role of the government's bureaucracy until they made it the main obstacle, rather than the main incentive, to progress, and together they produced, without consulting the electorate, almost 99 percent of the laws governing us." There are differences between leftand right-wing approaches: The left governs with an eye to redistributing wealth and well-being to the neediest groups, and the right tends to govern to serve foreign investors or national business interests. "Both, however, will do so with bad laws which explicitly benefit some and harm others. Although their aims may seem to differ, the result is that in Peru one wins or loses by political decisions. Of course, there is a big difference between a fox and a wolf but, for the rabbit, it is the similarity that counts."

DeSoto attacked the bureaucracy head-on when his private research center, the Institute for Liberty and Democracy, drafted legislation to abolish a collection of requirements built on the assumption that citizens are liars until proven otherwise. The law, which took effect in April 1989, reflected a growing rebellion against bureaucracy in Peru. Another law, which took effect in October 1989, radically simplified the process of gaining title to land. (DeSoto discovered that, to purchase a parcel of state-owned land in Peru, one had to invest 56 months of effort and 207 visits to 48 different offices). The legislation will have an important impact on the slum dwellers of Lima, for it will take much less time to regularize land titles as the result of invasions and seizures. Slum dwellers with land titles, according to DeSoto, invest in home improvements at a rate nine times greater than that of slum dwellers without titles. Slum dwellers who own property will be less inclined to turn to violent solutions to their problems.

The debureaucratization campaign has been paralleled by grassroots social movements that grew in response to a state that no longer could or would respond to the needs of its citizenry. Cataline Romero, director of the Bartolome de Las Cases Institute of Lima, said that "grass-roots social movements have blossomed into political participants that allow historically marginalized people to feel a sense of their own dignity and rights as citizens." Poor people have developed different strategies for survival as the government has failed to meet even their most basic needs. Most have entered the informal sector and have learned to work together through the formation of unions, mothers' clubs, and cooperatives. Concluded Romero: "As crisis tears institutions down, these communities are preparing the ground for building new institutions that are more responsive to the needs of the majority." DeSoto concurs and adds: "No one has ever considered that most poor Peruvians are a step ahead of the revolutionaries and are already changing the country's structures, and what politicians should be doing is guiding the change and giving it an appropriate institutional framework so that it can be properly used and governed."

DEMOCRACY AND THE "SELF-COUP"

In April 1992, President Fujimori, increasingly isolated and unable to effect economic and political reforms, suspended the Constitution, arrested a number of opposition leaders, shut down Congress, and openly challenged the power of the judiciary. The military, Fujimori's staunch ally, openly supported the *autogolpe*, or "self-coup," as did business leaders and about 80 percent of the Peruvian people. In the words of political scientist Cynthia McClintock, writing in *Current History*, "Fujimori emerged a new caudillo, destroying the conventional wisdom that institutions, whether civilian or military, had become more important than individual leaders in Peru and elsewhere in Latin America." In 1993, a constitutional amendment allowed Fujimori to run for a second consecutive term.

In April 1995, Fujimori won a comfortable victory, with 64 percent of the vote. This was attributable to his successful economic policies, which saw the Peruvian economy grow by 12 percent—the highest in the world for 1994—and the campaign against Sendero Luminoso.

This represented the high point of Fujimori's administration. Increasingly dicta-

torial behavior and a fraudulent election in 2000, coupled with a severe economic slump precipitated by the crisis in Asian financial markets and the chaos wreaked on the infrastructure, coastal agriculture, and fishing industry by the weather phenomenon known as El Niño, undermined Fujimori's popularity. Rampant corruption was symbolized by one woman who, according to *The Christian Science Monitor*, "became so disgusted with her country's electoral fraud and corruption … that she undertook a simple but memorable political protest: handwashing the Peruvian flag in a public square for months on end."

Fujimori's decision to run for a third term, despite a constitutional prohibition, was followed by an election in April 2000 that observers characterized as "rife with fraud." Prodemocracy forces led by Alejandro Toledo, a one-time shoeshine boy, boycotted the run-off election and helped to organize a massive national protest march against Fujimori's swearing-in ceremony in July. Violence in the streets, press censorship, and revelations of massive corruption by Fujimori's intelligence chief, Vladomiro Montesinos, forced Fujimori to resign from office and flee the country. Interim president Valentin Paniagua began the process of national reconstruction and created several commissions to investigate corruption and human-rights abuses.

His task has been daunting. Despite modest economic growth, which stood at 5 percent in 2001, Toledo has seen his popularity tumble from 60 percent in 2001 to 10 percent in 2004. Persistent corruption and scandal in government, his failure to de-

liver on campaign promises of jobs, prosperity, and a return to democracy have hamstrung his administration. Indeed, former President Fujimori, whose supporters fondly refer to him as "El Chino"—and whose detractors call him "Chinochet"—still retains a large measure of popularity despite outstanding criminal charges. Many people support Fujimori because he is perceived as strong and decisive. Toledo is considered weak and indecisive. Troubling also is the reappearance of Sendero Luminoso in 2003. Although small in number they have attacked security personnel, taken hostages, and initiated a rural campaign to win peasant support. They are well-financed because of their ties to Colombian cocaine traffickers. A number of observers feel that Toledo will not complete his term.

Timeline: PAST

1500
The Inca Empire is at its height

1535
The Spanish found Lima

1821
Independence is proclaimed

1955
Women gain the right to vote

1968
A military coup: far-reaching reforms are pursued

1989
Debureaucratization campaign begins

1990s
El Niño spreads economic havoc and human misery; privatization

PRESENT

2000s
President Alberto Fujimori resigns

Reappearance of Sendero Luminoso in 2003

Suriname (Republic of Suriname)

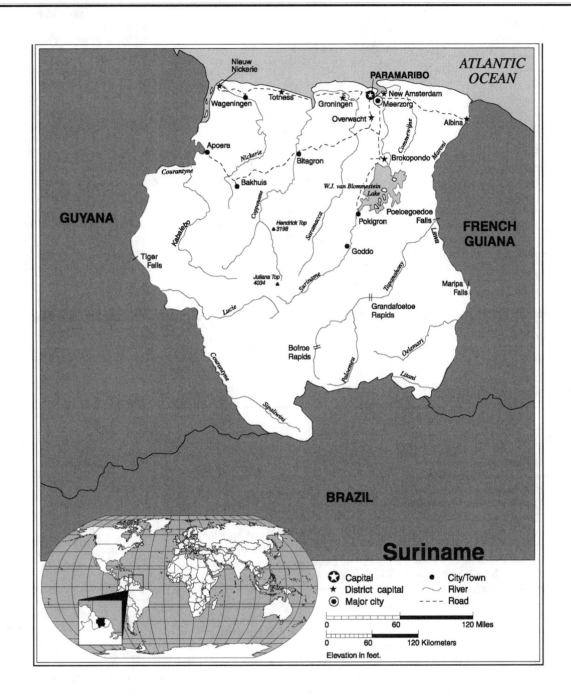

Suriname Statistics

GEOGRAPHY

Area in Square Miles (Kilometers): 63,037 (163,265) (about the size of Georgia)

Capital (Population): Paramaribo (216,000)

Environmental Concerns: deforestation; water pollution; threatened wildlife populations

Geographical Features: mostly rolling hills; a narrow coastal plain with swamps; mostly tropical rain forest

Climate: tropical

PEOPLE

Population

Total: 436,935

Annual Growth Rate: 0.31%

Rural/Urban Population Ratio: 50/50

Major Languages: Dutch; Sranantonga; English; Hindustani

Ethnic Makeup: 37% Hindustani (locally called East Indian); 31% Creole; 15% Javanese; 10% Bush Negro; 3% Amerindian; 3% Chinese

Religions: 27% Hindu; 25% Protestant; 23% Roman Catholic; 20% Muslim; 5% others

Health

Life Expectancy at Birth: 66 years (male); 71 years (female)

Infant Mortality Rate (Ratio): 25/1,000

Physicians Available (Ratio): 1/1,348

Education

Adult Literacy Rate: 93%

Compulsory (Ages): 6–16; free

COMMUNICATION

Telephones: 78,700 main lines

Daily Newspaper Circulation: 107 per 1,000 people

Televisions: 146 per 1,000 people

Internet Users: 20,000 (2002)

TRANSPORTATION

Highways in Miles (Kilometers): 2,813 (4,530)

Railroads in Miles (Kilometers): 103 (166)

Usable Airfields: 46

Motor Vehicles in Use: 66,000

GOVERNMENT

Type: constitutional democracy

Independence Date: November 25, 1975 (from the Netherlands)

Head of State/Government: President Runaldo Ronald Venetiaan is both head of state and head of government

Political Parties: New Front; Progressive Reform Party; National Democratic Party; National Party; others

Suffrage: universal at 18

MILITARY

Military Expenditures (% of GDP): 0.7%

Current Disputes: territorial disputes with Guyana and French Guiana

ECONOMY

Currency ($ U.S. Equivalent): 2,346 guilders = $1

Per Capita Income/GDP: $3,500/$1.53 billion

GDP Growth Rate: 1.5%

Inflation Rate: 17%

Unemployment Rate: 17%

Labor Force: 100,000

Natural Resources: timber; hydropower; fish; kaolin; shrimp; bauxite; gold; nickel; copper; platinum; iron ore

Agriculture: paddy rice; bananas; palm kernels; coconuts; plantains; peanuts; livestock; forest products; shrimp

Industry: bauxite and gold mining; alumina and aluminum production; lumbering; food processing; fishing

Exports: $495 million (primary partners Norway, Netherlands, United States)

Imports: $604 million (primary partners United States, Netherlands, Trinidad and Tobago)

SUGGESTED WEBSITE

http://www.cia.gov/cia/
publications/factbook/index.html

Suriname Country Report

SURINAME: A SMALL-TOWN STATE

Settled by the British in 1651, Suriname, a small colony on the coast of Guiana, prospered with a plantation economy based on cocoa, sugar, coffee, and cotton. The colony came under Dutch control in 1667; in exchange, the British were given New Amsterdam (Manhattan, New York). The colony was often in turmoil because of Indian and slave uprisings, which took advantage of a weak Dutch power. When slavery was finally abolished, in 1863, plantation owners brought contract workers from China, India, and Java.

DEVELOPMENT

The bauxite industry, which had been in decline for 2 decades, now accounts for 15 percent of GDP and 70 percent of export earnings.

On the eve of independence of the Netherlands in 1975, Suriname was a complex, multiracial society. Although existing ethnic tensions were heightened as communal groups jockeyed for power in the new state, other factors cut across racial lines. Even though Creoles (native-born whites) were dominant in the bureaucracy as well as in the mining and industrial sectors, there was sufficient economic opportunity for all ethnic groups, so acute socioeconomic conflict was avoided.

THE POLITICAL FABRIC

Until 1980, Suriname enjoyed a parliamentary democracy that, because of the size of the nation, more closely resembled a small town or extended family in terms of its organization and operation. The various ethnic, political, and economic groups that comprised Surinamese society were united in what sociologist Rob Kroes describes as an "oligarchic web of patron-client relations" that found its expression in government. Through the interplay of the various groups, integration in the political process and accommodation of their needs were achieved. Despite the fact that most interests had access to the center of power, and despite the spirit of accommodation and cooperation, the military seized power early in 1980.

THE ROOTS OF MILITARY RULE

In Kroes's opinion, the coup originated in the army among noncommissioned officers, because they were essentially outside the established social and political system—they were denied their "rightful" place in the patronage network. The officers had a high opinion of themselves and resented what they perceived as discrimination by a wasteful and corrupt government. Their demands for reforms, including recognition of an officers' union, were ignored. In January 1980, one government official talked of disbanding the army altogether.

FREEDOM

The Venetiaan government successfully brought to an end the Maroon insurgency of 8 years' duration. Under the auspices of the Organization of American States, the rebels turned in their weapons, and an amnesty for both sides in the conflict was declared.

The coup, masterminded and led by Sergeant Desire Bouterse, had a vague, undefined ideology. It claimed to be nationalist; and it revealed itself to be puritanical, in that it lashed out at corruption and demanded that citizens embrace civic duty and a work ethic. Ideological purity was maintained by government control or censorship of a once-free media. Wavering between left-wing radicalism and middle-of-the-road moderation, the rapid shifts in

Bouterse's ideological declarations suggest that this was a policy designed to keep the opposition off guard and to appease factions within the military.

The military rule of Bouterse seemed to come to an end early in 1988, when President Ramsewak Shankar was inaugurated. However, in December 1990, Bouterse masterminded another coup. The military and Bouterse remained above the rule of law, and the judiciary was not able to investigate or prosecute serious cases involving military personnel.

HEALTH/WELFARE

Amerindians and Maroons (the descendants of escaped African slaves) who live in the interior have suffered from the lack of educational and social services, partly from their isolation and partly from insurgency. With peace, however, it is hoped that the health, education, and general welfare of these peoples will improve.

With regard to Suriname's economic policy, most politicians see integration into Latin American and Caribbean markets as critical. The Dutch, who suspended economic aid after the 1990 coup, restored their assistance with the election of President Ronald Venetiaan in 1991. But civilian authorities were well aware of the roots

of military rule and pragmatically allowed officers a role in government befitting their self-perceived status.

ACHIEVEMENTS

Suriname, unlike most other developing countries, has a small foreign debt and a relatively strong repayment capacity. This is substantially due to its export industry.

In 1993, Venetiaan confronted the military when it refused to accept his choice of officers to command the army. Army reform was still high on the agenda in 1995 and was identified by President Venetiaan as one of his government's three great tasks. The others were economic reform necessary to ensure Dutch aid and establish the country's eligibility for international credit; and the need to reestablish ties with the interior to consolidate an Organization of American States–brokered peace, after almost a decade of insurgency.

A loan negotiated with the Dutch in 2001 will help Suriname to develop agriculture "bauxite" and the gold-mining industry. Unfortunately the development policy also threatens deforestation, because of timber exports, and the pollution of waterways as a result of careless mining practices. Housing and health care also rank highly on the government's list of pri-

orities, under President Jules Wijdenbosch. The government realizes that it cannot forever depend on the largesse of the Netherlands. De Ware Tijd notes that the planning and development minister stated that aid must be sought from other countries and that Suriname must increasingly rely on its own resources.

Timeline: PAST

1651
British colonization efforts

1667
The Dutch receive Suriname from the British in exchange for New Amsterdam

1975
Independence of the Netherlands

1980s
A military coup

1990s
A huge drug scandal implicates high-level government officials

PRESENT

2000s
The Netherlands extends loan aid

Uruguay (Oriental Republic of Uruguay)

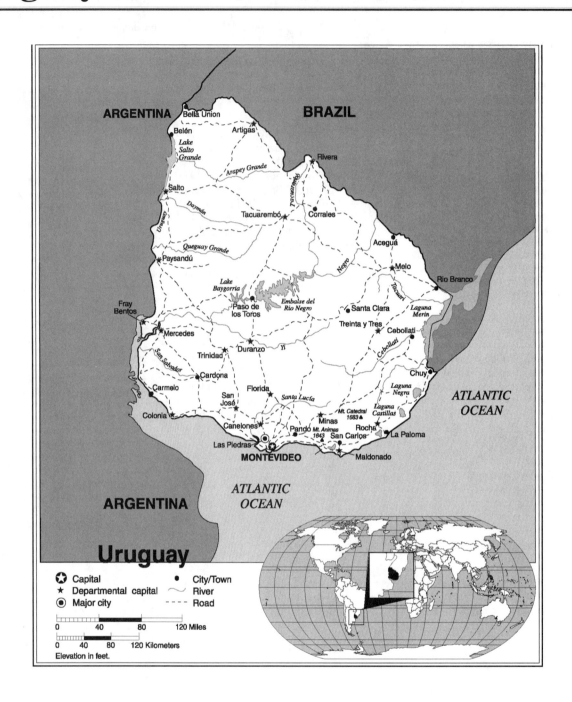

Uruguay Statistics

GEOGRAPHY

Area in Square Miles (Kilometers): 68,037 (176,215) (about the size of Washington State)

Capital (Population): Montevideo (1,304,000)

Environmental Concerns: transboundary pollution from Brazilian power plant; water pollution; waste disposal

Geographical Features: mostly rolling plains and low hills; fertile coastal lowland

Climate: warm temperate

PEOPLE

Population

Total: 3,399,237

Annual Growth Rate: 0.51%

Rural/Urban Population Ratio: 9/91

Major Languages: Spanish; Portunol: Brazilero

Ethnic Makeup: 88% white; 8% Mestizo; 4% black

Religions: 66% Roman Catholic; 2% Protestant; 2% Jewish; 30% nonprofessing or others

Health

Life Expectancy at Birth: 72 years (male); 79 years (female)

Infant Mortality Rate (Ratio): 15.2/1,000

Physicians Available (Ratio): 1/282

Education

Adult Literacy Rate: 97.3%

Compulsory (Ages): for 6 years between 6 and 14; free

COMMUNICATION

Telephones: 946,500 main lines

Daily Newspaper Circulation: 241 per 1,000 people

Televisions: 191 per 1,000

Internet Users: 400,000 (2002)

TRANSPORTATION

Highways in Miles (Kilometers): 5,390 (8,983)

Railroads in Miles (Kilometers): 1,243 (2,073)

Usable Airfields: 65

Motor Vehicles in Use: 525,000

GOVERNMENT

Type: republic

Independence Date: August 25, 1828 (from Brazil)

Head of State/Government: President Jorge Battle is both head of state and head of government

Political Parties: National (Blanco) Party factions; Colorado Party factions; Broad Front Coalition; others

Suffrage: universal and compulsory at 18

MILITARY

Military Expenditures (% of GDP): 2%

Current Disputes: boundary disputes with Brazil

ECONOMY

Currency ($ U.S. Equivalent): 28.21 pesos = $1

Per Capita Income/GDP: $12,600/$42.94 billion

GDP Growth Rate: 0.3%

Inflation Rate: 10.2%

Unemployment Rate: 16.1%

Labor Force: 1,380,000

Natural Resources: arable land; hydropower; minor minerals; fisheries

Agriculture: wheat; rice; corn; sorghum; livestock; fish

Industry: food processing; textiles; chemicals; beverages; transportation equipment; petroleum products

Exports: $2.1 billion (primary partners Mercosur, European Union, United States)

Imports: $1.9 billion (primary partners Mercosur, European Union, United States)

SUGGESTED WEBSITES

http://www.cia.gov/cia/
publications/factbook/index.html

Uruguay Country Report

URUGUAY: ONCE A PARADISE

The modern history of Uruguay begins with the administration of President José Batlle y Ordoñez. Between 1903 and 1929, Batlle's Uruguay became one of the world's foremost testing grounds for social change, and it eventually became known as the "Switzerland of Latin America." Batlle's Colorado Party supported a progressive role for organized labor and formed coalitions with the workers to challenge the traditional elite and win benefits. Other reforms included the formal separation of church and state, nationalization of key sectors of the economy, and the emergence of mass-based political parties. Batlle's masterful leadership was facilitated by a nation that was compact in size; had a small, educated, and homogeneous population; and had rich soil and a geography that facilitated easy communication and national integration.

DEVELOPMENT

Uruguay's people, in an assertion of their independence, in 2003 rejected recommendations made by the IMF with regard to economic reform.

Although the spirit of Batllismo eventually faded after his death in 1929, Batlle's legacy is still reflected in many ways. Reports on income distribution reveal an evenness that is uncommon in developing countries. Extreme poverty is unusual in Uruguay, and most of the population enjoy an adequate diet and minimal standards of living. Health care is within the reach of all citizens. And women in Uruguay are granted equality before the law, are present in large numbers at the national university, and have access to professional careers.

FREEDOM

Uruguay's military is constitutionally prohibited from involvement in issues of domestic security unless ordered to do so by civilian authorities. The press is free and unrestricted, as is speech. The political process is open, and academic freedom is the norm in the national university.

But this model state fell on bad times beginning in the 1960s. Runaway inflation, declining agricultural production, a swollen bureaucracy, official corruption, and bleak prospects for the future led to the appearance of youthful middle-class urban

guerrillas. Known as Tupamaros, they first attempted to jar the nation to its senses with a Robin Hood–style approach to reform. When that failed, they turned increasingly to terrorism in an effort to destroy a state that resisted reform. The Uruguayan government was unable to quell the rising violence. It eventually called on the military, which crushed the Tupamaros and then drove the civilians from power in 1973.

RETURN TO CIVILIAN RULE

In 1980, the military held a referendum to try to gain approval for a new constitution. Despite extensive propaganda, 60 percent of Uruguay's population rejected the military's proposals and forced the armed forces to move toward a return to civilian government. Elections in 1984 returned the Colorado Party to power, with Julio Maria Sanguinetti as president.

By 1989, Uruguay was again a country of laws, and its citizens were anxious to heal the wounds of the 1970s. A test of the nation's democratic will involved the highly controversial 1986 Law of Expiration, which effectively exempted military and police personnel from prosecution for alleged human-rights abuses committed under orders during the military regime.

Many Uruguayans objected and created a pro-referendum commission. They invoked a provision in the Constitution that is unique to Latin America: *Article 79* states that if 25 percent of eligible voters sign a petition, it will initiate a referendum, which, if passed, will implicitly annul the Law of Expiration. Despite official pressure, the signatures were gathered. The referendum was held on April 16, 1989. It was defeated by a margin of 57 to 43 percent.

HEALTH/WELFARE

Uruguay compares favorably with all of Latin America in terms of health and welfare. Medical care is outstanding, and the quality of public sanitation equals or exceeds that of other developing countries. Women, however, still experience discrimination in the workplace.

The winds of free-market enterprise and privatization are starting to blow through the country. When Sanguinetti regained the presidency in 1994, he was expected, as the leader of the Colorado Party—the party of José Batlle—to maintain the economic status quo. But in 1995, he said that his first priority would be to reform the social-security system, which cannot pay for itself, in large part because people in Uruguay are allowed to retire years earlier than in other countries. Reform was also begun in other sectors of the economy. Government employees were laid off, tariffs were reduced, and a program to privatize state industries was inaugurated. The new policies, according to officials, would produce "a change of mentality and culture" in public administration.

ACHIEVEMENTS

Of all the small countries in Latin America, Uruguay has been the most successful in creating a distinct culture. High levels of literacy and a large middle class have allowed Uruguay an intellectual climate that is superior to many much-larger nations.

In his first two years in power, Sanguinetti's successor, Jorge Batlle, has been unable to bring recession to an end. Low prices for agricultural exports, Argentina's economic malaise, and a public debt that stands at 45 percent of gross domestic product present the government with difficult policy decisions. To add to these woes, the appearance of hoof-and-mouth disease in southern Brazil in mid-2001 threatened Uruguay's important beef and wool industries. Once again there was talk of privatization, but a referendum held in December 2003 on the future of ANCAP, the national oil company showed that 62 percent of the electorate wanted no change. Interestingly, these same respondents also oppose monopolies. The failure of the referendum has been seen by some political observers as a signal that Batlle may not win re-election in October 2004.

Timeline: PAST

1624
Jesuits and Franciscans establish missions in the region

1828
Uruguay is established as a buffer state between Argentina and Brazil

1903–1929
The era of President José Batlle y Ordoñez; social reform

1932
Women win the right to vote

1963–1973
Tupamaro guerrillas wage war against the government

1990s
The government endorses sweeping economic and social reforms

PRESENT

2000s
President Battle struggles with the economy

Presidential elections scheduled for October 2004

Venezuela (Bolivarian Republic of Venezuela)

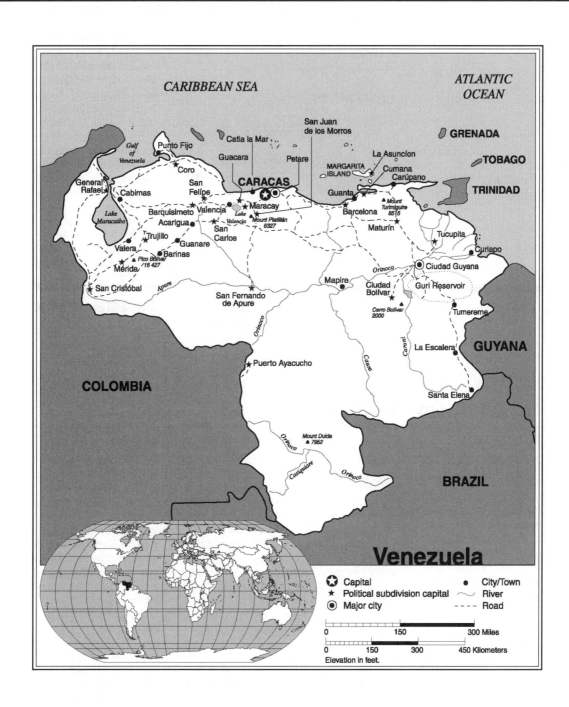

Venezuela

⊛ Capital	● City/Town
★ Political subdivision capital	〜 River
◉ Major city	- - - - Road

0   150   300 Miles
0   150   300   450 Kilometers
Elevation in feet.

Venezuela Statistics

GEOGRAPHY

Area in Square Miles (Kilometers):
352,143 (912,050) (about twice the size of California)

Capital (Population): Caracas (3,700,000)

Environmental Concerns: water, sewage, air, oil, and urban pollution; deforestation; soil degradation

Geographical Features: a flat coastal plain and the Orinoco Delta are bordered by

Andes Mountains and hills; plains (llanos) extend between the mountains and the Orinoco; Guyana Highlands and plains are south of the Orinoco

Climate: varies from tropical to temperate

PEOPLE

Population

Total: 25,017,387
Annual Growth Rate: 1.44%
Rural/Urban Population Ratio: 14/86
Major Languages: Spanish; indigenous
dialects
Ethnic Makeup: 67% Mestizo; 21% white;
10% black; 2% Indian
Religions: 96% Roman Catholic; 4%
Protestant and others

Health

Life Expectancy at Birth: 70 years (male);
76 years (female)
Infant Mortality Rate (Ratio): 23/1,000
Physicians Available (Ratio): 1/576

Education

Adult Literacy Rate: 93%
Compulsory (Ages): 5–15; free

COMMUNICATION

Telephones: 2,841,800 main lines
Daily Newspaper Circulation: 215 per
1,000 people
Televisions: 183 per 1,000

Internet Users: 1,274,400 (2002)

TRANSPORTATION

Highways in Miles (Kilometers): 57,693
(96,155)
Railroads in Miles (Kilometers): 363 (584)
Usable Airfields: 366
Motor Vehicles in Use: 2,025,000

GOVERNMENT

Type: republic
Independence Date: July 5, 1811 (from
Spain)
Head of State/Government: President
Hugo Chavez Frias is both head of state
and head of government
Political Parties: National Convergence;
Social Christian Party; Democratic
Action; Movement Toward Socialism;
Radical Cause; Homeland for All
Suffrage: universal at 18

MILITARY

Military Expenditures (% of GDP): 1.3%
Current Disputes: territorial disputes with
Guyana and Colombia

ECONOMY

Currency ($ U.S. Equivalent): 1,607
bolívars = $1
Per Capita Income/GDP: $4,800/$117.9
billion
GDP Growth Rate: –9.2%
Inflation Rate: 31.1%
Unemployment Rate: 18%
Labor Force: 9,900,000
Natural Resources: petroleum; natural
gas; iron ore; gold; bauxite; other
minerals; hydropower; diamonds
Agriculture: corn; sorghum; sugarcane;
rice; bananas; vegetables; coffee;
livestock; fish
Industry: petroleum; mining; construction
materials; food processing; textiles;
steel; aluminum; motor-vehicle
assembly
Exports: $25.86 billion (primary partners
United States, Colombia, Brazil)
Imports: $10.71 billion (primary partners
United States, Japan, Colombia)

SUGGESTED WEBSITE

http://www.cia.org/cia/
publications/factbook/index.html

Venezuela Country Report

VENEZUELA: CHANGING TIMES

Venezuela is a country in transition. After decades of rule by a succession of *caudillos* (strong, authoritarian rulers), national leaders can now point to four decades of unbroken civilian rule and peaceful transfers of presidential power. Economic growth—stimulated by mining, industry, and petroleum—has, until recently, been steady and, at times, stunning. With the availability of better transportation; access to radio, television, newspapers, and material goods; and the presence of the national government in once-isolated towns, regional diversity is less striking now than a decade ago. Fresh lifestyles and perspectives, dress and music, and literacy and health care are changing the face of rural Venezuela.

THE PROBLEMS OF CHANGE

Such changes have not been without problems—significant ones. Venezuela, despite its petroleum-generated wealth, remains a nation plagued by huge imbalances, inequalities, contradictions, and often bitter debate over the meaning and direction of national development. Some critics note the danger of the massive rural-to-urban

population shift and the influx of illegal immigrants (from Colombia and other countries), both the result of Venezuela's rapid economic development. Others warn of the excessive dependence on petroleum as the means of development and are concerned about the agricultural output at levels insufficient to satisfy domestic requirements. Venezuela, once a food exporter, periodically has had to import large amounts of basic commodities—such as milk, eggs, and meat—to feed the expanding urban populations. Years of easy, abundant money also promoted undisciplined borrowing abroad to promote industrial expansion and has saddled the nation with a serious foreign-debt problem. Government corruption is rampant and, in fact, led to the impeachment of President Carlos Andrés Pérez in 1993.

THE CHARACTER OF MODERNIZATION

The rapid changes in Venezuelan society have produced a host of generalizations as to the nature of modernization in this Andean republic. Commentators who speak of a revolutionary break with the past—of a "new" Venezuela completely severed from its historic roots reaching back to the

sixteenth century—ignore what is enduring about Venezuela's Hispanic culture.

Even before it began producing petroleum, Venezuela was not a sleepy backwater. Its Andean region was always the most prosperous area in the South American continent and was a refuge from the civil wars that swept other parts of the country. There were both opportunity and wealth in the coffee-growing trade. With the oil boom and the collapse of coffee prices in 1929, the Andean region experienced depopulation as migrants left the farms for other regions or for the growing Andean cities. In short, Venezuela's rural economy should not be seen as a static point from which change began but as a part of a dynamic process of continuing change, which now has the production of petroleum as its focus.

CULTURAL IDENTITY

Historian John Lombardi identifies language, culture, and an urban network centered on the capital city of Caracas as primary forces in the consolidation of the nation. "Across the discontinuities of civil war and political transformation, agricultural and industrial economies, rural life styles and urban agglomerations, Venezuela

(United Nations photo/H. Null)

When oil was discovered in Venezuela, rapid economic growth caused many problems in national development. By depending on petroleum as the major source of wealth, Venezuela was at the mercy of the fickle world energy market.

has functioned through the stable network of towns and cities whose interconnections defined the patterns of control, the directions of resource distribution, and the country's identity."

One example of the country's cultural continuity can be seen by looking into one dimension of Venezuelan politics. Political parties are not organized along class lines but tend to cut across class divisions. This is not to deny the existence of class consciousness—which is certainly ubiquitous in Venezuela—but it is not a major *political* force. Surprisingly, popular support for elections and strong party affiliations are more characteristic of rural areas than of cities. The phenomenon cannot be explained as a by-product of modernization. Party membership and electoral participation are closely linked to party organization, personal ties and loyalties, and charismatic leadership. The party, in a sense, becomes a surrogate *patrón* that has power and is able to deliver benefits to the party faithful.

IMPACTS OF URBANIZATION

Another insight into Hispanic political culture can be found in the rural-to-urban shift in population that has often resulted in large-scale seizures of land in urban areas by peasants. Despite the illegality of the seizures, such actions are frequently encouraged by officials because, they argue, it provides the poor with enough land to maintain political stability and to prevent peasants from encroaching on richer neighborhoods. Pressure by the new urban dwellers at election time usually results in their receiving essential services from government officials. In other words, municipal governments channel resources in return for expected electoral support from the migrants. Here is a classic Hispanic response to challenge from below—to bend, to cooperate.

DEVELOPMENT

The policies of President Chavez have led to a good deal of uncertainty in foreign and domestic economic, financial, and commercial circles. The fiscally responsible policies he announced just after taking office—including, bringing about a sharp reduction in the budget deficit by raising new taxes, cutting spending, restructuring debt, and imposing austerity, are in limbo.

Cultural values also underlie both the phenomenon of internal migration and the difficulty of providing adequate skilled labor for Venezuela's increasingly technological economy. While the attraction of the city and its many opportunities is one reason for the movement of population out of rural areas, so too is the Venezuelan culture, which belittles the peasant and rural life in general. Similarly, the shortage of skilled labor is the result not only of inadequate training but also of social values that neither reward nor dignify skilled labor.

THE SOCIETY

The rapid pace of change has contributed to a reexamination of the roles and rights of women in Venezuela. In recent years, women have occupied positions in the cabinet and in the Chamber of Deputies; several women deputies have held important posts in political parties.

Yet while educated women are becoming more prominent in the professions, there is a reluctance to employ women in traditional "men's" jobs, and blatant inequality still blemishes the workplace. Women, for example, are paid less than men for similar work. And although modern feminist goals have become somewhat of a social and economic force, at least in urban centers, the traditional roles of wife and mother continue to hold the most prestige, and physical beauty is still often viewed as a woman's most precious asset. In addition, many men seek deference from women rather than embracing social equality. Nevertheless, the younger generations

113

(Photo Lisa Clyde)

Caracas, Venezuela, an ultra-modern city of 3.7 million exemplifying the extremes of poverty and wealth that exist in Latin America, sprawls for miles over mountains and valleys.

of Venezuelans are experiencing the social and cultural changes that have tended to follow women's liberation in Western industrialized nations: higher levels of education and career skills; broadened intellectualism; increasing freedom and equality for both men and women; relaxed social mores; and the accompanying personal turmoil, such as rising divorce and single-parenthood rates.

FREEDOM

Venezuela has a free and vigorous daily press, numerous weekly news magazines, 3 nationwide television networks, and nearly 200 radio stations. Censorship or interference with the media on political grounds is rare. Venezuela has traditionally been a haven for refugees and displaced persons. "Justice" in the justice system remains elusive for the poor.

Venezuelans generally enjoy a high degree of individual liberty. Civil, personal, and political rights are protected by a strong and independent judiciary. Citizens generally enjoy a free press. There exists the potential for governmental abuse of press freedom, however. Several laws leave journalists vulnerable to criminal charges, especially in the area of libel. Journalists must be certified to

work, and certification may be withdrawn by the government if journalists are perceived to stray from the "truth," misquote sources, or refuse to correct "errors." But as a rule, radio, television, and newspapers are free and are often highly critical of the government.

The civil and human rights enjoyed by most Venezuelans have not necessarily extended to the nation's Indian population in the Orinoco Basin. For years, extra-regional forces—in the form of rubber gatherers, missionaries, and developers—have to varying degrees undermined the economic self-sufficiency, demographic viability, and tribal integrity of indigenous peoples. A government policy that stresses the existence of only one Venezuelan culture poses additional problems for Indians.

In 1991, however, President Pérez signed a decree granting a permanent homeland, encompassing some 32,000 square miles in the Venezuelan Amazon forest, to the country's 14,000 Yanomamö Indians. Venezuela will permit no mining or farming in the territory and will impose controls on existing religious missions. President Pérez stated that "the primary use will be to preserve and to learn the traditional ways of the Indians." As James Brooke reported in *The New York Times,* "Venezuela's move has left anthropologists euphoric."

Race relations are outwardly tranquil in Venezuela, but there exists an underlying racism in nearly all arenas. People are commonly categorized by the color of their skin, with white being the most prized. Indeed, race, not economic level, is still the major social-level determinant. This unfortunate reality imparts a sense of frustration and a measure of hopelessness to many of Venezuela's people, in that even those who acquire a good education and career training may be discriminated against in the workplace because they are "of color." Considering that only one fifth of the population are of white extraction, with 67 percent Mestizos and 10 percent blacks, this is indeed a widespread and debilitating problem.

A VIGOROUS FOREIGN POLICY

Venezuela has always pursued a vigorous foreign policy. In the words of former president Luis Herrera Campins: "Effective action by Venezuela in the area of international affairs must take certain key facts into account: economics—we are a producer-exporter of oil; politics—we have a stable, consolidated democracy; and geopolitics—we are at one and the same time a Caribbean, Andean, Atlantic, and Amazonian country." Venezuela has long assumed that it should be the guardian of Simón Bolívar's ideal of creating an independent and united Latin America. The nation's memory of its continental leadership, which developed during the Wars for Independence (1810–1826), has been rekindled in Venezuela's desire to promote the political and economic integration of both the continent and the Caribbean. Venezuela's foreign policy remains true to the Bolivarian ideal of an independent Latin America. It also suggests a prominent role for Venezuela in Central America. In the Caribbean, Venezuela has emerged as a source of revenue for the many microstates in the region; the United States is not without competitors for its Caribbean policy.

PROMISING PROSPECTS TURN TO DISILLUSIONMENT

The 1980s brought severe turmoil to Venezuela's economy. The boom times of the 1970s turned to hard times as world oil prices dropped. Venezuela became unable to service its massive foreign debt (currently $34 billion) and to subsidize the "common good," in the form of low gas and transportation prices and other amenities. In 1983, the currency, the bolívar, which had remained stable and strong for many years at 4.3 to the U.S. dollar, was devalued, to an official rate of 14.5 bolívars to the dollar. This was a boon to foreign visitors to the country, which became known as one of the world's greatest travel bargains, but a catastrophe for Venezuelans. (In June 2004, the exchange rate was about 1,116 bolívars to the dollar on the free market.)

President Jaime Lusinchi of the Democratic Action Party, who took office early in 1984, had the unenviable job of trying to cope with the results of the preceding years of free spending, high expectations, dependence on oil, and spiraling foreign debt. Although the country's gross national product grew during his tenure (agriculture growth contributed significantly, rising from 0.4 percent of gross national product in 1983 to 6.8 percent in 1986), austerity measures were in order. The Lusinchi government was not up to the challenge. Indeed, his major legacy was a corruption scandal at the government agency Recadi, which was responsible for allocating foreign currency to importers at the official rate of 14.5 bolívars to the dollar. It was alleged that billions of dollars were skimmed, with a number of high-level government officials, including three finance ministers, implicated. Meanwhile, distraught Venezuelans watched inflation and the devalued bolívar eat up their savings; the once-blooming middle class started getting squeezed out.

In the December 1988 national elections, another Democratic Action president, Carlos Andrés Pérez, was elected. Pérez, who had served as president from 1974 to 1979 (presidents may not serve consecutive terms), was widely rumored to have stolen liberally from Venezuela's coffers during that tenure. Venezuelans joked at first that "Carlos Andrés is coming back to get what he left behind," but as the campaign wore on, some political observers were dismayed to hear the preponderance of the naive sentiment that "now he has enough and will really work for Venezuela this time."

One of Pérez's first acts upon re-entering office was to raise the prices of government-subsidized gasoline and public transportation. Although he had warned that tough austerity measures would be implemented, the much-beleaguered and disgruntled urban populace took to the streets in February 1989 in the most serious rioting to have occurred in Venezuela since it became a democracy. Army tanks rolled down the major thoroughfares of Caracas, the capital; skirmishes between the residents and police and military forces were common; looting was widespread. The government announced that 287 people had been killed. Unofficial hospital sources charge that the death toll was closer to 2,000. A stunned Venezuela quickly settled down in the face of the violence, mortified that such a debacle, widely reported in the international press, should take place in this advanced and peaceable country. But tourism, a newly vigorous and promising industry as a result of favorable currency-exchange rates, subsided immediately; it has yet to recover fully.

On February 4, 1992, another ominous event highlighted Venezuela's continuing political and economic weaknesses. Rebel military paratroopers, led by Hugo Chavez, attacked the presidential palace in Caracas and government sites in several other major cities. The coup attempt, the first in Venezuela since 1962, was rapidly put down by forces loyal to President Pérez, who escaped what he described as an as-

sassination attempt. Reaction within Venezuela was mixed, reflecting widespread discontent with Pérez's tough economic policies, government corruption, and declining living standards. A second unsuccessful coup attempt, on November 27, 1992, followed months of public demonstrations against Pérez's government.

Perhaps the low point was reached in May 1993, when Pérez was suspended from office and impeachment proceedings initiated. Allegedly the president had embezzled more than $17 million and had facilitated other irregularities. Against a backdrop of military unrest, Ramón José Velásquez was named interim president.

In December 1993, Venezuelans elected Rafael Caldera, who had been president in a more prosperous and promising era (1969–1974). Caldera's presidency too was fraught with problems. In his first year, he had to confront widespread corruption in official circles, the devaluation of the bolívar, drug trafficking, a banking structure in disarray, and a high rate of violent crime in Caracas. Indeed, in 1997, a relative of President Caldera was mugged and a Spanish diplomat who had traveled to Caracas to negotiate a trade agreement with Venezuela was robbed in broad daylight.

In an attempt to restore order from chaos, President Caldera inaugurated his "Agenda Venezuela" program to address the difficult problems created by deep recession, financial instability, deregulation, privatization, and market reforms. The plan was showing signs of progress when it was undercut by the collapse of petroleum prices.

The stage was thus set for the emergence of a "hero" who would promise to solve all of Venezuela's ills. In the presidential election of 1998, the old parties were swept from power and a populist— the same Hugo Chavez who had attempted a coup in 1992—won with 55 percent of the popular vote. Those who expected change were not disappointed, although some of Chavez's actions have raised concerns about the future of democracy in Venezuela. A populist and a pragmatist, it is difficult to ascertain where Chavez's often contradictory policies will lead. Since taking power in February 1999, he has placed the army in control of the operation of medical clinics and has put soldiers to work on road and sewer repairs and in school and hospital construction. He has talked about the need to cut costs and uproot what he perceives as a deeply corrupt public sector—but he has refused to downsize the bureaucracy. Chavez supports privatization of the nation's pension fund and electric utilities, but he wants to maintain state control over health care and the petroleum industry. He wants more free-trade initiatives with foreign nations but at the same time threatens to prohibit some agricultural imports.

Perhaps of greater concern is Chavez's successful bid to redraft Venezuela's Constitution, to provide "a better version." He claimed that the document had eroded democracy by allowing a political elite to rule without restraint for decades. Chavez's "democratic" vision demands special powers to revamp the economy without congressional approval. Through clever manipulation of the people by means of his own radio and television shows, and newspaper, Chavez intimidated Congress into granting him almost all the power he wanted to enact financial and economic legislation by decree. A referendum in April 1999 gave him a huge majority supporting the creation of an assembly to redraft the Constitution. A draft was completed in November. The political opposition was convinced that the new document would allow Chavez to seek a second consecutive term in office, which had been prohibited in Venezuela, and that he was doing nothing less than creating a dictatorship under the cover of democracy and the law. Their fears have been realized, as the new Constitution allows for consecutive six-year terms.

The trend towards more centralized executive authority continued in 2000 and 2001. When the new 1999 Constitution was "reprinted" in March 2000, critics noted substantial changes from the original—changes that enhanced presidential power. In the same month, a group of retired military officers called on President Chavez to halt the politicization of the armed forces. The president's response was to appoint active-duty officers to a range of important positions in the government, including the state-owned petroleum company and foreign ministry. Organized labor complained that Chavez has attempted to transform the labor movement into an appendage of the ruling political party and has ignored union leadership in direct appeals by the government to rank-and-file workers. He has alienated the Catholic hierarchy over abortion and education issues; and the media, while legally free to criticize the government, have felt the need to exercise self-censorship. Perhaps most ominously, Chavez asked for and received from a compliant Legislature permission to rule by decree on a broad spectrum of issues, from the economy to public security. The *Ley Habilitante* allows him to enact legislation without parliamentary debate or even approval.

Equally radical and unpredictable is Chavez's policy toward neighboring Colombia. Recently Chavez opened a dialogue with Colombia's guerrillas and dismissed Colombia's protests with the statement that the guerrillas held effective power.

Chavez clearly sees himself as a major player in the region and seems to enjoy annoying the United States. He is friends with Fidel Castro, met with Saddam Hussein in Baghdad, and has strengthened relations with a number of Caribbean and Central American states. Venezuela's longstanding boundary dispute with neighboring Guyana has also been resurrected.

Venezuela's future is wholly unpredictable in large measure because its current government is unpredictable. It is not a formula that seems likely to assure long-term success and well-being for all citizens.

Growing dissatisfaction with Chavez's strong-arm rule precipitated street violence in early April 2002. For four days he was apparently forced from power by elements in the military, but demonstrations by Chavez's supporters resulted in his return to office. The political opposition mounted a campaign to gather the signatures necessary to force a recall vote in August 2004. Chavez is likely to meet the challenge by packing the Supreme Court with supporters who could determine the victor in the recall referendum if it is close.

Timeline: PAST

1520
The first Spanish settlement at Cumaná

1822–1829
Venezuela is part of Gran Colombia

1829
Venezuela achieves independence as a separate country

1922
The first productive oil well

1947
Women win the right to vote

1976
Foreign oil companies are nationalized

1980s
Booming public investment fuels inflation; Venezuela seeks renegotiation of foreign debt

1990s
Social and economic crisis grips the nation; Hugo Chavez wins the presidency and sets about to redraft the Constitution Chavez's government is challenged by massive flooding that leaves more than 30,000 people dead and many more homeless

PRESENT

2000s
A new Constitution is approved

Chavez strengthens executive power; Chavez is reelected

Recall vote scheduled for August 2004

The Caribbean

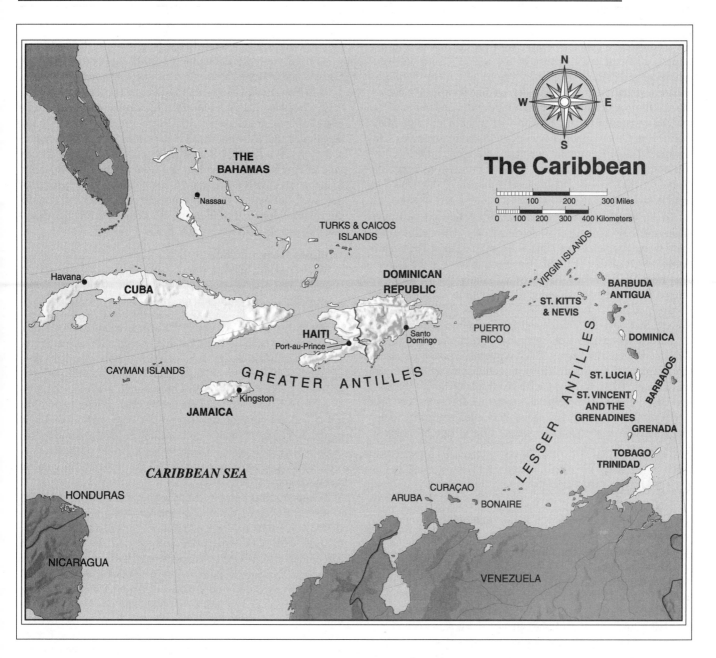

The Caribbean region consists of hundreds of islands stretching from northern South America to the southern part of Florida. Many of the islands cover just a few square miles and are dominated by a central range of mountains; only Cuba has any extensive lowlands. Almost every island has a ring of coral, making approaches very dangerous for ships. The land that can be used for agriculture is extremely fertile; but many islands grow only a single crop, making them vulnerable to fluctuations in the world market in that particular commodity.

The Caribbean: Sea of Diversity

To construct a coherent overview of the Caribbean is an extremely difficult task because of the region's profound geographical and cultural diversity. "The history of the Caribbean is the examination of fragments, which, like looking at a broken vase, still provides clues to the form, beauty, and value of the past." So writes historian Franklin W. Knight in his study of the Caribbean. Other authors have drawn different analogies: Geographer David Lowenthal and anthropologist Lambros Comitas note that the West Indies "is a set of mirrors in which the lives of black, brown, and white, of American Indian and East Indian, and a score of other minorities continually interact."

For the geographer, the pieces fall into a different pattern, consisting of four distinct geographical regions. The first contains the Bahamas as well as the Turks and Caicos Islands. The Greater Antilles—consisting of Cuba, Hispaniola (Haiti and the Dominican Republic), Jamaica, the Cayman Islands, Puerto Rico, and the Virgin Islands—make up the second region. Comprising the third region are the Lesser Antilles—Antigua and Barbuda, Dominica, St. Lucia, St. Vincent and the Grenadines, Grenada, and St. Kitts and Nevis as well as various French departments and British and Dutch territories. The fourth group consists of islands that are part of the South American continental shelf: Trinidad and Tobago, Barbados, and the Dutch islands of Aruba, Curaçao, and Bonaire. Within these broad geographical regions, each nation is different. Yet on each island there often is a firmly rooted parochialism—a devotion to a parish or a village, a mountain valley or a coastal lowland.

CULTURAL DIVERSITY

To break down the Caribbean region into culture groups presents its own set of problems. The term "West Indian" inadequately describes the culturally Hispanic nations of Cuba and the Dominican Republic. On the other hand, "West Indian" does capture the essence of the cultures of Belize, the Caribbean coast of Central America, and Guyana, Suriname, and Cayenne (French Guiana). In Lowenthal's view: "Alike in not being Iberian [Hispanic], the West Indies are not North American either, nor indeed do they fit any ordinary regional pattern. Not so much undeveloped as overdeveloped, exotic without being traditional, they are part of the Third World yet ardent emulators of the West."

EFFORTS AT INTEGRATION

To complicate matters further, few West Indians would identify themselves as such. They are Jamaicans, or Bajans (people from Barbados), or Grenadans. Their economic, political, and social worlds are usually confined to the islands on which they live and work. In the eyes of its inhabitants, each island, no matter how small, is—or should be—sovereign. Communications by air, sea, and telephone with the rest of the world are ordinarily better than communications within the Caribbean region itself. Trade, even between neighboring islands, has always been minimal. Economic ties with the United States or Europe, and in some cases with Venezuela, are more important.

A British attempt to create a "West Indies Federation" in 1958 was reduced to a shambles by 1962. Member states had the same historical background; spoke the same languages; had similar economies; and were interested in the same kinds of food, music, and sports. But their spirit of independence triumphed over any kind of regional federation that "threatened" their individuality. In the words of a former Bajan prime minister, "We live together very well, but we don't like to live together together." A Trinidadian explanation for the failure of the federation is found in a popular calypso verse from the early 1960s:

Plans was moving fine
When Jamaica stab we from behind
Federation bust wide open
But they want Trinidad to bear the burden.

Recently, however, the Windward Islands (Dominica, Grenada, St. Lucia, and St. Vincent and the Grenadines) have discussed political union. While each jealously guards its sovereignty, leaders are nevertheless aware that some integration is necessary if they are to survive in a changing world. The division of the world into giant economic blocs points to political union and the creation of a Caribbean state with a combined population of nearly half a million. Antigua and Barbuda resist because they believe that, in the words of former prime minister Vere Bird, "political union would be a new form of colonialism and undermine sovereignty."

While political union remains problematic, the 15 members of the Caribbean Community and Common Market (CARICOM, a regional body created in 1973) began long-term negotiations with Cuba in 1995 with regard to a free-trade agreement. CARICOM leaders informed Cuba that "it needs to open up its economy more." The free-market economies of CARICOM are profoundly different from Cuba's rigid state controls. "We need to assure that trade and investment will be mutually beneficial." Caribbean leaders have pursued trade with Cuba in the face of strong opposition from the United States. In general, CARICOM countries are convinced that "constructive engagement" rather than a policy of isolation is the best way to transform Cuba.

Political problems also plague the Dutch Caribbean. Caribbean specialist Aaron Segal notes that the six-island Netherlands Antilles Federation has encountered severe internal difficulties. Aruba never had a good relationship with the larger island of Curaçao and, in 1986, became a self-governing entity, with its own flag, Parliament, and currency, but still within the Netherlands. "The other Netherlands Antillean states have few complaints about their largely autonomous relations with the Netherlands but find it hard to get along with one another."

Interestingly, islands that are still colonial possessions generally have a better relationship with their "mother" countries than with one another. Over the past few decades, smaller islands—

populations of about 50,000 or less—have learned that there are advantages to a continued colonial connection. The extensive subsidies paid by Great Britain, France, or the Netherlands have turned dependency into an asset. Serving as tax-free offshore sites for banks and companies as well as encouraging tourism and hotel investments have led to modest economic growth.

CULTURAL IDENTIFICATION

Despite the local focus of the islanders, there do exist some broad cultural similarities. To the horror of nationalists, who are in search of a Caribbean identity that is distinct from Western civilization, most West Indians identify themselves as English or French in terms of culture. Bajans, for example, take a special pride in identifying their country as the "Little England of the Caribbean." English or French dialects are the languages spoken in common.

Nationalists argue that the islands will not be wholly free until they shatter the European connection. In the nationalists' eyes, that connection is a bitter reminder of slavery. After World War II, several Caribbean intellectuals attacked the strong European orientation of the islands and urged the islanders to be proud of their black African heritage. The shift in focus was most noticeable in the French Caribbean, although

this new ethnic consciousness was echoed in the English-speaking islands as well in the form of a black-power movement during the 1960s and 1970s. It was during those years, when the islands were in transition from colonies to associated states to independent nations, that the Caribbean's black majorities seized political power by utilizing the power of their votes.

It is interesting to note that at the height of the black-power and black-awareness movements, sugar production was actually halted on the islands of St. Vincent, Antigua, and Barbuda—not because world-market prices were low, but because sugar cultivation was associated with the slavery of the past.

African Influences

The peoples of the West Indies are predominantly black, with lesser numbers of people of "mixed blood" and small numbers of whites. Culturally, the blacks fall into a number of groups. Throughout the nineteenth century, in Haiti, blacks strove to realize an African-Caribbean identity. African influences have remained strong on the island, although they have been blended with European Christianity and French civilization. Mulattos, traditionally the elite in Haiti, have strongly identified with French culture in an obvious attempt to distance themselves from the black majority, who comprise about 95 percent of the population. African-Creoles, as blacks of the English-speaking

(United Nations photo/King)

These Jamaican agricultural workers, who reflect the strong African heritage of the Caribbean, contribute to the ethnic and cultural diversity of the region.

islands prefer to be called, are manifestly less "African" than the mass of Haitians. An exception to this generalization is the Rastafarians, common in Jamaica and found in lesser numbers on some of the other islands. Convinced that they are Ethiopians, the Rastafarians hope to return to Africa.

Racial Tension

The Caribbean has for years presented an image of racial harmony to the outside world. Yet, in actuality, racial tensions are not only present but also have become sharper during the past few decades. Racial unrest broke to the surface in Jamaica in 1960 with riots in the capital city of Kingston. Tensions heightened again in 1980–1981 and in 1984, to the point that the nation's tourist industry drastically declined. A recent slogan of the Jamaican tourist industry, "Make It Jamaica Again," was a conscious attempt to downplay racial antagonism. The black-power movement in the 1960s on most of the islands also put to the test notions of racial harmony.

Most people of the Caribbean, however, believe in the myth of racial harmony. It is essential to the development of nationalism, which must embrace all citizens. Much racial tension is officially explained as class difference rather than racial prejudice. There is some merit to the class argument. A black politician on Barbuda, for example, enjoys much more status and prestige than a poor white "Redleg" from the island's interior. Yet if a black and a Redleg competed for the job of plantation manager, the white would likely win out over the black. In sum, race does make a difference, but so too does one's economic or political status.

East Indians

The race issue is more complex in Trinidad and Tobago, where there is a large East Indian (i.e., originally from India) minority. The East Indians, for the most part, are agricultural workers. They were originally introduced by the British between 1845 and 1916 to replace slave labor on the plantations. While numbers of East Indians have moved to the cities, they still feel that they have little in common with urban blacks. Because of their large numbers, East Indians are able to preserve a distinctive, healthy culture and community and to compete with other groups for political office and status.

East Indian culture has also adapted, but not yielded, to the West Indian world. In the words of Trinidadian-East Indian author V. S. Naipaul: "We were steadily adopting the food styles of others: The Portuguese stew of tomato and onions … the Negro way with yams, plantains, breadfruit, and bananas," but "everything we adopted became our own; the outside was still to be dreaded…." The East Indians in Jamaica, who make up about 3 percent of the population, have made even more accommodations to the cultures around them. Most Jamaican-East Indians have become Protestant (the East Indians of Trinidad have maintained their Hindu or Islamic faith).

East Indian conformity and internalization, and their strong cultural identification, have often made them the targets of the black majority. Black stereotypes of the East Indians describe them in the following terms: "secretive," "greedy," and "stingy." And East Indian stereotypes describing blacks as

(Photo Lisa Clyde)

The weekly open-air market in St. Lucia provides a variety of local produce.

"childish," "vain," "pompous," and "promiscuous" certainly do not help to ease ethnic tensions.

REVOLUTIONARY CUBA

In terms of culture, the Commonwealth Caribbean (former British possessions) has little in common with Cuba or the Dominican Republic. But Cuba has made its presence felt in other ways. The Cuban Revolution, with the social progress that it entailed for many of its people and the strong sense of nationalism that it stimulated, impressed many West Indians. For new nation-states still in search of an identity, Cuba offered some clues as to how to proceed. For a time, Jamaica experimented with Cuban models of mass mobilization and programs designed to bring social services to the majority of the population. Between 1979 and 1983, Grenada saw merit in the Cuban approach to social and economic problems. The message that Cuba seemed to represent was that a small Caribbean state could shape its own destiny and make life better for its people.

The Cuba of Fidel Castro, while revolutionary, is also traditional. Hispanic culture is largely intact. The politics are authoritarian and personality-driven, and Castro himself easily fits into the mold of the Latin American leader, or caudillo, whose charisma and benevolent paternalism win him the widespread support of his people. Castro's relationship with the Roman Catholic Church is also traditional and corresponds to notions of a dualistic culture that has its roots in the Middle Ages. In Castro's words: "The same respect that the Revolution ought to

(United Nations photo/M. Hopp)

Certain crops in Caribbean countries generate a disproportionate amount of the nation's foreign incomes—so much so that their entire economies are vulnerable to changes in world demand. This harvest of bananas in Dominica is ready for shipment to a fickle world market.

have for religious beliefs, ought also to be had by those who talk in the name of religion for the political beliefs of others. And, above all, to have present that which Christ said: 'My kingdom is not of this world.' What are those who are said to be the interpreters of Christian thought doing meddling in the problems of this world?" Castro's comments should not be interpreted as a Communist assault on religion. Rather, they express a time-honored Hispanic belief that religious life and everyday life exist in two separate spheres.

The social reforms that have been implemented in Cuba are well within the powers of all Latin American governments to enact. Those governments, in theory, are duty-bound to provide for the welfare of their peoples. Constitutionally, the state is infallible and all-powerful. Castro has chosen to identify with the needs of the majority of Cubans, to be a "father" to his people. Again, his actions are not so much Communistic as Hispanic.

Where Castro has run against the grain is in his assault on Cuba's middle class. In a sense, he has reversed a trend that is evident in much of the rest of Latin America—the slow, steady progress of a middle class that is intent on acquiring a share of the power and prestige traditionally accorded to elites. Cuba's middle class was effectively shattered—people were deprived of much of their property; their livelihood; and, for those who fled into exile, their citizenship. Many expatriate Cubans remain bitter toward what they perceive as Castro's betrayal of the Revolution and the middle class.

EMIGRATION AND MIGRATION

Throughout the Caribbean, emigration and migration are a fact of life for hundreds of thousands of people. These are not new phenomena; their roots extend to the earliest days of European settlement. The flow of people looking for work is deeply rooted in history, in contemporary political economy, and even in Caribbean island culture. The Garifuna (black–Indian mixture) who settled in Belize and coastal parts of Mexico, Guatemala, Honduras, and Nicaragua originally came from St. Vincent. There, as escaped slaves, they intermixed with remnants of Indian tribes who had once peopled the islands, and they adopted many of their cultural traits. Most of the Garifuna (or Black Caribs, as they are also known) were deported from St. Vincent to the Caribbean coast of Central America at the end of the eighteenth century.

From the 1880s onward, patois-speaking (French dialect) Dominicans and St. Lucians migrated to Cayenne (French Guiana) to work in the gold fields. The strong identification with Europe has drawn thousands more to what many consider their cultural homes.

High birth rates and lack of economic opportunity have forced others to seek their fortunes elsewhere. Many citizens of the Dominican Republic have moved to New York, and Haitian refugees have thrown themselves on the coast of Florida by the thousands. Other Haitians seek seasonal employment in the Dominican Republic or the Bahamas. There are sizable Jamaican communities in the Dominican Republic, Haiti, the Bahamas, and Belize.

(United Nations photo/J. Viesti)

Economic hardship in parts of the Caribbean region is exemplified by this settlement in Port-au-Prince, Haiti. Such grinding poverty causes large numbers of people to migrate in search of a better life.

On the smaller islands, stable populations are the exception rather than the rule. The people are constantly migrating to larger places in search of higher pay and a better life. Such emigrants moved to Panama when the canal was being cut in the early 1900s or sought work on the Dutch islands of Curaçao and Aruba when oil refineries were built there in the 1920s. They provided much of the labor for the banana plantations in Central America.

The greatest number of people by far have left the Caribbean region altogether and emigrated to the United States, Canada, and Europe. Added to those who have left because of economic or population pressures are political refugees. The majority of these are Cubans, most of whom have resettled in Florida.

Some have argued that the prime mover of migration from the Caribbean lies in the *ideology* of migration—that is, the expectation that all nonelite males will migrate abroad. Sugarcane slave plantations left a legacy that included little possibility of island subsistence; and so there grew the need to migrate to survive, a reality that was absorbed into the culture of lower-class blacks. But for these blacks, there has also existed the expectation to return. (In contrast, middle- and upper-class migrants have historically departed permanently.) Historian Bonham Richardson writes: "By traveling away and returning the people have been able to cope more successfully with the vagaries of man and nature than they would have by staying at home. The small islands of the region are the most vulnerable to environmental and economic uncertainty. Time and again in the Lesser Antilles, droughts, hurricanes, and economic depressions have diminished wages, desiccated provision grounds, and destroyed livestock, and there has been no local recourse to disease or starvation." Hence men and women of the small West Indian islands have been obliged to migrate. "And like migrants everywhere, they have usually considered their travels temporary, partly because they have never been greeted cordially in host communities."

On the smaller islands, such as St. Kitts and Nevis, family and community ceremonies traditionally reinforce and sustain the importance of emigration and return. Funerals reunite families separated by vast distances; Christmas parties and carnival celebrations are also occasions to welcome returning family and friends.

Monetary remittances from relatives in the United States, England, Canada, or the larger islands are a constant reminder of the importance of migration. According to Richardson: "Old men who have earned local prestige by migrating and returning exhort younger men to follow in their footsteps…. Learned cultural responses thereby maintain a migration ethos … that is not only valuable in coping with contemporary problems, but also provides continuity with the past."

The Haitian diaspora (dispersion) offers some significant differences. While Haitian migration is also a part of the nation's history, a return flow is noticeably absent. One of every six Haitians now lives abroad—primarily in Cuba, the Dominican Republic, Venezuela, Colombia, Mexico, and the Bahamas. In French Guiana, Haitians comprise more than 25 percent of the population. They are also found in large numbers in urban areas of the United States, Canada, and France. The typical Haitian emigrant is poor, has little education, and has few skills or job qualifications.

Scholar Christian A. Girault remarks that although "ordinary Haitian migrants are clearly less educated than the Cubans, Dominicans, Puerto Ricans and even Jamaicans, they are not Haiti's most miserable; the latter could never hope to buy an air ticket or boat passage, or to pay an agent." Those who establish new roots in host countries tend to remain, even though they experience severe discrimination and are stereotyped as "undesirable" because they are perceived as bringing with them "misery, magic and disease," particularly AIDS.

There is also some seasonal movement of population on the island itself. Agricultural workers by the tens of thousands are found in neighboring Dominican Republic. *Madames sara*, or peddlers, buy and sell consumer goods abroad and provide "an essential provisioning function for the national market."

AN ENVIRONMENT IN DANGER

When one speaks of soil erosion and deforestation in a Caribbean context, Haiti is the example that usually springs to mind. While that image is accurate, it is also too limiting, for much of the Caribbean is threatened with ecological disaster. Part of the problem is historical, for deforestation began with the development of sugarcane cultivation in the seventeenth century. But now, soil erosion and depletion as well as the exploitation of marginal lands by growing populations perpetuate a vicious cycle between inhabitants and the land on which they live. Cultivation of sloping hillsides, or denuding the slopes in the search of wood to make charcoal, creates a situation in which erosion is constant and an ecological and human disaster likely. In 2004 days of heavy rain on the island of Hispaniola generated thousands of mudslides and killed an estimated 2,000 people in Haiti and the Dominican Republic.

A 1959 report on soil conditions in Jamaica noted that, in one district of the Blue Mountains, on the eastern end of that island, the topsoil had vanished, a victim of rapid erosion. The problem is not unique to the large islands, however. Bonham Richardson observes that ecological degradation on the smallest islands is acute. Thorn scrub and grasses have replaced native forest. "A regional drought in 1977, leading to starvation in Haiti and producing crop and livestock loss south to Trinidad, was severe only partly because of the lack of rain. Grasses and shrubs afford little protection against the sun and thus cannot help the soil to retain moisture in the face of periodic drought. Neither do they inhibit soil loss."

Migration of the islands' inhabitants has at times exacerbated the situation. In times of peak migration, a depleted labor force on some of the islands has resulted in landowners resorting to the raising of livestock, which is not labor-intensive. But livestock contribute to further ecological destruction. "Emigration itself has thus indirectly fed the ongoing devastation of island environments, and some of the changes seem irreversible. Parts of the smaller islands already resemble moonscapes. They seem simply unable to sustain their local resident populations, not to mention future generations or those working abroad who may someday be forced to return for good."

MUSIC, DANCE, FOLKLORE, AND FOOD

Travel accounts of the Caribbean tend to focus on local music, dances, and foods. Calypso, the limbo, steel bands, reggae, and African–Cuban rhythms are well known. Much of the music derives from Amerindian and African roots.

Calypso music apparently originated in Trinidad and spread to the other islands. Calypso singers improvise on any theme; they are particularly adept at poking fun at politicians and their shortcomings. Indeed, governments are as attentive to the lyrics of a politically inspired calypso tune as they are to the opposition press. On a broader scale, calypso is a mirror of Caribbean society.

Some traditional folkways, such as storytelling and other forms of oral history, are in danger of being replaced by electronic media, particularly radio, tape recorders, and jukeboxes. The new entertainment is both popular and readily available.

(Photo Lisa Clyde)

These lush mountain peaks in St. Lucia are volcanic in origin.

Scholar Laura Tanna has gathered much of Kingston, Jamaica's, oral history. Her quest for storyteller Adina Henry took her to one of the city's worst slums, the Dungle, and was reprinted in *Caribbean Review*: "We walked down the tracks to a Jewish cemetery, with gravestones dating back to the 1600s. It, too, was covered in litter, decaying amid the rubble of broken stones. Four of the tombs bear the emblem of the skull and crossbones. Popular belief has it that Spanish gold is buried in the tombs, and several of them have been desecrated by treasure seekers. We passed the East Indian shacks, and completed our tour of Majesty Pen amidst greetings of 'Love' and 'Peace' and with the fragrance of ganja [marijuana] wafting across the way. Everywhere, people were warm and friendly, shaking hands, chatting, drinking beer, or playing dominos. One of the shacks had a small bar and jukebox inside. There, in the midst of pigs grunting at one's feet in the mud and slime, in the dirt and dust, people had their own jukeboxes, tape recorders, and radios, all blaring out reggae, the voice of the ghetto." Tanna found Miss Adina, whose stories revealed the significant African contribution to West Indian folk culture.

In recent years, Caribbean foods have become more accepted, and even celebrated, within the region as well as internationally. Part of the search for an identity involves a new attention to traditional recipes. French, Spanish, and English recipes have been adapted to local foods—iguana, frogs, seafood, fruits, and vegetables. Cassava, guava, and mangos figure prominently in the islanders' diets.

The diversity of the Caribbean is awesome, with its potpourri of peoples and cultures. Its roots lie in Spain, Portugal, England, France, the Netherlands, Africa, India, China, and Japan. There has emerged no distinct West Indian culture, and the Caribbean peoples' identities are determined by the island—no matter how small—on which they live. For the Commonwealth Caribbean, nationalist stirrings are still weak and lacking in focus; while people in Cuba and the Dominican Republic have a much surer grasp on who they are. Nationalism is a strong integrating force in both of these nations. The Caribbean is a fascinating and diverse corner of the world that is far more complex than the travel posters imply.

Antigua and Barbuda

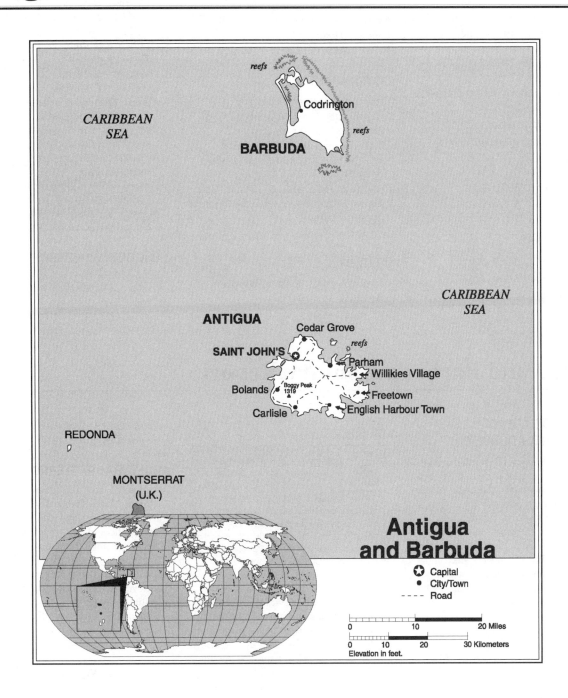

reefs

CARIBBEAN
SEA

Codrington

BARBUDA

CARIBBEAN
SEA

ANTIGUA

Cedar Grove

SAINT JOHN'S

reefs

Parham

Willikies Village

Bolands

Boggy Peak
1319

Freetown

Carlisle

English Harbour Town

REDONDA

MONTSERRAT
(U.K.)

**Antigua
and Barbuda**

☆ Capital
● City/Town
- - - - Road

| 0 | 10 | 20 Miles |
| 0 | 10 | 20 | 30 Kilometers |

Elevation in feet.

Antigua and Barbuda Statistics

GEOGRAPHY

Area in Square Miles (Kilometers): 171
(442) (about 21?2 times the size of
Washington, D.C.)

Capital (Population): Saint John's
(35,700)

Environmental Concerns: water
management; clearing of trees

Geographical Features: mostly low-lying
limestone and coral islands, with some
higher volcanic areas

Climate: tropical marine

PEOPLE

Population

Total: 68,320

Annual Growth Rate: 0.06%

Rural/Urban Population Ratio: 64/36

Major Languages: English; Creole

Ethnic Makeup: almost entirely black African origin; some of British, Portuguese, Lebanese, or Syrian origin

Religions: predominantly Anglican; other Protestant sects; some Roman Catholic

Health

Life Expectancy at Birth: 69 years (male); 74 years (female)

Infant Mortality Rate (Ratio): 20/1,000

Physicians Available (Ratio): 1/1,083

Education

Adult Literacy Rate: 89%

Compulsory (Ages): 5–16

COMMUNICATION

Telephones: 38,000 main lines

Televisions: 435 per 1,000 people

Internet Users: 10,000 (2002)

TRANSPORTATION

Highways in Miles (Kilometers): 150 (240)

Railroads in Miles (Kilometers): 48 (77)

Usable Airfields: 3

Motor Vehicles in Use: 14,700

GOVERNMENT

Type: parliamentary democracy

Independence Date: November 1, 1981 (from the United Kingdom)

Head of State/Government: Queen Elizabeth II; Prime Minister Baldwin Spencer

Political Parties: Antigua Labour Party; United Progressive Party; a coalition of opposing parties

Suffrage: universal at 18

MILITARY

Current Disputes: tensions between Antiguans and Barbudans

ECONOMY

Currency ($ U.S. Equivalent): 2.67 East Caribbean dollars = $1

Per Capita Income/GDP: $11,000/$750 million

GDP Growth Rate: 3%

Inflation Rate: 0.4%

Unemployment Rate: 11%

Labor Force: 30,000

Natural Resources: negligible; the pleasant climate fosters to union

Agriculture: cotton; fruits; vegetables; sugarcane; livestock

Industry: tourism; construction; light manufacturing

Exports: $689 million (primary partners Caribbean, Guyana, United States)

Imports: $692 million (primary partners United States, United Kingdom, Canada)

SUGGESTED WEBSITE

http://www.cia.gov/cia/
publications/factbook/geos/
ac.html

Antigua and Barbuda Country Report

ANTIGUA AND BARBUDA: A STRAINED RELATIONSHIP

The nation of Antigua and Barbuda gained its independence from Great Britain on November 1, 1981. Both islands, tenuously linked since 1967, illustrate perfectly the degree of localism characteristic of the West Indies. Barbudans—who number approximately 1,200—culturally and politically believe that they are not Antiguans; indeed, since independence of Britain, they have been intent on secession. Barbudans view Antiguans as little more than colonial masters.

MEMORIES OF SLAVERY

Antigua was a sugar island for most of its history. This image changed radically in the 1960s, when the black-power movement then sweeping the Caribbean convinced Antiguans that work on the sugar plantations was "submissive" and carried the psychological and social stigma of historic slave labor. In response to the clamor, the government gradually phased out sugar production, which ended entirely in 1972. The decline of agriculture resulted in a strong rural-to-urban flow of people. To replace lost revenue from the earnings of sugar, the government promoted tourism.

Tourism produced the unexpected result of greater freedom for women, in that they

gained access to previously unavailable employment opportunities. Anthropologist W. Penn Handwerker has shown that a combination of jobs and education for women has resulted in a marked decline in fertility. Between 1965 and the 1980s, real wages doubled, infant mortality fell dramatically, and the proportion of women ages 20 to 24 who completed secondary school rose from 3 percent to about 50 percent. "Women were freed from dependency on their children" as well as their men and created "conditions for a revolution in gender relations." Men outmigrated as the economy shifted, and women took the new jobs in tourism. Many of the jobs demanded higher skills, which in turn resulted in more education for women, followed by even better jobs. And notes Handwerker: "Women empowered by edu-

cation and good jobs are less likely to suffer abuse from partners."

CULTURAL PATTERNS

Antiguans and Barbudans are culturally similar. Many islanders still have a strong affinity for England and English culture, while others identify more with what they hold to be their African–Creole roots. On Antigua, for example, Creole, which is spoken by virtually the entire population, is believed to reflect what is genuine and "natural" about the island and its culture. Standard English, even though it is the official language, carries in the popular mind an aura of falseness.

FOREIGN RELATIONS

Despite the small size of the country, Antigua and Barbuda are actively courted by regional powers. The United States maintains a satellite-tracking station on Antigua, and Brazil has provided loans and other assistance. A small oil refinery, jointly supported by Venezuela and Mexico, began operations in 1982.

FAMILY POLITICS

From 1951 to 2004, with one interruption, one interruption, Antiguan politics has been dominated by the family of Vere Bird

FREEDOM

The Bird's control of the electronic media is an area of concern for the new Baldwin government. It must be allowed to air opinions divergent from those of government ministries.

and his Antigua Labour Party (ALP). Charges of nepotism, corruption, drug smuggling, and money laundering dogged the Vere Bird administration for years. Still, in 1994, Lester Bird managed to succeed his 84-year-old father as prime minister, and the ALP won 11 of 17 seats in elections. Lester admitted that his father had been guilty of some "misjudgments" and quickly pledged that the ALP would improve education, better the status of women, and increase the presence of young people in government.

HEALTH/WELFARE

The government has initiated programs to enhance educational opportunities for men and women and to assist in family planning. The new Directorate of Women's Affairs helps women to advance in government and in the professions. It has also sponsored educational programs for women in health, crafts, and business skills.

The younger Bird, in his State of the Nation address early in 1995, challenged Antiguans to transform their country on their own terms, rather than those dictated by the International Monetary Fund. His government would take "tough and unpopular" measures to avoid the humiliation of going "cap in hand" to foreign financial institutions. Those tough measures have included increases in contributions for medical benefits, property and personal taxes, and business and motor-vehicle licenses.

ACHIEVEMENTS

Antigua has preserved its rich historical heritage, from the dockyard named for Admiral Lord Nelson to the Ebenezer Methodist Church. Built in 1839, the latter was the "mother church" for Methodism in the Caribbean.

In 2003, however, the government angered public employees, who constitute one-third of the labor force, when it failed to pay salaries on time. Tourism was stagnant and the public debt was a very high 140 percent of GDP. Economic difficulties when coupled with persistent scandal and corruption brought 90 percent of the electorate to the polls in 2004 and Bird's ALP was soundly defeated. Prime Minister Baldwin Spencer's government must now live up to the expectations of the electorate.

Tourism, the earnings from which slipped from 80 percent of GDP in 1994 to only 50 percent in 2004, is a major area of concern. Agricultural production, consisting primarily of fruits and vegetables, concentrates on the domestic market and does not generate significant foreign exchange earnings. Spencer's administration must make some difficult choices in the near future.

Timeline: PAST

1632
The English settle Antigua

1834
Antigua abolishes slavery

1958–1962
Antigua becomes part of the West Indies Federation

1981
Independence from Great Britain

1990s
Barbuda talks of secession; Hurricane Luis devastates the islands

PRESENT

2000s
The Bird government announces a "zero tolerance" drug policy
Bird political dynasty comes to an end in 2004

The Bahamas (Commonwealth of the Bahamas)

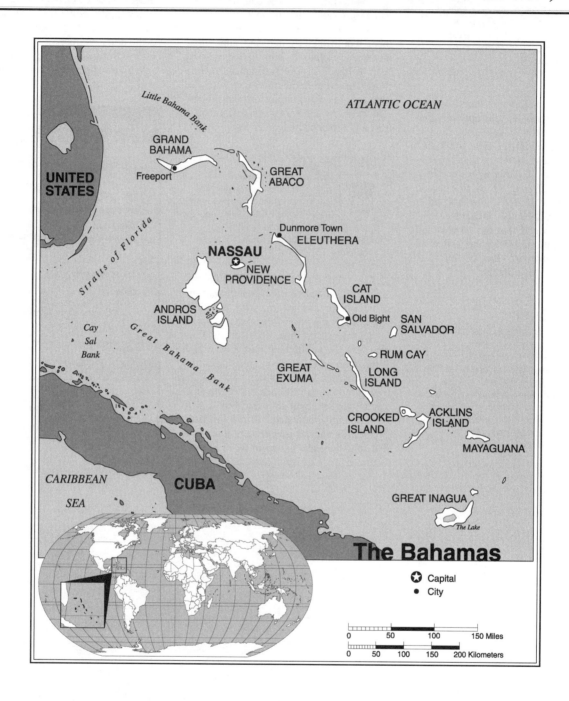

The Bahamas Statistics

GEOGRAPHY

Area in Square Miles (Kilometers): 5,380 (13,934) (about the size of Connecticut)

Capital (Population): Nassau (172,300)

Environmental Concerns: coral-reef decay; waste disposal

Geographical Features: long, flat coral formations with some low, rounded hills

Climate: tropical marine

PEOPLE

Population

Total: 299,697

Annual Growth Rate: 0.72%

Rural/Urban Population Ratio: 13/87

Ethnic Makeup: 85% black; 15% white

Major Language: English

Religions: 32% Baptist; 22% Protestant; 20% Anglican; 19% Roman Catholic; 7% unaffiliated or unknown

Health

Life Expectancy at Birth: 62 years (male); 69 years (female)
Infant Mortality Rate (Ratio): 25/1,000
Physicians Available (Ratio): 1/709

Education

Adult Literacy Rate: 95.6%
Compulsory (Ages): 5–14; free

COMMUNICATION

Telephones: 126,600 main lines
Daily Newspaper Circulation: 126 per 1,000 people
Televisions: 179 per 1,000 people
Internet Users: 60,000 (2002)

TRANSPORTATION

Highways in Miles (Kilometers): 1,672 (2,693)
Railroads in Miles (Kilometers): none

Usable Airfields: 62
Motor Vehicles in Use: 83,000

GOVERNMENT

Type: constitutional parliamentary democracy
Independence Date: July 10, 1973 (from the United Kingdom)
Head of State/Government: Queen Elizabeth II; Prime Minister Perry Christie
Political Parties: Free National Movement; Progressive Liberal Party
Suffrage: universal at 18

MILITARY

Current Disputes: none

ECONOMY

Currency ($ U.S. Equivalent): 1.00 Bahamian dollar = $1

Per Capita Income/GDP: $16,800/$5.09 billion
GDP Growth Rate: 1%
Inflation Rate: 1.7%
Unemployment Rate: 7%
Labor Force: 156,000
Natural Resources: salt; aragonite; timber
Agriculture: citrus fruits; vegetables; poultry
Industry: tourism; banking; cement; oil refining and transshipment; salt production; rum; aragonite; pharmaceuticals; steel pipe
Exports: $361 million (primary partners United States, Switzerland, United Kingdom)
Imports: $1.61 billion (primary partners United States, Italy, Japan)

SUGGESTED WEBSITE

http://www.cia.gov/cia/
publications/factbook/index.html

The Bahamas Country Report

BAHAMAS: A NATION OF ISLANDS

Christopher Columbus made his first landfall in the Bahamas in 1492, when he touched ashore on the island of San Salvador. Permanent settlements on the islands were not established by the British until 1647, when the Eleutheran Adventurers, a group of English and Bermudan religious dissidents, landed. The island was privately governed until 1717, when it became a British Crown colony. During the U.S. Civil War, Confederate blockade runners used the Bahamas as a base. The tradition continued in the years after World War I, when Prohibition rum runners used the islands as a base. Today, drug traffickers utilize the isolation of the out-islands for their illicit operations.

DEVELOPMENT

Together manufacturing and agriculture account for only 10 % of GDP. There has been little growth in either sector. Offshore banking is important.

Although the Bahamas are made up of almost 700 islands, only 10 have populations of any significant size. Of these, New Providence and Grand Bahama contain more than 75 percent of the Bahamian population. Because most economic and cultural activities take place on the larger islands, other islands—particularly those in the southern region—have suffered de-

population over the years as young men and women have moved to the two major centers of activity.

FREEDOM

Women participate actively in all levels of government and business. The Constitution does, however, make some distinctions between males and females with regard to citizenship and permanent-resident status.

Migrants from Haiti and Jamaica have also caused problems for the Bahamian government. There are an estimated 60,000 illegal Haitians now resident in the Bahamas—equivalent to nearly one fifth of the total Bahamian population of 300,000. The Bahamian response was tolerance until late 1994, when the government established tough new policies that reflected a fear that the country would be "overwhelmed" by Haitian immigrants. In the words of one official, the large numbers of Haitians would "result in a very fundamental economic and social transformation that even the very naïve would understand to be undesirable." Imprisonment, marginalization, no legal right to work, and even the denial of access to schools and hospitals are now endured by the immigrants.

Bahamian problems with Jamaicans are rooted differently. The jealous isolation of

each of the new nations is reflected in the peoples' fears and suspicions of the activities of their neighbors. As a result, interisland freedom of movement is subject to strict scrutiny.

The Bahamas were granted their independence from Great Britain in 1973 and established a constitutional parliamentary democracy governed by a freely elected prime minister and Parliament. Upon independence, there was a transfer of political power from a small white elite to the black majority, who comprise 85 percent of the population. Whites continue to play a role in the political process, however, and several hold high-level civil-service and political posts.

The country has enjoyed a marked improvement in health conditions over the past few decades. Life expectancy has risen, and infant mortality has declined. Virtually all people living in urban areas have access to good drinking water, although the age and dilapidated condition of the capital's (Nassau) water system could present problems in the near future.

HEALTH/WELFARE

Cases of child abuse and neglect in the Bahamas rose in the 1990s. The Government and Women's Crisis Centre focused on the need to fight child abuse through a public-awareness program that had as its theme: "It shouldn't hurt to be a child."

The government has begun a program to restructure education on the islands. The authorities have placed a new emphasis on technical and vocational training so that skilled jobs in the economy now held by foreigners will be performed by Bahamians. But while the literacy rate has remained high, there is a shortage of teachers, equipment, and supplies.

ACHIEVEMENTS

The natural beauty of the islands has had a lasting effect on those who have visited them. As a result of his experiences in the waters off Bimini, Ernest Hemingway wrote his classic *The Old Man and the Sea.*

The government of Prime Minister Hubert A. Ingraham and his Free National Movement won a clear mandate in 1997 over the opposition Progressive Liberal Party to continue the policies and programs it initiated in 1992. *The Miami Herald* reported that the election "marked a watershed in Bahamian politics, with many new faces on the ballot and both parties facing leadership succession struggles before the next vote is due in 2002."

Ideologically, the two contending political parties were similar; thus, voters made their decisions on the basis of who they felt would provide jobs and bring crime under control. In 2002 voters decided that the Progressive Liberal Party would do a better job and elected Perry Christie as Prime Minister.

Honest government and a history of working effectively with the private sector to improve the national economy have dramatically increased foreign investment in the Bahamas. Rapid growth in the service sector of the economy has stimulated the migration of people from fishing and farming villages to the commercial tourist centers in New Providence Island, Grand Bahama, and Great Abaco. It is estimated that tourism now accounts for 60 percent of GDP and absorbs half of the labor force. Importantly, today there are more companies owned by Bahamians than ever before.

Despite new investments, many young Bahamians out-migrate. The thousands of illegal Haitian immigrants have added pressure to the job market and still worry some Bahamians that their own sense of identity may be threatened. But in general, there is a sense of optimism in the islands.

Timeline: PAST

1492
Christopher Columbus first sights the New World at San Salvador Island

1647
The first English settlement in the Bahamas

1967
Black-power controversy

1973
Independence from Great Britain

1980s
Violent crime, drug trafficking, and narcotics addiction become serious social problems

1990s
New investments create jobs and cut the unemployment rate

PRESENT

2000s
Employment is up, and so is many Bahamians' sense of optimism

Perry Christie elected prime minister in 2002

Barbados

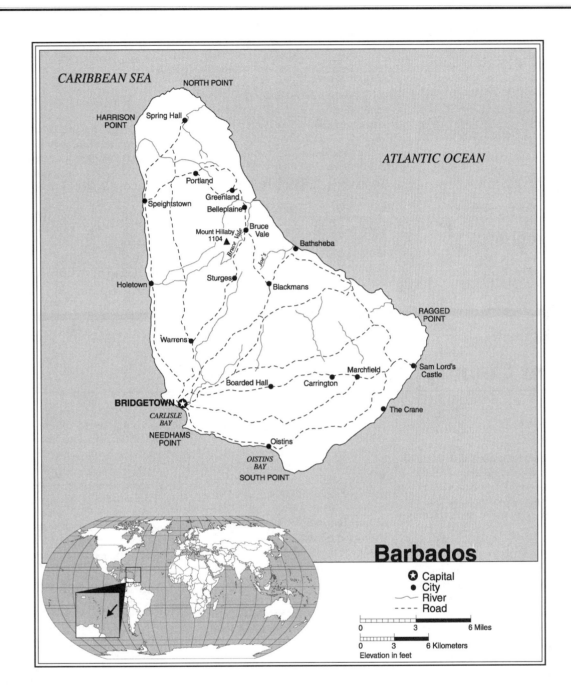

Barbados Statistics

GEOGRAPHY

Area in Square Miles (Kilometers): 166
(431) (about 2½ times the size of
Washington, DC.)

Capital (Population): Bridgetown (6,000)

Environmental Concerns: pollution of
coastal waters from waste disposal by
ships; soil erosion; illegal solid-waste
disposal

Geographical Features: relatively flat;
rises gently to central highland region

Climate: tropical marine

PEOPLE

Population

Total: 278,289
Annual Growth Rate: 0.36%
Rural/Urban Population Ratio: 52/48
Major Language: English

Ethnic Makeup: 80% black; 16% mixed or other; 4% white

Religions: 67% Protestant (Anglican, Pentecostal, Methodist, others); 4% Roman Catholic; 17% unaffiliated; 12% others or unknown

Health

Life Expectancy at Birth: 70 years (male); 73 years (female)
Infant Mortality Rate (Ratio): 12.3/1,000
Physicians Available (Ratio): 1/842

Education

Adult Literacy Rate: 97.4%
Compulsory (Ages): 5–16

COMMUNICATION

Telephones: 133,000 main lines
Daily Newspaper Circulation: 157 per 1,000 people
Televisions: 287 per 1,000 people
Internet Users: 30,000 (2002)

TRANSPORTATION

Highways in Miles (Kilometers): 1,025 (1,650)
Railroads in Miles (Kilometers): none
Usable Airfields: 1
Motor Vehicles in Use: 48,500

GOVERNMENT

Type: parliamentary democracy; independent sovereign state within Commonwealth
Independence Date: November 30, 1966 (from the United Kingdom)
Head of State/Government: Queen Elizabeth II; Prime Minister Owen Seymour Arthur
Political Parties: Democratic Labour Party; Barbados Labour Party; National Democratic Party
Suffrage: universal at 18

MILITARY

Current Disputes: none

ECONOMY

Currency ($ U.S. Equivalent): 2.00 Bajan dollars = $1
Per Capita Income/GDP: $16,200/$4.49 billion
GDP Growth Rate: -0.6%
Inflation Rate: -0.5%
Unemployment Rate: 10.7%
Labor Force: 128,500
Natural Resources: petroleum; fish; natural gas
Agriculture: sugarcane; vegetables; cotton
Industry: tourism; light manufacturing; sugar; component assembly
Exports: $206 million (primary partners United Kingdom, United States, Trinidad and Tobago)
Imports: $1.01 billion (primary partners United States, Trinidad and Tobago, Japan)

SUGGESTED WEBSITE

http://www.cia.gov/cia/
publications/factbook/index.html

Barbados Country Report

THE LITTLE ENGLAND OF THE CARIBBEAN

A parliamentary democracy that won its independence from Britain in 1966, Barbados boasts a House of Assembly that is the third oldest in the Western Hemisphere, after Bermuda's and Virginia's. A statement of the rights and privileges of Bajans (as Barbadians are called), known as the Charter of Barbados, was proclaimed in 1652 and has been upheld by those governing the island. The press is free, labor is strong and well organized, and human rights are respected.

DEVELOPMENT

Between 1971 and 1999, there was an approximate 30% decrease in the amount of land used for agriculture. Formerly agricultural land has been transformed into golf courses, residential areas, commercial developments, tourist facilities, or abandoned.

While the majority of the populations of the English-speaking West Indies still admire the British, this admiration is carried to extremes in Barbados. In 1969, for example, Bajan soccer teams chose English names and colors—Arsenal, Tottenham Hotspurs, Liverpool, and Coventry City.

Among the primary religions are Anglican and Methodist Protestantism.

Unlike most of the other islands of the Caribbean, European sailors initially found Barbados uninhabited. It has since been determined that the island's original inhabitants, the Arawak Indians, were destroyed by Carib Indians who overran the region and then abandoned the islands. Settled by the English, Barbados was always under British control until its independence.

FREEDOM

Barbados has maintained an excellent human-rights record. The government officially advocates strengthening the human-rights machinery of the United Nations and the Organization of American States. Women are active participants in the country's economic, political, and social life.

A DIVERSIFYING ECONOMY

In terms of wealth, as compared to other West Indian nations, Barbados is well off. One important factor is that Barbados has been able to diversify its economy; thus, the country is no longer dependent solely on sugar and its by-products rum and molasses. Manufacturing and high-technology industries now contribute to economic growth, and tourism has overtaken agriculture as a generator of foreign exhange. Offshore finance and information services have also become important.

The Constitution of 1966 authorized the government to promote the general welfare of the citizens of the island through equitable distribution of wealth. While governments have made a sincere effort to wipe out pockets of poverty, a great disparity in wealth still exists.

RACE AND CLASS

Barbados is a class- and race-conscious society. One authority noted that there are three classes (elite, middle class, and masses) and two colors (white/light and black). Land is highly concentrated in the hands of a few; 10 percent of the population own 95 percent of the land. Most of the nation's landed estates and businesses are owned by whites, even though they comprise a very small percentage of the population (4 percent).

HEALTH/WELFARE

By 2000, unemployment had dropped to 9%, from the 1993 high of 26%. Although prices have risen, a sound economy has given people more money to spend on consumer goods, durables, and housing.

While discrimination based on color is legally prohibited, color distinctions continue to correlate with class differences and dominate most personal associations. Although whites have been displaced politically, they still comprise more than half of the group considered "influential" in the country.

ACHIEVEMENTS

Bajan George Lamming has won attention from the world's literary community for his novels, each of which explores a stage in or an aspect of the colonial experience. Through his works, he explains what it is to be simultaneously a citizen of one's island and a West Indian.

Even though Barbados's class structure is more rigid than that of other West Indian states, there is upward social mobility for all people, and the middle class has been growing steadily in size. Poor whites, known as "Redlegs," have frequently moved into managerial positions on the estates. The middle class also includes a fairly large percentage of blacks and mulattos. Bajans have long enjoyed access to public and private educational systems, which have been the object of a good deal of national pride. Adequate medical care is available to all residents through local clinics and hospitals under a government health program. All Bajans are covered under government health insurance programs.

SEEKING A LEADERSHIP ROLE

Given the nation's relative wealth and its dynamism, Bajans have been inclined to seek a strong role in the region. In terms of Caribbean politics, economic development, and defense, Bajans feel that they have a right and a duty to lead.

The Labour Party has continued to push privatization policies. In 1993, an important step was taken toward the greater diversification of the nation's economic base with the creation of offshore financial services. By 1995, the new industry had created many new jobs for Bajans and had significantly reduced the high unemployment rate. Recent discussions on the Free Trade Area of the Americas (FTAA) has stimulated much debate among the smaller Caribbean states. While the Barbados government sees the possibilities of tying into a market of 800 million consumers, it also feels that the larger states must afford smaller nations special and differential treatment. Others are concerned about maintaining the "Bajan way of life" and worry that "the world is falling in on us." Critics charge that any new wealth would be skewed saying, it has the attributes of "fancy molasses." "Very little trickles down, the rich get richer while the poor become marginalized." Other concerns have been expressed about pollution and the possible loss, because of FTAA, of offshore financial privileges.

Timeline: PAST

1625
Barbados is occupied by the English

1647
The first sugar from Barbados is sent to England

1832
Full citizenship is granted to nonwhites

1951
Universal suffrage

1966
Independence from Great Britain

1990s
Barbados develops an offshore banking industry

PRESENT

2000s
The Arthur government continues its policy of economic diversification

Inflation falls to 2.5% in 2003

Cuba (Republic of Cuba)

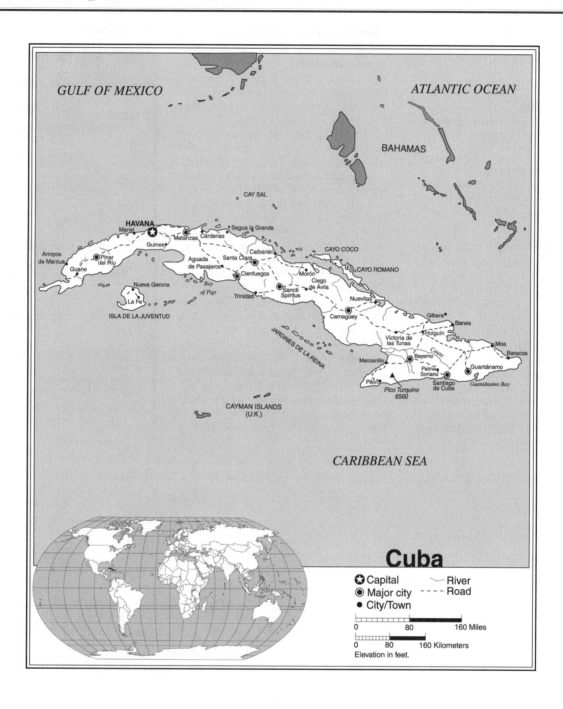

Cuba

- ⊛ Capital
- ◉ Major city
- ● City/Town
- River
- Road

0   80   160 Miles

0   80   160 Kilometers

Elevation in feet.

Cuba Statistics

GEOGRAPHY

Area in Square Miles (Kilometers): 44,200 (114,471) (about the size of Pennsylvania)

Capital (Population): Havana (2,185,000)

Environmental Concerns: pollution of Havana Bay; threatened wildlife populations; deforestation

Geographical Features: mostly flat to rolling plains; rugged hills and mountains in the southeast

Climate: tropical

PEOPLE

Population

Total: 11,308,764

Annual Growth Rate: 0.34%

Rural/Urban Population Ratio: 24/76

Ethnic Makeup: 51% mulatto; 37% white; 11% black; 1% Chinese
Major Language: Spanish
Religion: 85% Roman Catholic before Castro assumed power

Health

Life Expectancy at Birth: 74 years (male); 79 years (female)
Infant Mortality Rate (Ratio): 7.5/1,000
Physicians Available (Ratio): 1/231

Education

Adult Literacy Rate: 95.7%
Compulsory (Ages): 6–11; free

COMMUNICATION

Telephones: 574,400 main lines
Daily Newspaper Circulation: 122 per 1,000 people
Televisions: 200 per 1,000 people
Internet Users: 120,000

TRANSPORTATION

Highways—Miles (Kilometers): 37,793 (60,858)

Railroads—Miles (Kilometers): 2,985 (4,807)
Usable Airfields: 170

GOVERNMENT

Type: Communist state
Independence Date: May 20, 1902 (from Spain)
Head of State/Government: President Fidel Castro Ruz is both head of state and head of government
Political Parties: Cuban Communist Party
Suffrage: universal at 16

MILITARY

Military Expenditures (% of GDP): 1.8% (est.)
Current Disputes: U.S. Naval Base at Guantanamo Bay is leased to the United States

ECONOMY

Currency ($ U.S. Equivalent): 1.000 Cuban pesos = $1 (official rate)

Per Capita Income/GDP: $2,800/$31.59 billion
GDP Growth Rate: 1.3%
Inflation Rate: 5%
Unemployment Rate: 3.2%
Labor Force: 4,500,000
Natural Resources: cobalt; nickel; iron ore; copper; manganese; salt; timber; silica; petroleum; arable land
Agriculture: sugarcane; tobacco; citrus fruits; coffee; rice; potatoes; beans; livestock
Industry: sugar; petroleum; food; textiles; tobacco; chemicals; paper and wood products; metals; cement; fertilizers; consumer goods; agricultural machinery
Exports: $1.4 billion (primary partners Russia, the Netherlands, Canada)
Imports: $4.5 billion (primary partners Spain, Venezuela, Mexico)

SUGGESTED WEBSITE

http://www.cia.gov/cia/
publications/factbook/index.html

Cuba Country Report

REFLECTIONS ON A REVOLUTION

Cuba, which contains about half the land area of the West Indies, has held the attention of the world since 1959. In that year, Fidel Castro led his victorious rebels into the capital city of Havana and began a revolution that has profoundly affected Cuban society. The Cuban Revolution had its roots in the struggle for independence of Spain in the late nineteenth century, in the aborted Nationalist Revolution of 1933, and in the Constitution of 1940. It grew from Cuba's history and must be understood as a Cuban phenomenon.

The Revolution in some respects represents the fulfillment of the goals of the Cuban Constitution of 1940, a radically nationalist document that was never fully implemented. It banned *latifundia* (the ownership of vast landed estates) and discouraged foreign ownership of the land. It permitted the confiscation of property in the public or social interest. The state was authorized to provide full employment for its people and to direct the course of the national economy. Finally, the Constitution of 1940 gave the Cuban state control of the sugar industry, which at the time was controlled by U.S. companies.

The current Constitution, written in 1976, incorporates 36 percent of the articles of the 1940 Constitution. In other words, many of Castro's policies and programs are founded in Cuban history and the aspirations of the Cuban people. Revolutionary Cuba—at least in its earlier years—was very successful in solving the nation's most pressing problems of poverty. But those successes must be balanced against the loss of basic freedoms imposed by a strong authoritarian state.

ACHIEVEMENTS OF THE REVOLUTION

Education

One of the Revolution's most impressive successes has been in the area of education. In 1960, the Castro regime decided to place emphasis on raising the minimum level of education for the whole population. To accomplish this, some 200,000 Cubans were mobilized in 1961 under the slogan "Let those who know more teach those who know less." In a single year, the literacy rate rose from 76 to 96 percent. Free education was made available to all Cubans. The literacy campaign involved many Cu-

bans in an attempt to recognize and attack the problems of rural impoverishment. It was the first taste of active public life for many women who were students or teachers and because of their involvement, they began to redefine sex roles and attitudes.

While the literacy campaign was a resounding triumph, long-term educational policy was less satisfactory. Officials blamed the high dropout rate in elementary and junior high schools on poor school facilities and inadequate teacher training. Students also apparently lacked enthusiasm, and Castro himself acknowledged that students needed systematic, constant, daily work and discipline.

"Scholarship students and students in general," in Castro's words, "are willing to do anything, except to study hard."

Health Care

The Revolution took great strides forward in improving the health of the Cuban population, especially in rural regions. Success in this area is all the more impressive when one considers that between one third and one half of all doctors left the country between 1959 and 1962. Health care initially declined sharply, and the infant mortality rate rose rapidly. But with the training of new health-care professionals, the gaps were filled. The infant mortality rate in Cuba is now at a level comparable to that of developed countries.

DEVELOPMENT

Vladimir Putin, president of Russia, visited Cuba in 2000 and promised a stronger economic relationship between the two countries. After Brazil, Cuba is Russia's largest trading partner in the region.

From the outset, the government decided to concentrate on rural areas, where the need was the greatest. Medical treatment was free, and newly graduated doctors had to serve in the countryside for at least two years. The Cuban health service was founded on the principle that good health for all, without discrimination, is a birthright of Cubans. All Cubans were included under a national health plan.

The first national health standards were developed between 1961 and 1965, and eight priority areas were identified: infant and maternal care, adult health care, care for the elderly, environmental health, nutrition, dentistry, school health programs, and occupational health. A program of insect spraying and immunization eradicated malaria and poliomyelitis. Cuban life expectancy became one of the highest in the world, and Cuba's leading causes of death became the same as in the United States—heart disease, cancer, and stroke.

Before the Revolution of 1959, there was very little health and safety regulation for workers. Afterward, however, important advances were made in the training of specialized inspectors and occupational physicians. In 1978, a Work Safety and Health Law was enacted, which defined the rights and responsibilities of government agencies, workplace administrators, unions, and workers.

Cuba also exported its health-care expertise. It has had medical teams in countries from Nicaragua to Yemen and more doctors overseas than the World Health Organization. In 2003 and 2004 Cuban medical personnel provided health care to Venezuela, which in turn provided Cuba with cheap petroleum.

Redistribution of Wealth

The third great area of change presided over by the Revolution was income redistribution. The Revolution changed the lives of rural poor and agricultural workers. They gained the most in comparison to other groups in Cuban society—especially

urban groups. From 1962 to 1973, for example, agricultural workers saw their wages rise from less than 60 percent to 93 percent of the national average.

Still, Cuba's minimum wage was inadequate for most families. Many families needed two wage earners to make ends meet. All wages were enhanced by the so-called social wage, which consisted of free medical care and education, subsidized housing, and low food prices. Yet persistent shortages and tight rationing of food undermined a good portion of the social wage. Newly married couples found it necessary to live with relatives, sometimes for years, before they could obtain their own housing, which was in short supply. Food supplies, especially those provided by the informal sector, were adversely affected by a 1986 decision to eliminate independent producers because an informal private sector was deemed antithetical to "socialist morality" and promoted materialism.

FREEDOM

The Committee to Protect Journalists noted that those who try to work outside the confines of the state media face tremendous obstacles. "The problems of a lack of basic supplies ... are dwarfed by Fidel Castro's campaign of harassment and intimidation against the fledgling free press."

Women in Cuba

From the outset of the Revolution, Fidel Castro appealed to women as active participants in the movement and redefined their political roles. Women's interests were protected by the Federation of Cuban Women, an integral part of the ruling party. The Family Code of 1975 equalized pay scales, reversed sexual discrimination against promotions, provided generous maternity leave, and gave employed women preferential access to goods and services. Although women comprised approximately 30 percent of the Cuban workforce, most were still employed in traditionally female occupations; the Third Congress of the Cuban Communist Party admitted in 1988 that both racial minorities and women were underrepresented in responsible government and party positions at all levels. This continues to be a problem.

SHORTCOMINGS

Even at its best, the new Cuba had significant shortcomings. Wayne Smith, a former chief of the U.S. Interest Section in Havana who was sympathetic to the Revolution, wrote: "There is little freedom of expres-

sion and no freedom of the press at all. It is a command society, which still holds political prisoners, some of them under deplorable conditions. Further, while the Revolution has provided the basic needs of all, it has not fulfilled its promise of a higher standard of living for the society as a whole. Cuba was, after all, an urban middle-class society with a relatively high standard of living even before the Revolution.... The majority of Cubans are less well off materially."

Castro, to win support for his programs, did not hesitate to take his revolutionary message to the people. Indeed, the key reason why Castro enjoyed such widespread support in Cuba was because the people had the sense of being involved in a great historical process.

Alienation

Not all Cubans identified with the Revolution, and many felt a deep sense of betrayal and alienation. The elite and most of the middle class strongly resisted the changes that robbed them of influence, prestige, and property. Some were particularly bitter, for at its outset, the Revolution had been largely a middle-class movement. For them, Castro was a traitor to his class. Thousands fled Cuba, and some formed the core of an anti-Castro guerrilla movement based in South Florida.

(United Nations photo)

Fidel Castro has been the prime minister of Cuba since he seized power in 1959. Pictured above is Castro at the United nations, as he looked in 1960.

There are many signs that Castro's government, while still popular among many people, has lost the widespread acceptance it enjoyed in the 1960s and 1970s. While Castro still has the support of the older generation and those in rural areas who benefited from the social transformation of the island, limited economic growth has led to dissatisfaction among urban workers and youth, who are less interested in Castro as a revolutionary hero and more interested in economic gains.

HEALTH/WELFARE

In August 1997, the Cuban government reported 1,649 HIV cases, 595 cases of full-blown AIDS, and 429 deaths, a significant increase over figures for 1996. Cuban medical personnel are working on an AIDS vaccine. AIDS has been spread in part because of an economic climate that has driven more women to prostitution.

More serious disaffection may exist in the army. Journalist Georgie Anne Geyer, writing in *World Monitor*, suggests that the 1989 execution of General Arnaldo Ochoa, ostensibly for drug trafficking, was actually motivated by Castro's fears of an emerging competitor for power. "The 1930s-style show trial effectively revealed the presence of an 'Angola generation' in the Cuban military.... That generation, which fought in Angola between 1974 and 1989, is the competitor generation to Castro's own Sierra Maestra generation." The condemned officers argued that their dealings with drug traffickers were not for personal enrichment but were designed to earn desperately needed hard currency for the state. Some analysts are convinced that Castro knew about drug trafficking and condoned it; others claim that it took place without his knowledge. But the bottom line is that the regime had been shaken at the highest levels, and the purge was the most far-reaching since the 1959 Revolution.

The Economy

The state of the Cuban economy and the future of the Cuban Revolution are inextricably linked. Writing in *World Today*, James J. Guy predicted that, given the economic collapse of the former Soviet Union and its satellites, "Cuba is destined to face serious structural unemployment: its agrarian economy cannot generate the white-collar, technical jobs demanded by a swelling army of graduates.... The entire system is deteriorating—the simplest services take months to deliver, water and electricity are constantly interrupted...," and there is widespread corruption and black-marketeering.

Oil is particularly nettlesome. For years after the collapse of the Soviet Union Cuba had no access to affordable petroleum, at great cost to the economy. That changed in 2003 when Venezuela provided Cuba with discounted oil in exchange for Cuban expertise in the areas of health and sports.

Although Castro prides Cuba on being one of the last bulwarks of untainted Marxism-Leninism, in April 1991 he said: "We are not dogmatic ... we are realistic ... Under the special conditions of this extraordinary period we are also aware that different forms of international cooperation may be useful." He noted that Cuba had contacted foreign capitalists about the possibility of establishing joint enterprises and remarked that more than 49 percent foreign participation in state businesses was a possibility.

In 1993, Castro called for economic realism. Using the rhetoric of the Revolution, he urged the Legislative Assembly to think seriously about the poor condition of the Cuban economy: "It is painful, but we must be sensible.... It is not only with decisiveness, courage and heroism that one saves the Revolution, but also with intelligence. And we have the right to invent ways to survive in these conditions without ever ceasing to be revolutionaries."

ACHIEVEMENTS

A unique cultural contribution of Cuba to the world was the Afro-Cuban movement, with its celebration of black song and dance rhythms. The work of contemporary prize-winning Cuban authors such as Alejo Carpentier and Edmundo Desnoes has been translated into many languages.

A government decree in September 1993 allowed Cubans to establish private businesses; today, Cubans in some 140 professions can work on their own for a profit. At about the same time, the use of dollars was decriminalized, the Cuban currency became convertible, and, in the agricultural sector, the government began to transform state farms into cooperatives. Farmers are now allowed to sell some of their produce in private markets and, increasingly, market forces set the prices of many consumer goods. Managers in state-owned enterprises have been given unprecedented autonomy; and foreign investment, in contrast with past practice, is now encouraged.

Still, the Cuban economy has continued its decline. Mirta Ojito, writing in *The New York Times*, sees older revolutionaries "coming to terms with the failure of their dreams." Cuba now resembles most other underdeveloped countries, with "many needy, unhappy, sad people." The Revolution was supposed to make Cuba prosperous, "not merely survive," and end the country's dependence on the U.S. dollar. By 1999, dollars in circulation in Cuba had created a parallel speculative economy. With the new millenium, Cuba's infrastructure continued to crumble. In 2001, salaries averaged just $15 per month, and the weekly ration card given to each family provides one chicken, just over three pounds of rice and beans, sugar, and two pints of cooking oil The peso is virtually worthless, and bartering or U.S. dollars are needed to acquire all of the luxuries and many of the necessities of life. With rising prices, it is not surprising that prostitution, moonlighting, black-marketeering, and begging have rapidly increased. Castro has talked with CARICOM states about the possibilities of free trade, but the stifling bureaucracy makes it much easier to export *from* rather than export *to* Cuba.

Timeline: PAST

1492
The island is discovered by Christopher Columbus

1511
The founding of Havana

1868–1878
The Ten Years' War in Cuba

1895–1898
The Cuban War of Independence

1902
The Republic of Cuba is established

1940
Cuba writes a new, progressive Constitution

1959
Fidel Castro seizes power

1961
An abortive U.S.–sponsored invasion at the Bay of Pigs

1980s
Mass exodus from Cuba; trial and execution of top military officials for alleged dealing in drugs

1990s
The economy rapidly deteriorates; Castro pursues Economic Liberalization Tensions flare between Cuba and the United Sates over the disposition of a young Cuban refugee, Elian González

PRESENT

2000s
The U.S. trade embargo, supported only by Israel, continues to make life difficult for the Cuban people

Freedom Issues

Soon after the Revolution, the government assumed total control of the media. No independent news organization is allowed, and all printed publications are censored by the government or the Communist Party. The arts are subject to strict censorship, and even sports must serve the purposes of the Revolution. As Castro noted: "Within the Revolution everything is possible. Outside it, nothing."

In many respects, there is less freedom now in Cuba than there was before the Revolution. Cuba's human-rights record is not good. There are thousands of political prisoners, and rough treatment and torture—physical and psychological—occur. The Constitution of 1976 allows the repression of all freedoms for all those who oppose the Revolution. U.S. political scientist William LeGrande, who was sympathetic to the Revolution, nevertheless noted that "Cuba is a closed society. The Cuban Communist Party does not allow dissenting views on fundamental policy. It does not allow people to challenge the basic leadership of the regime." But here, too, there are signs of change. In 1995, municipal elections were held under a new system that provides for run-offs if none of the candidates gains a clear majority. In an indication of a new competitiveness in Cuban politics, 326 out of 14,229 positions were subject to the run-off rule.

THE FUTURE

It will be difficult for Castro to maintain the unquestioned support of the Cuban population. There must be continued positive accomplishments in the economy. Health and education programs are successful and will continue to be so. "Cubans get free health care, free education and free admission to sports and cultural events [and] 80% of all Cubans live in rent-free apartments, and those who do pay rent pay only between 6 and 10% of their salaries," according to James J. Guy.

But there must be a recovery of basic political and human freedoms. Criticism of the government must not be the occasion for jail terms or exile. The Revolution must be more inclusive and less exclusive.

Although Castro has never been effectively challenged, there are signs of unrest on the island. The military, as noted, is a case in point. Castro has also lost a good deal of luster internationally, as most countries have moved away from statism and toward free-market economies and more open forms of government.

Even though a similar trend is apparent on the island, in 1994 and 1995 many Cubans grew increasingly frustrated with their lives and took to the sea in an attempt to reach the United States. Thousands were intercepted by the U.S. Coast Guard and interned in U.S. military facilities at Guantanamo Bay and Fort Howard in the Panama Canal Zone.

The question is increasingly asked, What will happen once Fidel, through death or retirement, is gone from power? Castro's assumption is that the new Constitution, which institutionalizes the Revolution, will provide a mechanism for succession. Over the past few years, he has made some effort to depersonalize the Revolution; his public appearances are fewer and he does less traveling around the countryside. But there is no transition plan, and Castro continues to behave as if he is the embodiment of the Revolution. As for his staying power, a recent anecdote is revealing. When presented with a gift of a Galapagos tortoise, Castro asked how long they lived. The reply, "More than a hundred years," prompted Castro to say, "How sad it is to outlive one's pets."

Change must come to Cuba. More than half of all Cubans alive today were born after the Revolution. They are not particularly attuned to the rhetoric of revolution and seem more interested in the attainment of basic freedoms and consumer goods. In January 1999, *The Economist* asked: "What will follow Fidel?" The magazine suggested that Cubans could be faced with violence and political turmoil, for there were "no plausible political heirs in sight, no credible opposition, and an exile community eager not only for return but also revenge."

Dominica (Commonwealth of Dominica)

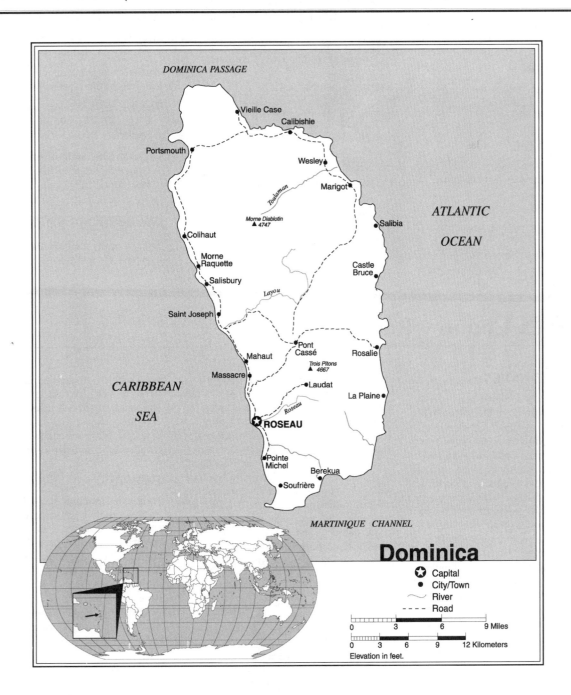

Dominica Statistics

GEOGRAPHY

Area in Square Miles (Kilometers): 289 (752) (about 4 times the size of Washington, D.C.)
Capital (Population): Roseau (16,000)
Geographical Features: rugged mountains of volcanic origin
Climate: tropical

PEOPLE

Population

Total: 69,278
Annual Growth Rate: –0.45%
Rural/Urban Population Ratio: 30/70
Major Languages: English; French Creole

Ethnic Makeup: mostly black; some Carib Indians
Religions: 77% Roman Catholic; 15% Protestant; 8% other or unaffiliated

Health

Life Expectancy at Birth: 71 years (male); 76 years (female)

Infant Mortality Rate (Ratio): 14/1,000
Physicians Available (Ratio): 1/2,112

Education
Adult Literacy Rate: 94%
Compulsory (Ages): 5–15; free

COMMUNICATION
Telephones: 23,700 main lines
Televisions: 1 per 13 people
Internet Users: 12,500

TRANSPORTATION
Highways in Miles (Kilometers): 484 (780)
Railroads in Miles (Kilometers): none
Usable Airfields: 2
Motor Vehicles in Use: 13,000

GOVERNMENT
Type: parliamentary democracy

Independence Date: November 3, 1978
(from the United Kingdom)
Head of State/Government: President
Nicholas J. O. Liverpool; Prime Minister
Roosevelt Skerrit
Political Parties: United Workers Party;
Dominica Freedom Party; Dominica
Labour Party
Suffrage: universal at 18

MILITARY
Current Disputes: none

ECONOMY
Currency ($ U.S. Equivalent): 2.67 East
Caribbean dollars = $1
Per Capita Income/GDP: $5,400/$380
million
GDP Growth Rate: -1%
Inflation Rate: 1.1%

Unemployment Rate: 23%

Labor Force: 25,000

Natural Resources: timber

Agriculture: fruits; cocoa; root crops;
forestry and fishing potential

Industry: soap; coconut oil; tourism;
copra; furniture; cement blocks; shoes

Exports: $39 million (primary partners
CARICOM, United Kingdom, United
States)

Imports: $98.2 million (primary partners
United States, CARICOM, United
Kingdom)

SUGGESTED WEBSITE
http://www.cia.gov/cia/
publications/factbook/index.html

Dominica Country Report

A FRAGMENTED NATION

Dominica is a small and poor country that gained its independence of Great Britain in 1978. Culturally, the island reflects a number of patterns. Ninety percent of the population speak French patois (dialect), and most are Roman Catholic, while only a small minority speak English and are Protestant. Yet English is the official language. There are also small groups of Indians who may have descended from the original Carib inhabitants; they are alternately revered and criticized. Many Dominicans perceive the Carib Indians as drunken, lazy, and dishonest. Others see them as symbolically important because they represent an ancient culture and fit into the larger Caribbean search for cultural and national identity. There is also a small number of Rastafarians, who identify with their black African roots.

DEVELOPMENT

Banana exports fell from $30 million in 1992 to $5 million in 2003. Only 6,000 people out of 70,000 earn enough to pay an income tax and the unemployment rate stands at 25%.

Christopher Columbus discovered the island of Dominica on his second voyage to the New World in 1493. Because of the presence of Carib Indians, who were

known for their ferocity, Spanish efforts to settle the island were rebuffed. It was not until 1635 that France took advantage of Spanish weakness and claimed Dominica as its own. French missionaries became the island's first European settlers. Because of continued Carib resistance, the French and English agreed in 1660 that both Dominica and St. Vincent should be declared neutral and left to the Indians. Definitive English settlement did not occur until the eighteenth century, and the island again became a bone of contention between the French and English. It became Britain's by treaty in 1783.

FREEDOM

Freedom House, an international human-rights organization, listed Dominica as "free." It also noted that "the rights of the native Caribs may not be fully respected." The example set by former prime minister Mary Eugenia Charles led to greater participation by women in the island's political life.

Today, Dominica's population is broken up into sharply differentiated regions. The early collapse of the plantation economy left pockets of settlements, which are still isolated from one another. A difficult topography and poor communications exaggerate the differences between these small communities. This contrasts with na-

tions such as Jamaica and Trinidad and Tobago, which have a greater sense of national awareness because there are good communications and mass media that reach most citizens and foster the development of a national perception.

EMIGRATION

Although Dominica has a high birth rate and its people's life expectancy has measurably increased over the past few years, the growth rate has been dropping due to significant out-migration. Out-migration is not a new phenomenon. From the 1880s until well into the 1900s, many Dominicans sought economic opportunity in the gold fields of French Guiana. Today, most move to the neighboring French departments of Guadeloupe and Martinique.

THE ECONOMY

Dominica's chief export, bananas, has suffered for some years from natural disasters and falling prices. Hurricanes blew down the banana trees in 1979, 1980, 1989, and 1995, and banana exports fell dramatically. A drop in banana prices in 1997 prompted the opposition Dominica Freedom Party to demand that Dominica become part of a single market in order to take advantage of set prices enjoyed by the producers of Martinique and Guadeloupe. Recent talks among producers, Windward Island governments, and the European Union focused on the need for radical changes in the banana industry. The head of the Windward Islands Banana Development and Exporting Company said that the industry should be "market-led rather than production-led." He also noted that the industry was too fragmented, with 10,000 growers all over the islands. This was one reason why costs were high and yields low. With new technology, an acre should produce 20 tons of fruit instead of the four tons now harvested. Together with other banana-producing small states in the Caribbean, Dominica has increasingly turned to non-traditional crops, including root crops, cucumbers, flowers, hot peppers, tomatoes, and nonbanana tropical fruits.

HEALTH/WELFARE

With the assistance of external donors, Dominica has rebuilt many primary schools destroyed in Hurricane Hugo in 1989. A major restructuring of the public health administration has improved the quality of health care, even in the previously neglected rural areas.

Hard-pressed for revenue, the *Economist* reports that Dominica has traded on its sovereignty for cash. It has sold passports to foreigners, hosted off-shore banks, and

voted with Japan in favor of commercial whaling. The "favorite local game" involves playing off China and Taiwan for economic gain.

ACHIEVEMENTS

Traditional handcrafts—especially intricately woven baskets, mats, and hats—have been preserved in Dominica. Schoolchildren are taught the techniques to pass on this dimension of Dominican culture.

POLITICAL FREEDOM

Despite economic difficulties and several attempted coups, Dominica still enjoys a parliamentary democracy patterned along British lines. The press is free and has not been subject to control—save for a brief state of emergency in 1981, which corresponded to a coup attempt by former prime minister Patrick John and unemployed members of the disbanded Defense Force. Political parties and trade unions are free to organize. Labor unions are small but enjoy the right to strike. Women have full rights under the law and are active in the political system; former prime minister Mary Eugenia Charles was the Caribbean's first woman to become a head of government.

Timeline: PAST

1493
Dominica is sighted on Christopher Columbus's second voyage

1783
Dominica is deeded to the British by France

1978
Independence of Great Britain

1979–1980
Hurricanes devastate Dominica's economy

1989
Hurricane Hugo devastates the island

1990s
Dominica seeks stronger tourism revenues, especially in ecotourism

The banana industry is in crisis

PRESENT

2000s
The banana industry looks for solutions to its problems

Roosevelt Skerrit elected prime minister in 2004

Dominican Republic

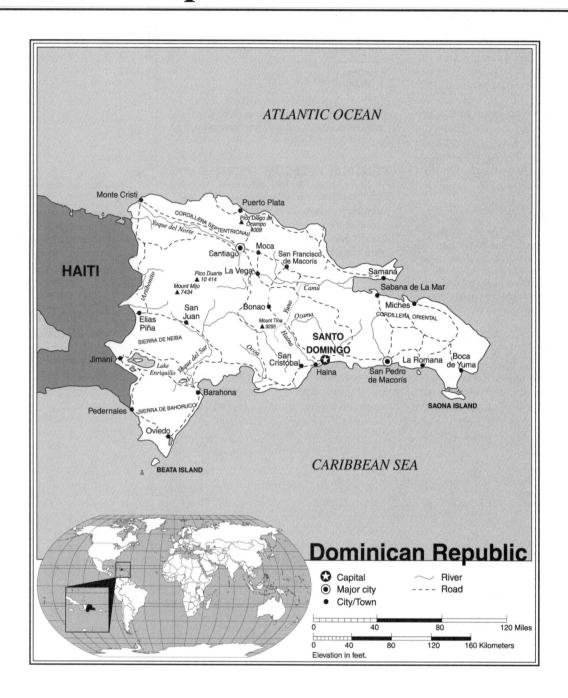

ATLANTIC OCEAN

HAITI

Monte Cristi
Puerto Plata
CORDILLERA SEPTENTRIONAL
Pico Diego de Ocampo ▲1009
Yaque del Norte
Moca
Santiago
San Francisco de Macorís
Samaná
Sabana de La Mar
Pico Duarte ▲10 414
La Vega
Mount Mijo ▲7434
Camú
Artibonito
San Juan
Bonao
Yuna
Miches
CORDILLERA ORIENTAL
Elias Piña
Mount Tina ▲9285
Ozama
SIERRA DE NEIBA
Haina
SANTO DOMINGO
Jimaní
Lake Enriquillo
Yaque del Sur
Ocoa
San Cristóbal
Haina
La Romana
Boca de Yuma
San Pedro de Macorís
Barahona
SAONA ISLAND
Pedernales
SIERRA DE BAHORUCO
Oviedo
BEATA ISLAND
CARIBBEAN SEA

Dominican Republic

⭐ Capital   ~ River
◉ Major city   ---- Road
• City/Town

0   40   80   120 Miles
0   40   80   120   160 Kilometers
Elevation in feet.

Dominican Republic Statistics

GEOGRAPHY

Area in Square Miles (Kilometers): 18,712
(48,464) (about twice the size of New
Hampshire)

Capital (Population): Santo Domingo
(3,166,000)

Environmental Concerns: water
shortages; soil erosion; damage to coral
reefs; deforestation; damage from
Hurricane Georges
Geographical Features: rugged highlands
and mountains with fertile valleys
interspersed
Climate: tropical maritime

PEOPLE

Population

Total: 8,833,634
Annual Growth Rate: 1.33%
Rural/Urban Population Ratio: 37/63
Major Language: Spanish

Ethnic Makeup: 73% mixed; 16% white; 11% black
Religions: 95% Roman Catholic; 5% others

Health

Life Expectancy at Birth: 65 years (male); 69 years (female)
Infant Mortality Rate (Ratio): 36/1,000
Physicians Available (Ratio): 1/1,076

Education

Adult Literacy Rate: 84.7%
Compulsory (Ages): 6–14

COMMUNICATION

Telephones: 909,000 main lines
Daily Newspaper Circulation: 35 per 1,000 people
Televisions: 97 per 1,000 people
Internet Users: 300,000

TRANSPORTATION

Highways in Miles (Kilometers): 7,825 (12,600)

Railroads in Miles (Kilometers): 470 (757)
Usable Airfields: 28
Motor Vehicles in Use: 206,000

GOVERNMENT

Type: republic
Independence Date: February 27, 1844 (from Haiti)
Head of State/Government: President Leonel Fernández is both head of state and head of government
Political Parties: Dominican Revolutionary Party; Social Christian Reformist Party; Dominican Liberation Party; Independent Revolutionary Party; others
Suffrage: universal and compulsory at 18, or at any age if married; members of the armed forces or the police cannot vote

MILITARY

Military Expenditures (% of GDP): 1.1%
Current Disputes: none

ECONOMY

Currency ($ U.S. Equivalent): 30.83 Dominican pesos = $1
Per Capita Income/GDP: $6,000/$52.16 billion
GDP Growth Rate: -1.8%
Inflation Rate: 21.2%
Unemployment Rate: 15.5%
Labor Force: 2,300,000–2,600,000
Natural Resources: nickel; bauxite; gold; silver; arable land
Agriculture: sugarcane; coffee; cotton; cocoa; tobacco; rice; beans; potatoes; corn; bananas; livestock
Industry: tourism; sugar processing; ferronickel and gold mining; textiles; cement; tobacco
Exports: $5.1 billion (primary partners United States, Belgium, Asia)
Imports: $7.9 billion (primary partners United States, Venezuela, Mexico)

SUGGESTED WEBSITE

http://www.cia.org/cia/
publications/factbook/index.html

Dominican Republic Country Report

DOMINICAN REPUBLIC: RACIAL STRIFE

Occupying the eastern two thirds of the island of Hispaniola (Haiti comprises the western third), the Dominican Republic historically has feared its neighbor to the west. Much of the fear has its origins in race. From 1822 until 1844, the Dominican Republic—currently 73 percent mixed, or mulatto—was ruled by a brutal black Haitian regime. One authority noted that the Dominican Republic's freedom from Haiti has always been precarious: "Fear of reconquest by the smaller but more heavily populated (and, one might add, black) neighbor has affected Dominican psychology more than any other factor."

DEVELOPMENT

The U.S. decision to grant the Dominican Republic NAFTA parity with regard to textiles will allow these goods to enter the U.S. market at prices close to those of Mexico. The Dominican government is generally supportive of recent FTAA initiatives because of market accessibility and job creation.

In the 1930s, for example, President Rafael Trujillo posed as the defender of Catholic values and European culture against the "barbarous" hordes of Haiti. Trujillo ordered the massacre of from 12,000 to 20,000 Haitians who had settled in the Dominican Republic in search of work. For years, the Dominican government had encouraged Haitian sugarcane cutters to cross the border to work on the U.S.–owned sugar plantations. But with the world economic depression in the 1930s and a fall in sugar prices and production, many Haitians did not return to their part of the island; in fact, additional thousands continued to stream across the border. The response of the Dominican government was wholesale slaughter.

Since 1952, a series of five-year agreements have been reached between the two governments to regularize the supply of Haitian cane cutters. An estimated 20,000 cross each year into the Dominican Republic legally, and an additional 60,000 enter illegally. Living and working conditions are very poor for these Haitians, and the migrants have no legal status and no rights. Planters prefer the Haitian workers because they are "cheaper and more docile" than Dominican laborers, who expect reasonable food, adequate housing, electric lights, and transportation to the fields. Today, as in the 1930s, economic troubles have gripped the Dominican Republic; the government has promised across-the-board sacrifices.

There is a subtle social discrimination against darker-skinned Dominicans, although this has not proved to be an insurmountable obstacle, as many hold elected political office. Discrimination is in part historical, in part cultural, and must be set against a backdrop of the sharp prejudice against Haitians. This prejudice is also directed against the minority in the Dominican population who are of Haitian descent. For example, during the contested presidential election of 1994, President Joaquín Balaguer Ricardo introduced the issue of race when questions were raised about his opponent's rumored Haitian origins. While in office, President Leonel Fernández worked hard for better relations with Haitians, but the bitter memories and policies of the past undercut his efforts.

FREEDOM

Dominican politics remain volatile even as the country returns to economic stability. The media are generally free, but from time to time the government reveals a degree of intolerance against its critics.

WOMEN'S RIGHTS

Women in the Dominican Republic have enjoyed political rights since 1941. While in office, President Balaguer, in an unprecedented move, named women governors for eight of the country's 29 provinces. Sexual discrimination is prohibited by law, but women have not shared equal social or economic status or opportunity with men. Divorce, however, is easily obtainable, and women can hold property in their own names. A 1996 profile of the nation's population and health noted that 27 percent of Dominican households were headed by women. In urban areas, the percentage rose to 31 percent.

HEALTH/WELFARE

Sociologist Laura Raynolds notes that a restructuring of labor that moved thousands of women into nontraditional agriculture and manufacturing for export has reduced them to a "cheap and disciplined" workforce. Their work is undervalued to enhance profits. In that the majority of these workers are mothers, there has been a redefinition of family identity and work.

AN AIR OF CHANGE

Progress toward a political scene free of corruption and racism has been fitful. The 1994 presidential election was marred by what multinational observers called massive fraud. The opposition claimed that Balaguer not only "stole the election" but also employed racist, anti-Haitian rhetoric that "inflamed stereotypes of Haitians in the Dominican Republic." Widespread unrest in the wake of the election, together with pressure from the Roman Catholic Church, the Organization of American States, and the United States, resulted in the "Pact for Democracy," which forced Balaguer to serve a shortened two-year term as president. New elections in 1996 returned Leonel Fernández to the presidency.

ACHIEVEMENTS

Some of the best baseball in the hemisphere is played in the Dominican Republic. Two of its citizens, pitcher Pedro Martínez and slugger Sammy Sosa, have become stars in major-league baseball in the United States. Both have raised awareness of their country and both have contributed to the welfare of Dominicans.

A brief economic recovery has been followed by sharp recession. Inflation soared to 10 percent a month in 2003, the slowdown in the global economy cut into the tourist industry, and assembly plants in the free trade zone were forced to cut back. Many of the problems were blamed on President Hipólito Mejía's economic policies, which included a $2.4 billion bail-out of the nation's third largest commercial bank—bankrupted by massive fraud. In elections in 2004 Mejía resorted to demagoguery, distributed motorcycles at cut rates, and promised a 30 percent raise to public employees. He lost the election to former president Le-onel Fernández, who has pledged to cut inflation, stabilize the exchange rate, and restore investor confidence.

Timeline: PAST

1496
The founding of Santo Domingo, the oldest European city in the Americas

1821
Independence from Spain is declared

1822–1844
Haitian control

1844
Independence as a separate state

1930–1961
The era of General Rafael Trujillo

1965
Civil war and U.S. intervention

1990s
Diplomatic relations are restored with Cuba

Hurricane Georges slams into the nation, killing many and causing $1.3 billion in damage

PRESENT

2000s

Leonel Fernández elected president in May 2004

Grenada

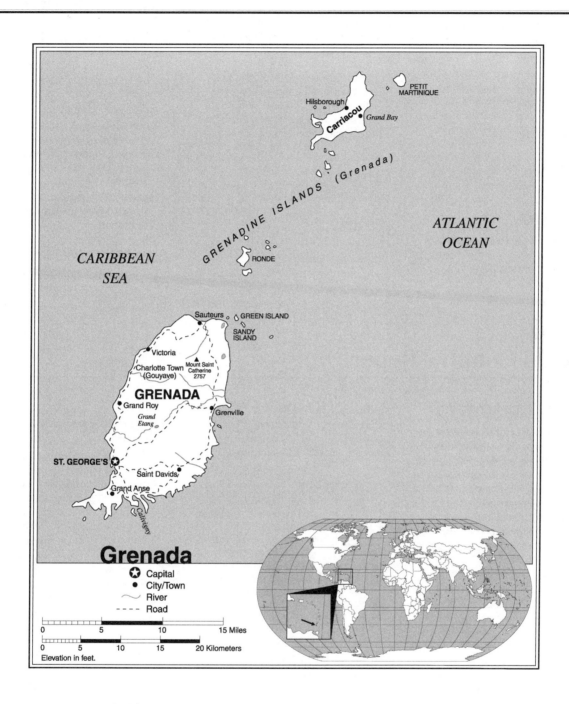

Grenada

- ⭐ Capital
- ● City/Town
- 〰 River
- - - - Road

| 0 | 5 | 10 | 15 Miles |

| 0 | 5 | 10 | 15 | 20 Kilometers |

Elevation in feet.

Grenada Statistics

GEOGRAPHY

Area in Square Miles (Kilometers): 133 (340) (about twice the size of Washington, D.C.)

Capital (Population): St. George's (4,500)

Geographical Features: volcanic in origin, with central mountains

Climate: tropical

PEOPLE

Population

Total: 89,357

Annual Growth Rate: 0.14%

Rural/Urban Population Ratio: 64/36

Major Languages: English; French patois

Ethnic Makeup: mainly black

Religions: largely Roman Catholic; Church of England; other Protestant sects

Health

Life Expectancy at Birth: 63 years (male);
66 years (female)
Infant Mortality Rate (Ratio): 14.6/1,000
Physicians Available (Ratio): 1/2,045

Education

Adult Literacy Rate: 98%
Compulsory (Ages): 5–6; free

COMMUNICATION

Telephones: 33,500 main lines
Televisions: 154 per 1,000 people
Internet Users: 15,000 (2002)

TRANSPORTATION

Highways in Miles (Kilometers): 646
(1,040)
Railroads in Miles (Kilometers): none
Usable Airfields: 3

GOVERNMENT

Type: parliamentary democracy
Independence Date: February 7, 1974
(from the United Kingdom)
Head of State/Government: Queen
Elizabeth II; Prime Minister Keith
Mitchell
Political Parties: New National Party;
Grenada United Labour Party; The
National Party; National Democratic
Congress; Maurice Bishop Patriotic
Movement; Democratic Labour Party
Suffrage: universal at 18

MILITARY

Current Disputes: none

ECONOMY

Currency ($ U.S. Equivalent): 2.67 East
Caribbean dollars = $1
Per Capita Income/GDP: $5,000/$440
million

GDP Growth Rate: 2.5%
Inflation Rate: 2.8%
Unemployment Rate: 12.5%
Labor Force: 42,300
Natural Resources: timber; tropical fruit;
deepwater harbors
Agriculture: bananas; cocoa; nutmeg;
mace; citrus fruits; avocados; root crops;
sugarcane; corn; vegetables
Industry: food and beverages; spice
processing; textiles; light assembly
operations; tourism; construction
Exports: $46 million (primary partners
CARICOM, United Kingdom, United
States)
Imports: $208 million (primary partners
United States, CARICOM, United
Kingdom)

SUGGESTED WEBSITE

http://www.cia.gov/cia/
publications/factbook/index.html

Grenada Country Report

GRENADA: A FRESH BEGINNING

On his third voyage to the New World in 1498, Christopher Columbus sighted Grenada, which he named Concepción. The origin of the name Grenada cannot be clearly established, although it is believed that the Spanish renamed the island for the Spanish city of Granada. Because of a fierce aboriginal population of Carib Indians, the island remained uncolonized for 100 years.

Grenada, like most of the Caribbean, is ethnically mixed. Its culture draws on several traditions. The island's French past is preserved among some people who still speak patois (a French dialect). There are few whites on the island, save for a small group of Portuguese who immigrated earlier in the century. The primary cultural identification is with Great Britain, from which Grenada won its independence in 1974.

Grenada's political history has been tumultuous. The corruption and violent tactics of Grenada's first prime minister, Eric Gairy, resulted in his removal in a bloodless coup in 1979. Even though this action marked the first extra-constitutional change of government in the Commonwealth Caribbean (former British colonies), most Grenadians supported the coup, led by Maurice Bishop and his New Joint Endeavor for Welfare, Education, and Liberation (JEWEL) movement. Prime Minister Bishop, like Jamaica's Michael Manley before him, attempted to break out of European cultural and institutional molds and mobilize Grenadians behind him.

DEVELOPMENT

Prime Minister Mitchell has moved to end what Grenadians call "barter trade" and the government calls "smuggling." For years, Grenadian fishermen have exchanged their fish in Martinique for beer, cigarettes, and appliances. The cash-strapped treasury desperately needs the tariff revenues and has used drug interdiction as a means to end the contraband trade.

Bishop's social policies laid the foundation for basic health care for all Grenadians. With the departure of Cuban medical doctors in 1983, however, the lack of trained personnel created a significant health-care problem. Moreover, although medical-care facilities exist, these are not always in good repair, and equipment is aging and not reliable. Methods of recording births, deaths, and diseases lack systemization in Grenada, so it is risky to rely on local statistics to estimate the health needs of the population. There has also been some erosion from Bishop's campaign to accord women equal pay, status, and treatment. Two women were elected to Parliament, but skilled employment for women tends to be concentrated in the lowest-paid sector.

On October 19, 1983, Bishop and several of his senior ministers were killed during the course of a military coup. Six days later, the United States, with the token assistance of soldiers and police from states of the Eastern Caribbean, invaded Grenada, restored the 1974 Constitution, and prepared the way for new elections (in 1984).

According to one scholar, the invasion was a "lesson in a peacemaker's role in rebuilding a nation. Although Grenada has a history of parliamentary democracy, an atmosphere of civility, fertile soil, clean drinking water, and no slums, continued aid has not appreciably raised the standard of living and the young are resentful and restless."

FREEDOM

Grenadians are guaranteed full freedom of the press and speech. Newspapers, most of which are published by political parties, freely criticize the government without penalty. The OAS reported ballot fraud in the elections of November 2003, thus giving PM Keith Mitchell yet another term.

Grenada's international airport, the focus of much controversy, has pumped new blood into the tourist industry. Moves have also been made by the Grenadian government to promote private-sector business and to diminish the role of the government in the economy. Large amounts of foreign aid, especially from the United States, have helped to repair the infrastructure.

In recent years, foreign governments such as Kuwait, attracted by the power of Grenada's vote in the United Nations, have committed millions of dollars to Grenada's infrastructure. Some of these partnerships, particularly that involving Japan's access to Caribbean fish stocks, may have severe consequences for Grenadians in the future.

HEALTH/WELFARE

Grenada still lacks effective legislation for regulation of working conditions, wages, and occupational-safety and health standards. Discrimination is prohibited by law, but women are often paid less than men for the same work.

Significant problems remain. Unemployment has not significantly decreased; it remains at 15 percent of the workforce. Not surprisingly, the island has experienced a rising crime rate.

Prime Minister Keith Mitchell of the New National Party has promised to create more jobs in the private sector and to cut taxes to stimulate investment in small,

high-technology businesses. He also stated that government would become smaller and leaner.

ACHIEVEMENTS

A series of public consultations have been held with respect to the reestablishment of local government in the villages. Some 52 village councils work with the government in an effort to set policies that are both responsive and equitable.

To ease his task, the Grenadian economy has experienced a modest recovery, which had begun in 1993. Privatization has continued, attracting foreign capital. As is the case in much of the Caribbean, tourism has become an important source of revenue and employment in Grenada, with a rapid expansion of the service sector. Despite the decline of agricultural exports, Grenada has maintained its position as the world's second-largest exporter of nutmeg. To protect forested areas and what remains of its agricultural base, in 2001 the government developed a "Land Bank" policy. Designed to promote the efficient use and management of all agricultural lands, the government helps those who want to engage in agricultural pursuits but lack access to land, and pressures landowners who have not maintained prime agricultural land in a productive state.

Timeline: PAST

1498
Grenada is discovered by Christopher Columbus

1763
England acquires the island from France by treaty

1834
Slavery is abolished

1958–1962
Member of the West Indies Federation

1974
Independence from Great Britain

1979
A coup brings Maurice Bishop to power

1983
Prime Minister Bishop is assassinated; U.S. troops land

1995
Former mathematics professor Keith Mitchell is elected prime minister

PRESENT

2000s

Venezuela experiments with new shipping routes to Grenada to expand markets in both nations

Haiti (Republic of Haiti)

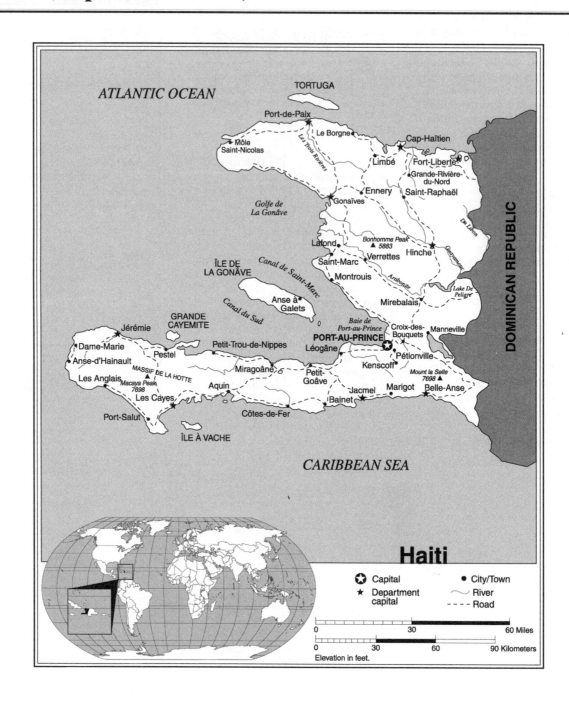

Haiti Statistics

GEOGRAPHY

Area in Square Miles (Kilometers):
10,714 (27,750) (about the size of Maryland)

Capital (Population): Port-au-Prince (845,000)

Environmental Concerns: extensive deforestation; soil erosion; inadequate potable water

Geographical Features: mostly rough and mountainous

Climate: tropical to semiarid

PEOPLE*

Population

Total: 7,656,166

*Note: Estimates explicitly take into account the effects of excess mortality due to AIDS.

148

Annual Growth Rate: 1.71%

Rural/Urban Population Ratio: 68/32

Major Languages: French; Creole

Ethnic Makeup: 95% black; 5% mulatto and white

Religions: 80% Roman Catholic (of which the majority also practice Vodun); 16% Protestant; 4% others

Health

Life Expectancy at Birth: 50 years (male); 53 years (female)

Infant Mortality Rate (Ratio): 95.2/1,000

Physicians Available (Ratio): 1/9,846

Education

Adult Literacy Rate: 52.9%

Compulsory (Ages): 6–12

COMMUNICATION

Telephones: 130,000 main lines

Daily Newspaper Circulation: 7 per 1,000 people

Televisions: 4 per 1,000 people

Internet Users: 80,000

TRANSPORTATION

Highways in Miles (Kilometers): 2,588 (4,160)

Railroads in Miles (Kilometers): privately owned industrial line

Usable Airfields: 13

Motor Vehicles in Use: 53,000

GOVERNMENT

Type: republic

Independence Date: January 1, 1804 (from France)

Head of State/Government: Interim President Boniface Alexander; Prime Minister Gerald Latortue

Political Parties: National Front for Change and Democracy; National Congress of Democratic Movements; Movement for the Installation of Democracy in Haiti; National Progressive Revolutionary Party; Lavalas Family; Haitian Christian Democratic Party; others

Suffrage: universal at 18

MILITARY

Current Disputes: claims U.S.–administered Navassa Island

ECONOMY

Currency ($ U.S. Equivalent): 40.5 gourde = $1

Per Capita Income/GDP: $1,600/$12.8 billion

GDP Growth Rate: -1%

Inflation Rate: 37.3%

Unemployment Rate: 65%

Labor Force: 3,600,000; unskilled labor abundant

Natural Resources: bauxite; copper; calcium carbonate; gold; marble; hydropower

Agriculture: coffee; mangoes; sugarcane; rice; corn; sorghum; wood

Industry: sugar refining; flour milling; textiles; cement; tourism; light assembly based on imported parts

Exports: $321 million (primary partners United States, European Union)

Imports: $1.2 billion (primary partners United States, European Union)

SUGGESTED WEBSITE

http://www.cia.gov/cia/
publications/factbook/geos/
ha.html

Haiti Country Report

HAITI

Haiti, which occupies the western third of the island of Hispaniola (the Dominican Republic comprises the other two thirds), was the first nation in Latin America to win independence from its mother country—in this instance, France. It is the poorest country in the Western Hemisphere and one of the least developed in the world. Agriculture, the main employer of the population, is pressed beyond the limits of the available land; the result has been catastrophic deforestation and erosion. While only roughly 30 percent of the land is suitable for planting, 50 percent is actually under cultivation. Haitians are woefully poor, suffer from poor health and lack of education, and seldom find work. Even when employment is found, wages are miserable, and there is no significant labor movement to intercede on behalf of the workers.

A persistent theme in Haiti's history has been a bitter rivalry between a small mulatto elite, consisting of 3 to 4 percent of the population, and the black majority. When François Duvalier, a black country doctor, was president (1957–1971), his avowed aim was to create a "new equilibrium" in the country—by which he meant a major shift in power from the established, predominantly mulatto, elite to a new, black middle class. Much of Haitian culture explicitly rejects Western civilization, which is identified with the mulattos. The Creole language of the masses and their practice of Vodun (voodoo), a combination of African spiritualism and Christianity, has not only insulated the population from the "culturally alien" regimes in power but has also given Haitians a common point of identity.

DEVELOPMENT

 Haiti's agricultural sector, where the vast majority of people earn a living, continues to suffer from massive soil erosion caused by deforestation, poor farming techniques, overpopulation, and low investment.

Haitian intellectuals have raised sharp questions about the nation's culture. Modernizers would like to see the triumph of the French language over Creole and Roman Catholicism over Vodun. Others argue that significant change in Haiti can come only from within, from what is authentically Haitian. The refusal of Haitian governments to recognize Creole as the official language has only added to the determination of the mulatto elite and the black middle class to exclude the rest of the population from effective participation in political life.

FREEDOM

 Demobilized soldiers and armed political factions are responsible for much of the violence in Haiti and have come to dominate drug trafficking. The rule of law cannot be maintained in the face of judicial corruption and a dysfunctional legal system.

For most of its history, Haiti has been run by a series of harsh authoritarian regimes. The ouster in 1986 of President-for-Life Jean-Claude Duvalier promised a more democratic opening as the new ruling National Governing Council announced as its primary goal the transition to a freely elected government. Political prisoners were freed; the dreaded secret police, the Tontons Macoute, were disbanded; and the press was unmuzzled.

HEALTH/WELFARE

Until 30 years ago, Haiti was self-sufficient in food production. It must now import about a third of its food needs. Nevertheless, the country has a rapidly expanding population, with a doubling time of 35 years overall, far faster than the Caribbean average of 52 years.

The vacuum left by Duvalier's departure was filled by a succession of governments that were either controlled or heavily influenced by the military. Significant change was heralded in 1990 with the election to power of an outspoken Roman Catholic priest, Jean-Bertrand Aristide. By the end of 1991, he had moved against the military and had formulated a foreign policy that sought to move Haiti closer to the nations of Latin America and the Caribbean. Aristide's promotion of the "church of the poor," which combined local beliefs with standard Catholic instruction, earned him the enmity of both conservative Church leaders and Vodun priests. The radical language of his Lavalas (Floodtide) movement, which promised sweeping economic and social changes, made business leaders and rural landowners uneasy.

Perhaps not surprisingly in this coup-ridden nation, the army ousted President Aristide in 1991. It took tough economic sanctions and the threat of an imminent U.S. invasion to force the junta to relinquish power. Aristide, with the support of U.S. troops, was returned to power in 1994. Once an uneasy stability was restored to the country, U.S. troops left the peacekeeping to UN soldiers.

Although there was a period of public euphoria over Aristide's return, the assessment of the *Guardian*, a British newspaper, was somber: Crime rates rose precipitously, political violence continued, and Aristide's enemies were still in Haiti—and armed. Haitians, "sensing a vacuum," took the law "into their own hands."

René Préval, who had served briefly as Aristide's prime minister, was himself elected to the presidency in 1996. According to *Caribbean Week*, Préval was caught between "a fiercely independent Parliament [and] an externally-imposed structural adjustment programme…." Préval, presiding over a divided party, was unable to have his choices for cabinet posts approved by the Legislature, which left Haiti without an effective government from 1997 to 1999.

ACHIEVEMENTS

In the late 1940s, Haitian "primitive" art created a sensation in Paris and other art centers. Although the force of the movement has now been spent, it still represents a unique, colorful, and imaginative art form.

Aristide was re-elected in 2000 in a vote characterized by irregularities and fraud. The result was parliamentary paralysis, as the opposition effectively boycotted Aristide's few initiatives, and a country where virtually every institution failed to function. Violent protests and equally violent government repression finally forced Aristide from power at the end of February, 2004. Meanwhile, the suffering of Haiti's people continued unabated, compounded by heavy spring rains and mudslides that killed perhaps 2,000 people.

Timeline: PAST

1492
The island is discovered by Christopher Columbus; named Hispaniola

1697
The western portion of Hispaniola is ceded to France

1804
Independence from France

1957–1971
The era of President François Duvalier

1971
Jean-Claude Duvalier is named president-for-life

1986
Jean-Claude Duvalier flees into exile

1991
A military coup ousts President Jean-Bertrand Aristide

PRESENT

2000s
The suffering of millions continues
Aristide ousted in February 2004

Jamaica

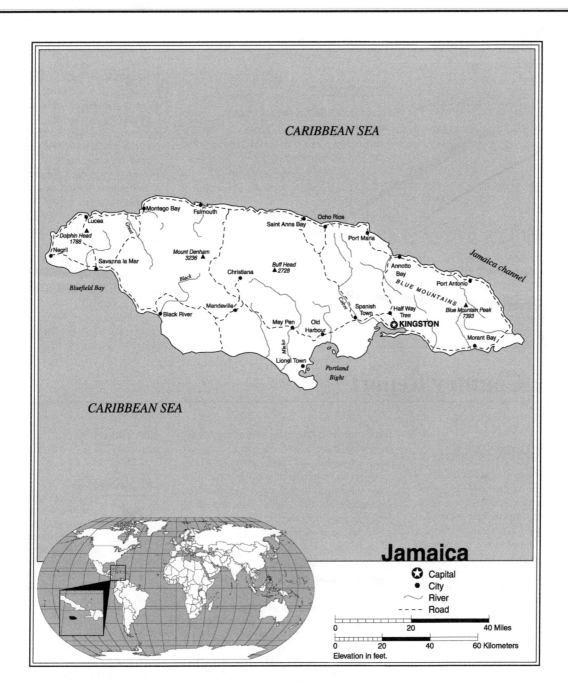

CARIBBEAN SEA

CARIBBEAN SEA

Jamaica channel

Montego Bay
Falmouth
Ocho Rios
Lucea
Saint Anns Bay
Port Maria
Dolphin Head
1788
Negril
Mount Denham
3236
Buff Head
2728
Annotto Bay
Port Antonio
Savanna la Mar
Christiana
BLUE MOUNTAINS
Black
Bluefield Bay
Mandeville
Cobre
Spanish Town
Half Way Tree
Blue Mountain Peak
7393
Black River
May Pen
Old Harbour
KINGSTON
Minho
Morant Bay
Lionel Town
Portland Bight

Jamaica

- ✪ Capital
- ● City
- ～ River
- - - Road

0   20   40 Miles
0   20   40   60 Kilometers
Elevation in feet.

Jamaica Statistics

GEOGRAPHY

Area in Square Miles (Kilometers): 4,244 (10,991) (slightly smaller than Connecticut)

Capital (Population): Kingston (104,000)

Environmental Concerns: deforestation; damage to coral reefs; water and air pollution

Geographical Features: Mostly mountains, with a narrow, discontinuous coastal plain

Climate: tropical; temperate Interior

PEOPLE

Population

Total: 2,713,130

Annual Growth Rate: 0.66%

Rural/Urban Population Ratio: 46/54

Major Languages: English; Jamaican
Creole
Ethnic Makeup: 90% black; 7% mixed; 3%
East Indian, white, Chinese and others
Religions: 56% Protestant; 5% Roman
Catholic; 39% others, including some
spiritualistic groups

Health

Life Expectancy at Birth: 73 years (male);
77 years (female)
Infant Mortality Rate (Ratio): 14/1,000
Physicians Available (Ratio): 1/6,043

Education

Adult Literacy Rate: 85%
Compulsory (Ages): 6–12; free

COMMUNICATION

Telephones: 444,400 main lines
Daily Newspaper Circulation: 65 per
1,000 people.
Televisions: 306 per 1,000
Internet Users: 600,000

TRANSPORTATION

Highways in Miles (Kilometers): 11,613
(18,700)
Railroads in Miles (Kilometers): 230 (370)
Usable Airfields: 36
Motor Vehicles in Use: 59,000

GOVERNMENT

Type: constitutional parliamentary
democracy
Independence Date: August 6, 1962 (from
the United Kingdom)
Head of State/Government: Queen
Elizabeth II; Prime Minister Percival J.
Patterson
Political Parties: People's National Party;
Jamaica Labour Party; National
Democratic Movement
Suffrage: universal at 18

MILITARY

Current Disputes: none

ECONOMY

Currency ($ U.S. Equivalent): 57.74
Jamaican dollars = $1
Per Capita Income/GDP: $3,800/$10.21
billion
GDP Growth Rate: 1.9%
Inflation Rate: 14.1%
Unemployment Rate: 16%
Labor Force: 1,130,000
Natural Resources: bauxite; gypsum;
limestone
Agriculture: sugarcane; bananas; coffee;
citrus fruits; potatoes; vegetables;
poultry; goats; milk
Industry: tourism; bauxite; textiles; food
processing; light manufactures; rum;
cement; metal
Exports: $1.7 billion (primary partners
United States, European Union, Canada)
Imports: $3 billion (primary partners United
States, CARICOM; European Union)

SUGGESTED WEBSITES

http://www.cia.gov/cia/
publications/factbook/index.html

Jamaica Country Report

JAMAICA:
"OUT OF MANY, ONE PEOPLE"

In 1962, Jamaica and Trinidad and Tobago
were the first of the English-speaking Caribbean islands to gain their independence.
A central problem since that time has been
the limited ability of Jamaicans to forge a
sense of nation. "Out of many, one people"
is a popular slogan in Jamaica, but it belies
an essential division of the population
along lines of both race and class. The elite,
consisting of a small white population and
Creoles (Afro-Europeans), still think of
themselves as "English." Local loyalties
notwithstanding, Englishness permeates
much of Jamaican life, from language to
sports. According to former prime minister
Michael Manley: "The problem in Jamaica
is how do you get the Jamaican to divorce
his mind from the paralysis of his history,
which was all bitter colonial frustration, so
that he sees his society in terms of this is
what crippled me?"

Manley's first government (1975–1980)
was one of the few in the Caribbean to incorporate the masses of the people into a political process. He was aware that in a
country such as Jamaica—where the majority of the population were poor, ill educated,
and lacked essential services—the promise
to provide basic needs would win him wide-

spread support. Programs to provide Jamaicans with basic health care and education
were expanded, as were services. Many
products were subjected to price controls or
were subsidized to make them available to
the majority of the people. Cuban medical
teams and teachers were brought to Jamaica
to fill the manpower gaps until local people
could be trained.

DEVELOPMENT

Prime Minister Patterson visited
Cuba, where agreements were
signed for closer cooperation
between the two nations in the
medical sphere and with a focus on
biotechnology. Agreement was also reached
on tourism issues and stressed cooperation
rather than competition

However, Jamaica's fragile economy
could not support Manley's policies, and he
was eventually opposed by the entrenched
elite and voted out of office. But Manley
was returned to office in 1989, with a new
image as a moderate, willing to compromise
and aware of the need for foreign-capital investment. Manley retired in 1992 and was
replaced as prime minister by Percival J.
Patterson, who promised to accelerate Jamaica's transition to a free-market econ-

omy. The government instituted a policy of
divestment of state-owned enterprises.

FREEDOM

Despite the repeal of the
controversial Suppression of
Crime Act of 1974, the
Parliament, in the face of
persistent high levels of crime, provided for
emergency police powers. Some critics
charge that the Parliament in essence recreated the repealed legislation in a different
guise.

The challenges remain. Crime and violence continue to be major social problems
in Jamaica. The high crime rate threatens
not only the lucrative tourist industry but the
very foundations of Jamaican society. Prime
Minister Patterson has called for a moral re-
awakening: "All our programs and strategies for economic progress are doomed to
failure unless there is a drastic change in social attitudes…." A stagnant economy, persistent inflation, and unemployment and
underemployment combine to lessen respect for authority and contribute to the
crime problem. In 2001, Amnesty International noted that in proportion to population,
more people are killed by police in Jamaica
than anywhere else in the world. Many of
the deaths are the result of clashes with

gangs of drug dealers, who usually outgun the police. Jamaica counted 1,000 murders in 2002, more, proportionately, than in South Africa, and less than Colombia or El Salvador. It remains a violent society, and the nation continues to walk the narrow line between liberty and license.

HEALTH/WELFARE

Jamaica's "Operation Pride" was designed to combine a dynamic program of land divestment by the state with provisions to meet demands for housing. Squatter colonies would be replaced by "proud home owners.

As is the case in many developing-world countries where unemployment and disaffection are common, drug use is high in Jamaica. The government is reluctant to enforce drug control, however, for approximately 8,000 rural families depend on the cultivation of ganja (marijuana) to supplement their already marginal incomes.

Some of Jamaica's violence is politically motivated and tends to be associated with election campaigns. Both major parties have supporters who employ violence for political purposes. The legal system has been unable to contain the violence or bring the guilty to justice, because of a pervasive code of silence enforced at the local level.

The Patterson government has moved deliberately in the direction of electoral re-

ACHIEVEMENTS

Marcus Garvey was posthumously declared Jamaica's first National Hero in 1964 because of his leading role in the international movement against racism. He called passionately for the recognition of the equal dignity of human beings regardless of race, religion, or national origin. Garvey died in London in 1940.

form in an attempt to reduce both violence and fraud. Until those reforms are in place, however, the opposition Jamaica Labour Party has decided to boycott by-elections.

Re-elected to fourth term in 2002, PM Patterson hopes to match Jamaica's political stability with improvements in the nation's social and economic sectors. He has successfully addressed inflation through tight monetary and fiscal policies and is redressing Jamaica's debt by privatizing inefficient state enterprises.

On the positive side, human rights are generally respected, and Jamaica's press is basically free. Press freedom is observed in practice within the broad limits of libel laws and the State Secrets Act. Opposition parties publish newspapers and magazines that are highly critical of government policies, and foreign publications are widely available.

Jamaica's labor-union movement is strong and well organized, and it has contributed many leaders to the political process. Unions are among the strongest and best organizations in the country and are closely tied to political parties.

Timeline: PAST

1509
The first Spanish settlement

1655
Jamaica is seized by the English

1692
An earthquake destroys Port Royal

1944
Universal suffrage is proclaimed

1962
Independence from Great Britain

1990s
Violent crime and strong-armed police responses plague the island

Percival J. Patterson is elected prime minister

PRESENT

2000s
Gun battles break out in Kingston

Patterson re-elected to a fourth term in 2002

St. Kitts–Nevis (Federation of St. Kitts and Nevis)

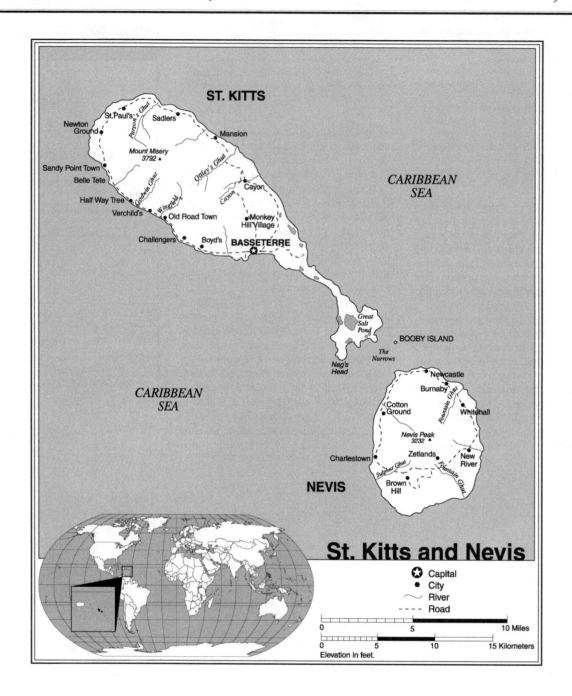

St. Kits–Nevis Statistics

GEOGRAPHY

Area in Square Miles (Kilometers): 101
(261) (about 1½ times the size of
Washington, D.C.)
Capital (Population): Basseterre (12,600)
Geographical Features: volcanic, with
mountainous interiors
Climate: subtropical

PEOPLE

Population

Total: 38,836
Annual Growth Rate: 0.25%
Rural/Urban Population Ratio: 66/34
Major Language: English
Ethnic Makeup: mainly of black African
descent

Religions: Anglican; other Protestant
sects; Roman Catholic

Health

Life Expectancy at Birth: 68 years (male);
74 years (female)
Infant Mortality Rate (Ratio): 16.3/1,000
Physicians Available (Ratio): 1/1,057

Education

Adult Literacy Rate: 97%
Compulsory (Ages): for 12 years between ages 5 and 18

COMMUNICATION

Telephones: 23,500 main lines
Televisions: 241 per 1,000
Internet Users: 10,000

TRANSPORTATION

Highways in Miles (Kilometers): 199 (320)
Railroads in Miles (Kilometers): 36 (58)
Usable Airfields: 2

GOVERNMENT

Type: constitutional monarchy within Commonwealth

Independence Date: September 19, 1983 (from the United Kingdom)
Head of State/Government: Queen Elizabeth II; Prime Minister Denzil Douglas
Political Parties: St. Kitts and Nevis Labour Party; People's Action Movement; Nevis Reformation Party; Concerned Citizens Movement
Suffrage: universal at 18

MILITARY

Current Disputes: Nevis has threatened to secede

ECONOMY

Currency ($ U.S. Equivalent): 2.67 East Caribbean dollars = $1
Per Capita Income/GDP: $8,800/$339 million

GDP Growth Rate: -1.9%
Inflation Rate: 1.7%
Unemployment Rate: 4.5%
Labor Force: 18,200
Natural Resources: negligible
Agriculture: sugarcane; rice; yams; vegetables; bananas; fish
Industry: sugar processing; tourism; cotton; salt; copra; clothing; footwear; beverages
Exports: $70 million (primary partners United States, United Kingdom, CARICOM)
Imports: $195 million (primary partners United States, CARICOM, United Kingdom)

SUGGESTED WEBSITES

http://www.cia.gov/cia/
publications/factbook/index.html

St. Kits–Nevis Country Report

ST. KITTS–NEVIS: ESTRANGED NEIGHBORS

On September 19, 1983, the twin-island state of St. Kitts–Nevis became an independent nation. The country had been a British colony since 1623, when Captain Thomas Warner landed with his wife and eldest son, along with 13 other settlers. The colony fared well, and soon other Caribbean islands were being settled by colonists sent out from St. Kitts (also commonly known as St. Christopher).

DEVELOPMENT

In an attempt to improve an economy that is essentially stagnant, the government of St. Kitts–Nevis, under the auspices of CARICOM, supports the idea of a Caribbean free-trade area. Of particular interest is the inclusion of Cuba in the agreement.

The history of this small island nation is the story of the classic duel between the big sea powers of the period—Great Britain, France, and Spain—and the indigenous people—in this case, the Carib Indians. (Although much of the nation's history has centered around St. Kitts, the larger of the two islands, Nevis, only two miles away, has always been considered a part of St. Kitts, and its history is tied into that of the larger island.) The British were the first settlers on the island of St. Kitts but were followed that same year by the French. In a

unique compromise, considering the era, the British and French divided the territory in 1627 and lived in peace for a number of decades. A significant reason for this British–French cooperation was the constant pressure from their common enemies: the aggressive Spanish and the fierce Carib Indians.

FREEDOM

The election in 1984 of Constance Mitcham to Parliament signaled a new role for women. She was subsequently appointed minister of women's affairs. However, despite her conspicuous success, women still occupy a very small percentage of senior civil-service positions.

With the gradual elimination of the mutual threat, Anglo–French tensions again mounted, resulting in a sharp land battle at Frigate Bay on St. Kitts. The new round of hostilities, which reflected events in Europe, would disrupt the Caribbean for much of the next century. Events came to a climactic head in 1782, when the British garrison at Brimstone Hill, commonly known as the "Gibraltar of the West Indies," was overwhelmed by a superior French force. In honor of the bravery of the defenders, the French commander allowed the British to march from the fortress in full formation. (The expression "peace with honor" has its roots in this historic encounter.) Later in the year, however, the British again seized the upper hand. A naval battle

at Frigate Bay was won by British Admiral Hood following a series of brilliant maneuvers. The defeated French admiral, the Count de Grasse, was in turn granted "peace with honor." Thereafter, the islands remained under British rule until their independence in 1983.

AGRICULTURE

Before the British colonized the island, St. Kitts was called Liamiuga ("Fertile Isle") by the Carib Indians. The name was, and is, apt, because agriculture plays a big role in the economy of the islands. Tourism has finally replaced sugar exports as the largest generator of foreign exchange.

Because the sugar market is so unstable, the economy of St. Kitts–Nevis fluctuates considerably. Nearly one third of St. Kitts's land (some 16,000 acres) is under cultivation. In a good year, sugar production can exceed 50,000 tons. Although over the years growers have experimented with a number of other crops, they always have come back to sugarcane.

HEALTH/WELFARE

Since the economy is so dependent on sugarcane, the overall welfare of the country is at the mercy of the world sugar market. Although a minimum wage exists by law, the amount is less than what a person can reasonably be expected to live on.

ECONOMIC CHANGE

Unlike such islands as Barbados and Antigua, St. Kitts–Nevis for years chose not to use tourism as a buffer to offset any disastrous fluctuations in sugar prices. On St. Kitts, there was an antitourism attitude that can be traced back to the repressive administration of Prime Minister Robert Bradshaw, a black nationalist who worked to discourage tourism and threatened to nationalize all land holdings.

That changed under the moderate leadership of Kennedy Simmonds and his People's Action Movement, who remained in power from 1980 until ousted in elections in July 1995. The new administration of Denzil Douglas promised to address serious problems that had developed, including drug trafficking, money laundering, and a lack of respect for law and order. In 1997, a 50-man "army" was created to wage war against heavily armed drug traffickers operating in the region. Agriculture Minister Timothy Harris noted that the permanent defense force "was critical to the survival of the sovereignty of the nation." Simmond's promotion of tourism took root. By 2001, tourism had become a major growth industry in the islands. Major airlines refused to schedule landings in St. Kitts until there were an adequate number of hotel rooms. Accordingly, the government promoted the construction of 1,500 rooms. A positive side-effect are the jobs produced in the construction trades and service industry.

ACHIEVEMENTS

St. Kitts–Nevis was the first successful British settlement in the Caribbean. St. Kitts–Nevis was the birthplace of Alexander Hamilton, the first U.S. secretary of the Treasury Department and an American statesman.

The future of St. Kitts–Nevis will depend on its ability to broaden its economic base. PM Douglas's economic policies include the promotion of export-oriented manufactures and off-shore banking. A potential problem of some magnitude looms, however: The island of Nevis, long in the shadow of the more populous and prosperous St. Kitts, nearly voted to secede in a referendum held in August 1998. The Constitution requires a two-thirds majority for secession; 61.7 percent of the population of Nevis voted "Yes." Not surprisingly, the government is working to fashion a new federalism with "appropriate power sharing" between the islands. That has not diminished the move toward independence for Nevis' 10,000 people. Another referendum on secession is set for 2004.

Timeline: PAST

1493
The islands are discovered and named by Christopher Columbus

1623
The British colony is settled by Captain Thomas Warner

1689
A land battle at Frigate Bay disrupts a peaceful accord between France and England

1782
The English are expelled by French at the siege of Brimstone Hill

The French are beaten at the sea battle of Frigate Bay; the beginning of continuous British rule

1967
Self-government as an Associate State of the United Kingdom

1983
Full independence from Great Britain

1998
A referendum on Nevis secession is narrowly defeated

PRESENT

2000s
Another vote on the secession of Nevis scheduled for 2004

St. Lucia

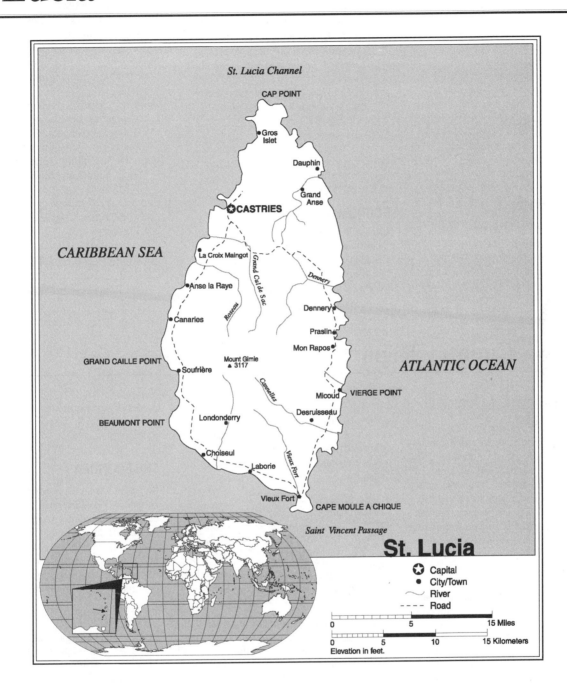

St. Lucia Channel

CAP POINT

Gros Islet

Dauphin

Grand Anse

⊗ CASTRIES

CARIBBEAN SEA

La Croix Maingot

Anse la Raye

Grand Cul de Sac

Roseau

Dennery

Canaries

Dennery

Praslin

Mon Rapos

GRAND CAILLE POINT

Mount Gimie
▲ 3117

Soufrière

Canvelles

ATLANTIC OCEAN

Micoud

VIERGE POINT

Desruisseau

BEAUMONT POINT

Londonderry

Vieux Fort

Choiseul

Laborie

Vieux Fort

CAPE MOULE A CHIQUE

Saint Vincent Passage

St. Lucia

⊗ Capital
• City/Town
⌇ River
--- Road

| 0 | 5 | 15 Miles |
| 0 | 5 | 10 | 15 Kilometers |

Elevation in feet.

St. Lucia Statistics

GEOGRAPHY

Area in Square Miles (Kilometers): 238 (619) (about 3 times the size of Washington, D.C.)

Capital (Population): Castries (13,600)

Environmental Concerns: deforestation; soil erosion

Geographical Features: volcanic and mountainous; some broad, fertile valleys

Climate: tropical maritime

PEOPLE

Population

Total: 164,213

Annual Growth Rate: 1.23%

Rural/Urban Population Ratio: 63/37

Major Languages: English; French patois

Ethnic Makeup: 90% black; 6% mixed; 3% East Indian; 1% white

Religions: 90% Roman Catholic; 3% Church of England; 7% other Protestant sects

Health

Life Expectancy at Birth: 69 years (male);
    76 years (female)
Infant Mortality Rate (Ratio): 13/1,000
Physicians Available (Ratio): 1/2,235

Education

Adult Literacy Rate: 67%
Compulsory (Ages): 5–15

COMMUNICATION

Telephones: 51,100 main lines
Televisions: 172 per 1,000 people
Internet Users: 13,000

TRANSPORTATION

Highways in Miles (Kilometers): 451
    (1,210)
Railroads in Miles (Kilometers): none
Usable Airfields: 2
Motor Vehicles in Use: 19,000

GOVERNMENT

Type: constitutional monarchy within
    Commonwealth
Independence Date: February 22, 1979
    (from the United Kingdom)
Head of State/Government: Queen
    Elizabeth II; Prime Minister Kenneth D.
    Anthony
Political Parties: United Workers Party;
    St. Lucia Labour Party; National
    Freedom Party
Suffrage: universal at 18

MILITARY

Current Disputes: none

ECONOMY

Currency ($ U.S. Equivalent): 2.67 East
    Caribbean dollars = $1
Per Capita Income/GDP: $5,400/$866
    million
GDP Growth Rate: 3.3%

Inflation Rate: 3%
Unemployment Rate: 16.5%
Labor Force: 43,800
Natural Resources: forests; sandy
    beaches; minerals (pumice); mineral
    springs; geothermal potential
Agriculture: bananas; coconuts;
    vegetables; citrus fruits; root crops;
    cocoa
Industry: clothing; assembly of electronic
    components; beverages; corrugated
    cardboard boxes; tourism; lime
    processing; coconut processing
Exports: $66 million (primary partners
    United Kingdom, United States,
    CARICOM)
Imports: $267 million (primary partners
    United States, CARICOM, United
    Kingdom)

SUGGESTED WEBSITES

http://www.cia.gov/cia/
    publications/factbook/index.html

St. Lucia Country Report

ST. LUCIA: ENGLISH POLITICS, FRENCH CULTURE

The history of St. Lucia gives striking testimony to the fact that the sugar economy, together with the contrasting cultures of various colonial masters, was crucial in shaping the land, social structures, and lifestyles of its people. The island changed hands between the French and the English at least seven times, and the influences of both cultures are still evident today. Ninety percent of the population speaks French patois (dialect), while the educated and the elite prefer English. Indeed, the educated perceive patois as suitable only for proverbs and curses. On St. Lucia and the other patois-speaking islands (Dominica, Grenada), some view the common language as the true reflection of their uniqueness. English, however, is the language of status and opportunity. In terms of religion, most St. Lucians are Roman Catholic.

DEVELOPMENT

The government negotiated a loan with France to construct new water pipelines. Provision of potable water is critical not only to agriculture but also to the tourist industry.

The original inhabitants of St. Lucia were Arawak Indians who had been forced off the South American mainland by the cannibalistic Carib Indians. Gradually, the Carib also moved onto the Caribbean islands and destroyed most of the Arawak culture. Evidence of that early civilization has been found in rich archaeological sites on St. Lucia..

FREEDOM

The St. Lucian political system is healthy, with opposition parties playing an active role in and out of Parliament. Women participate fully in government and hold prominent positions in the civil service.

The date of the European "discovery" of the island is uncertain; it may have occurred in 1499 or 1504 by the navigator and mapmaker Juan de la Cosa, who explored the Windward Islands during the early years of the sixteenth century. The Dutch, French, and English all established small settlements or trading posts on the island in the seventeenth century but were resisted by the Caribs. The first successful settlement dates from 1651, when the French were able to maintain a foothold.

The island's political culture is English. Upon independence from Great Britain in 1979, St. Lucians adopted the British parliamentary system, which includes specific safeguards for the preservation of human rights. Despite several years of political disruption, caused by the jockeying for power of several political parties and affiliated interests, St. Lucian politics is essentially stable.

THE ECONOMY

St. Lucia has an economy that is as diverse as any in the Caribbean. Essentially agricultural, the country has also developed a tourism industry, manufacturing, and related construction activity. A recent "mineral inventory" has located possible gold deposits, but exploitation must await the creation of appropriate mining legislation.

HEALTH/WELFARE

The minister of agriculture has linked marginal nutrition and malnutrition in St. Lucia with economic adjustment programs in the Caribbean. He noted that the success achieved earlier in raising standards of living was being eroded by "onerous debt burdens."

U.S. promises to the region made in the 1980s failed to live up to expectations. Although textiles, clothing, and nontraditional goods exported to the United States increased as a result of the Caribbean Basin Initiative, St. Lucia remained dependent on its exports of bananas. About a third of the island's workforce are involved in banana production, which accounts for 90 percent of St. Lucia's exports

St. Lucia's crucial banana industry suffered significant production losses in 1997 and 2001 in large part because of drought. Exports were half of the normal volume, and St. Lucia fell short of filling its quota for the European Union. A 1999 European Union decision to drop its import preferences for bananas from former colonial possessions in the Caribbean together with increased competition from Latin America growers have created an urgent demand to diversify St. Lucia's economy. Increased emphasis has been placed on exports of mangos and avocados. Tourism, light manufacturing, and offshore banking have also experienced growth. Despite these attempts unemployment, inflation, a high cost of living, and drug trafficking remain serious problems and have led to periodic unrest.

ACHIEVEMENTS

St. Lucians have won an impressive two Nobel prizes. Sir W. Arthur Lewis won the prize in 1979 for economics, and in 1993, poet Derek Walcott won the prize for literature. When asked how the island had produced two Nobel laureates, Wolcott replied: "It's the food."

St. Lucia, like several other islands, has also succeeded in trading on its sovereignty—a vote in the United Nations—to raise revenue. Since 1997, when banana exports reached crisis proportions, St. Lucia has supported the claims of China over the independence of Taiwan. St. Kitts–Nevis, St. Vincent, and Grenada still support Taiwan—in exchange benefits.

EDUCATION AND EMIGRATION

Education in St. Lucia has traditionally been brief and perfunctory. Few students attend secondary school, and very few (3 percent) ever attend a university. Although the government reports that 95 percent of those eligible attend elementary school, farm and related chores severely reduce attendance figures. In recent years, St. Lucia has channeled more than 20 percent of its expenditures into education and health care. Patient care in the general hospital was made free of charge in 1980.

Population growth is relatively low, but emigration off the island is a significant factor. For years, St. Lucians, together with Dominicans, traveled to French Guiana to work in the gold fields. More recently, however, they have crossed to neighboring

Martinique, a French department, in search of work. St. Lucians can also be found working on many other Caribbean islands.

Timeline: PAST

638
The English take possession of St. Lucia

1794
The English regain possession of St. Lucia from France

1908
Riots

1951
Universal adult suffrage

1979
Independence from Great Britain

1990s
Banana production suffers a serious decline

PRESENT

2000s
Economic diversification becomes a critical need

St. Vincent and the Grenadines

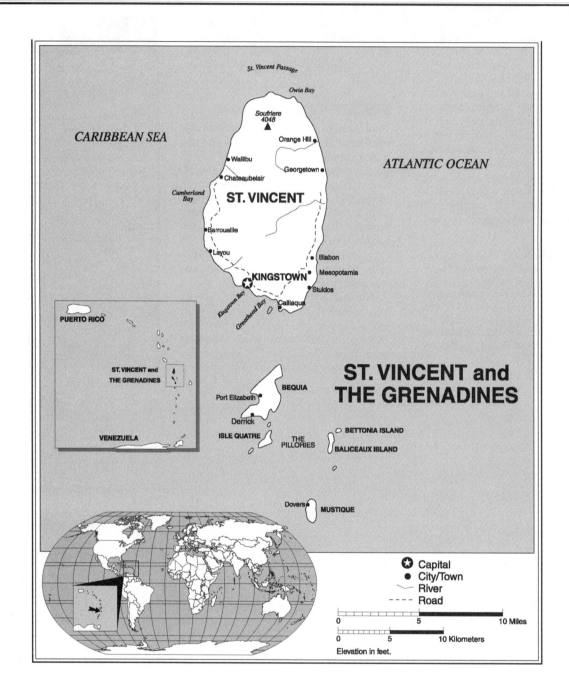

St. Vincent and the Grenadines Statistics

GEOGRAPHY

Area in Square Miles (Kilometers): 131 (340) (about twice the size of Washington, D.C.)

Capital (Population): Kingstown (16,000)

Environmental Concerns: pollution of coastal waters and shorelines by discharges from pleasure boats

Geographical Features: volcanic; mountainous

Climate: tropical

PEOPLE

Population

Total: 117,193
Annual Growth Rate: 0.31%
Rural/Urban Population Ratio: 50/50

Major Languages: English; French patois
Ethnic Makeup: mainly black African descent; remainder mixed, with some white, East Indian, and Carib Indian
Religions: Anglican; Methodist; Roman Catholic; Seventh-Day Adventist

Health

Life Expectancy at Birth: 71 years (male); 74 years (female)
Infant Mortality Rate (Ratio): 16.6/1,000
Physicians Available (Ratio): 1/2,708

Education

Adult Literacy Rate: 96%

COMMUNICATION

Telephones: 27,300 main lines
Televisions: 161 per 1,000 people
Internet Users: 7,000

TRANSPORTATION

Highways in Miles (Kilometers): 646 (1,040)

Railroads in Miles (Kilometers): none
Usable Airfields: 6
Motor Vehicles in Use: 8,200

GOVERNMENT

Type: constitutional monarchy within the British Commonwealth
Independence Date: October 27, 1979 (from the United Kingdom)
Head of State/Government: Queen Elizabeth II; Prime Minister Ralph Gonsalves
Political Parties: Unity Labour Party; New Democratic Party; United People's Movement; National Reform Party
Suffrage: universal at 18

MILITARY

Current Disputes: none

ECONOMY

Currency ($ U.S. Equivalent): 2.67 East Caribbean dollars = $1

Per Capita Income/GDP: $2,900/$339 million
GDP Growth Rate: -0.5%
Inflation Rate: -0.4%
Unemployment Rate: 22%
Labor Force: 67,000
Natural Resources: negligible
Agriculture: bananas; arrowroot; coconuts; sweet potatoes; spices; small amount of livestock; fish
Industry: food processing; cement; furniture; clothing; starch; tourism
Exports: $38 million (primary partners CARICOM, United Kingdom, United States)
Imports: $174 million (primary partners United States, CARICOM, United Kingdom)

SUGGESTED WEBSITES

http://www.cia.gov/cia/
publications/factbook/index.html

St. Vincent and the Grenadines Country Report

ST. VINCENT AND THE GRENADINES: POOR BUT FREE

Vincentians, like many other West Indians, either identify with or, as viewed from a different perspective, suffer from a deep-seated European orientation. Critics argue that it is an identification that is historical in origin, and that it is negative. For many, the European connection is nothing more than the continuing memory of a master–slave relationship.

DEVELOPMENT

A slump in banana production because of poor weather and low prices forced farmers to produce other crops, including marijuana. In a 1998 sweep, U.S. troops aided St. Vincentian soldiers in eradicating the crop, which spread animosity toward Washington.

St. Vincent is unique in that it was one of the few Caribbean islands where runaway black slaves intermarried with Carib Indians and produced a distinct racial type known as the Garifuna, or black Carib. Toward the end of the eighteenth century, the Garifuna and other native peoples mounted an assault on the island's white British planters. They were assisted by the French from Martinique but were defeated in 1796. As punishment, the Garifuna were

deported to what is today Belize, where they formed one of the bases of that nation's population.

In 1834, the black slaves were emancipated, which disrupted the island's economy by decreasing the labor supply. In order to fill this vacuum, Portuguese and East Indian laborers were imported to maintain the agrarian economy. This, however, was not done until later in the nineteenth century—not quickly enough to prevent a lasting blow to the island's economic base.

St. Vincent, along with Dominica, is one of the poorest islands in the West Indies. The current unemployment rate (2004) is estimated at between 33 percent. With more than half the population under age 15, unemployment will continue to be a major problem in the foreseeable future.

FREEDOM

The government took a great step forward in terms of wage scales for women by adopting a new minimum-wage law, which provided for equal pay for equal work done by men and women. Violence against women remains a significant problem.

Formerly one of the West Indian sugar islands, St. Vincent's main crops are now ba-

nanas and arrowroot. The sugar industry was a casualty of low world-market prices and a black-power movement in the 1960s that associated sugar production with memories of slavery. Limited sugar production has been renewed to meet local needs.

HEALTH/WELFARE

Minimum wages established in 1989 range from $3.85 *per day* in agriculture to $7.46 in industry. New minimums were to be presented to parliament in 2003. Clearly, the minimum is inadequate, although most workers earn significantly more than the minimum.

THE POLITICS OF POVERTY

Poverty affects everyone in St. Vincent and the Grenadines, except a very few who live in comfort. In the words of one Vincentian, for most people, "life is a study in poverty." In 1969, a report identified malnutrition and gastroenteritis as being responsible for 57 percent of the deaths of children under age five. Those problems persist.

Deep-seated poverty also has an impact on the island's political life. Living on the verge of starvation, Vincentians cannot appreciate an intellectual approach to politics. They find it difficult to wait for the effects of long-term trends or coordinated development. Bread-and-butter issues are what concern them. Accordingly, parties speak little of basic economic and social change, structural shifts in the economy, or the latest economic theories. Politics is reduced to personality contests and rabble-rousing.

Despite its severe economic problems, St. Vincent is a free society. Newspapers are uncensored. Some reports, however, have noted that the government has on occasion granted or withheld advertising on the basis of a paper's editorial position.

Unions enjoy the right of collective bargaining. They represent about 11 percent of the labor force. St. Vincent, which won its independence of Great Britain in 1979, is a parliamentary, constitutional democracy. Political parties have the right to organize.

ACHIEVEMENTS

A regional cultural organization was launched in 1982 in St. Vincent. Called the East Caribbean Popular Theatre Organisation, its membership extends to Dominica, Grenada, and St. Lucia.

POLITICS AND ECONOMICS

While the country's political life has been calm, relative to some of the other Caribbean islands, there are signs of voter unrest. Prime Minister James Mitchell was reelected in 1999 for an unprecedented fourth five-year term, but his New Democratic Party lost some ground in the Legislature. (Ralph Gonsalves took over the post in 2001.) In addition, two opposition parties, the St. Vincent Labour Party and the Movement for National Unity, merged to create the new Unity Labour Party. In an effort to recover the initiative, Mitchell promised major new investments in the crucial banana industry as well as improvements in St. Vincent's infrastructure and social services. Bananas accounted for 65 percent of St. Vincent's export earnings. As is the case with other Windward Islands, St. Vincent's economy has been hurt by the 1999 European Union decision to phase out preferential treatment to banana producers from former colonial possessions. Bananas now account for 50 percent of St. Vincent's export earnings, a decline of 15 percent in five years. Not surprisingly drug trafficking, marijuana cultivation, and money laundering have increased as a result of the general economic malaise.

In 2001, along with other islands in the Windwards, St. Vincent entered talks with European Union officials with an eye to improving both yields and quality of bananas. There was general agreement that the entire Windward Islands banana industry needed restructuring if the industry is to survive.

Timeline: PAST

1498
Christopher Columbus discovers and names St. Vincent

1763
Ceded to the British by France

1795
The Carib War

1902
St. Vincent's La Soufrière erupts and kills 2,000 people

1979
Independence from Great Britain

1990s
A new minimum-wage law takes effect

PRESENT

2000s
The country's financial problems remain severe

Ralph Gonsalves assumes the post of prime minister

Trinidad and Tobago (Republic of Trinidad and Tobago)

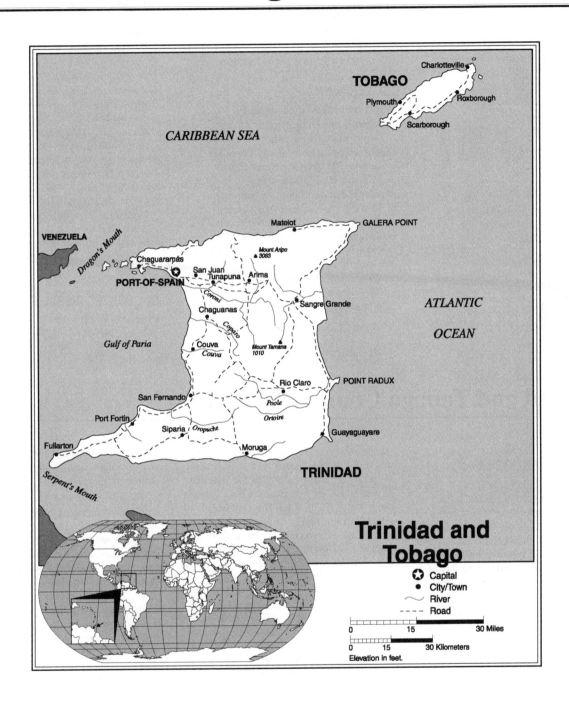

Trinidad and Tobago Statistics

GEOGRAPHY

Area in Square Miles (Kilometers): 1,980
  (5,128) (about the size of Delaware)

Capital (Population): Port-of-Spain
  (44,000)

Environmental Concerns: water pollution;
  oil pollution of beaches; deforestation;
  soil erosion
Geographical Features: mostly plains,
  with some hills and low mountains
Climate: tropical

PEOPLE

Population

Total: 1,096,585
Annual Growth Rate: -0.71%
Rural/Urban Population Ratio: 28/72

Major Language: English

Ethnic Makeup: 43% black; 40% East Indian; 14% mixed; 1% white; 1% Chinese; 1% others

Religions: 32% Roman Catholic; 24% Hindu; 14% Anglican; 14% other Protestant; 6% Muslim; 10% others

Health

Life Expectancy at Birth: 66 years (male); 71 years (female)

Infant Mortality Rate (Ratio): 25/1,000

Physicians Available (Ratio): 1/1,191

Education

Adult Literacy Rate: 98%

Compulsory (Ages): 5–12; free

COMMUNICATION

Telephones: 325,100 main lines

Daily Newspaper Circulation: 139 per 1,000 people

Televisions: 198 per 1,000 people

Internet Users: 138,000

TRANSPORTATION

Highways in Miles (Kilometers): 5,167 (8,320)

Railroads in Miles (Kilometers): minimal agricultural service

Usable Airfields: 6

Motor Vehicles in Use: 155,000

GOVERNMENT

Type: parliamentary democracy

Independence Date: August 31, 1962 (from United Kingdom)

Head of State/Government: President George Maxwell Richards; Prime Minister Patrick Manning

Political Parties: People's National Movement; National Alliance for Reconstruction; United National Congress; Movement for Social Transformation; National Joint Action Committee; others

Suffrage: universal at 18

MILITARY

Current Disputes: none

ECONOMY

Currency ($ U.S. Equivalent): 6.12 Trinidad/Tobago dollars = $1

Per Capita Income/GDP: $9,500/$10.6 billion

GDP Growth Rate: 4.5%

Inflation Rate: 3.7%

Unemployment Rate: 10.9%

Labor Force: 559,000

Natural Resources: petroleum; natural gas; asphalt

Agriculture: cocoa; sugarcane; rice; citrus fruits; coffee; vegetables; poultry

Industry: petroleum; chemicals; tourism; food processing; cement; beverages; textiles

Exports: $4.9 billion (primary partners United States, CARICOM, Latin America)

Imports: $3 billion (primary partners United States, Venezuela, European Union)

SUGGESTED WEBSITE

http://www.cia.gov/cia/ publications/factbook/index.html

Trinidad and Tobago Country Report

TRINIDAD AND TOBAGO: A MIDDLE-CLASS SOCIETY

The nation of Trinidad and Tobago, which became independent of Great Britain in 1962, differs sharply from other Caribbean countries in terms of both its wealth and its societal structure. More than one third of its revenues derive from the production of crude oil. Much of the oil wealth has been redistributed and has created a society that is essentially middle class. Health conditions are generally good, education is widely available, and the literacy rate is a very high 98 percent.

DEVELOPMENT

 The economy has experienced rapid growth thanks to vast reserves of natural gas, which has attracted investors from developed countries. The government has also promoted the use of natural gas, instituting a program to encourage consumers to switch from gasoline.

The country also enjoys an excellent human-rights record, although there is a good deal of tension between the ruling urban black majority and East Indians, who are rural. The divisions run deep and

parallel the situation in Guyana. East Indians feel that they are forced to submerge their culture and conform to the majority. In the words of one East Indian, "Where do Indians fit in when the culture of 40 percent of our people is denied its rightful place and recognition; when most of our people exist on the fringes of society and are considered as possessing nothing more than nuisance value?"

The lyrics of a black calypso artist that state the following are resented by East Indians:

> If you are an East Indian
> And you want to be an African
> Just shave your head just like me
> And nobody would guess your
> nationality.

The prosperity of the nation, however, tends to mute these tensions.

Freedom of expression and freedom of the press are constitutionally guaranteed as well as respected in practice. Opposition viewpoints are freely expressed in the nation's Parliament, which is modeled along British lines. There is no political censorship. Opposition parties are usually supported by rural Hindu East Indians; while they have freely participated in elections,

some East Indians feel that the government has gerrymandered electoral districts to favor the ruling party.

Violent crime and political unrest, including an attempted coup by black fundamentalist Muslim army officers in 1990, have become a way of life in the nation in recent years. Prime Minister Basdeo Panday, elected in 1996, noted that there were still agendas, "political and otherwise," that divided Trinidadian society. "How much better it will be," he stated, "if all in our society, and particularly those in a position to shape mass consciousness, will seize every opportunity to promote and mobilise the greater strength that comes out of our diversity...."

FREEDOM

 Freedom of expression on the islands is guaranteed by the Constitution. The independent judiciary, pluralistic political system, and independent and privately owned print media assure that free expression exists in practice as well as in theory.

Trade-union organization is the most extensive among Caribbean nations with ties to Britain and includes about 30 percent of the workforce. In contrast to other West Indian states, unions in Trinidad and Tobago are not government-controlled, nor are they generally affiliated with a political party.

Women are well represented in Parliament, serve as ministers, and hold other high-level civil-service positions. Several groups are vocal advocates for women's rights.

HEALTH/WELFARE

Legislation passed in 1991 greatly expanded the categories of workers covered by the minimum wage. The same legislation provided for 3 months' maternity leave for household and shop assistants as well as other benefits.

In an attempt to redress imbalances in the nation's agricultural structure, which is characterized by small landholdings—half of which are less than five acres each—the government has initiated a land-redistribution program using state-owned properties and estates sold to the government. The program is designed to establish more efficient medium-size family farms, of five to 20 acres, devoted to cash crops.

The islands' economic fortunes have tended to reflect the prices it can command from its exports of oil and liquefied natural gas. In 2004, with prices high, government revenues grew significantly. In 2001, British Petroleum began development of the Kapok gas field with a completion date of 2003. The project will give the nation one of the largest offshore gas-handling facilities in the world. Importantly, the company has indicated that the facility will conform to the most stringent environmental safeguards.

ACHIEVEMENTS

Eric Williams, historian, pamphleteer, and politician, left his mark on Caribbean culture with his scholarly books and his bitterly satirical *Massa Day Done.* V. S. Naipaul is an influential author born in Trinidad. Earl Lovelace is another well-known Trinidadian author. He won the 1997 Commonwealth Writers' Prize for his novel *Salt.*

TOBAGO

Residents of Tobago have come to believe that their small island is perceived as nothing more than a dependency of Trinidad. It has been variously described as a "weekend resort," a "desert island," and a "tree house"—in contrast to "thriving," "vibrant" Trinidad. Tobagans feel that they receive less than their share of the benefits generated by economic prosperity.

In 1989, the Constitution was reviewed with an eye to introducing language that would grant Tobago the right to secede. The chair of the Tobago House of Assembly argued that, "in any union, both partners should have the right to opt out if they so desire." Others warn that such a provision would ultimately snap the ties that bind two peoples into one. Trinidadian opposition leaders have observed that the areas that have historically supported the ruling party have more and better roads, telephones, and schools than those backing opposition parties.

Timeline: PAST

1498
The island now called Trinidad is discovered by Columbus and later colonized by Spain

1797
Trinidad is captured by the British

1889
Tobago is added to Trinidad as a colonial unit

1962
Independence from Great Britain

1980s
Oil-export earnings slump

1996
Basdeo Panday is elected prime minister

PRESENT

2000s
Further development of the natura-gas industry

Patrick Manning elected Prime minister, October 2002.

Annotated Table of Contents for Articles

Regional Articles

Mexico Articles

Central America Articles

South America Articles

Caribbean Articles

Latin America Losing Hope in Democracy, Report Says

By WARREN HOGE

UNITED NATIONS, April 21—A majority of Latin Americans say they would support the replacement of a democratic government with an "authoritarian" one if it could produce economic benefits, according to a United Nations report released Wednesday in Lima, Peru.

The report, a harsh self-analysis compiled by Latin Americans, says that the region, which has succeeded in freeing itself from a long history of military coups and dictatorships, is facing a new challenge to democratic rule because of popular disenchantment with its elected governments.

Created by the United Nations Development Program, the report looked at 18 nations and conducted opinion surveys of 18,643 citizens and lengthy interviews with 231 political, economic, social and cultural figures, including 41 current or former presidents and vice presidents.

Fifty-five percent of the people polled said they would support the replacement of a democratic government with an authoritarian one; 58 percent said they agreed that leaders should "go beyond the law" if they have to, and 56 percent said they felt that economic development was more important than maintaining democracy.

"This shows that democracy is not something that has taken hold of people's minds as strongly as we had thought it would," said Enrique Berruga Filloy, Mexico's ambassador to the United Nations.

The report says that while unhappiness with political leadership has a long history in Latin America, the people now complaining are faulting democracy itself.

Voter turnout is falling across the region, especially among the young, while civil unrest is on the rise.

Since 2000, four elected presidents in the 18 countries surveyed have been forced to step down because of plunges in public support, and others may now be in peril. The countries surveyed were Argentina, Bolivia, Brazil, Chile, Colombia, Costa Rica, the Dominican Republic, Ecuador, El Salvador, Guatemala, Honduras, Mexico, Nicaragua, Panama, Paraguay, Peru, Uruguay and Venezuela.

All of these countries have either introduced or consolidated electoral democracy over the past 25 years, emerging from unrepresentative one-party politics or harsh and repressive military rule. All of them hold regular elections that meet international standards of fairness and enjoy a free press and basic civil liberties.

The report acknowledges distinctive circumstances in individual countries, but it argues that there is a broadly shared political culture and social structure that transcends them. "The common denominators of this phenomenon outweigh the many national differences," it says.

The report attributes the erosion of confidence in elected governments to slow economic growth, social inequality and ineffective legal systems and social services. Despite gains in human rights from the days of dictatorship, most Latin Americans, it says, still cannot expect equal treatment before the law because of abusive police practices, politicized judiciaries and widespread corruption.

Drug Economies of the Americas

$$$$$$$

By JoAnn Kawell

In the Americas, the production and sale of illicit drugs generates tens of billions of dollars a year—perhaps far more. Though exact figures are impossible to come by, for reasons we discuss below, it's clear that many Latin American nations now earn as much, or more, from the drug trade than they do from any other single legal commodity or industry. Perhaps the most important thing about the drug industry, however, is not its size, but how deeply it has been woven into the region's political and cultural—as well as economic—life. But while officials, and nonofficials, have blamed the drug for many of the region's ills, from rampant violence, corruption and civil strife to economic disintegration, the rise of the industry is arguably more the result than the cause of these phenomena. What's more, despite our desire to divide our economies and societies into "legitimate" and "illegitimate" spheres for the purpose of analysis—and perhaps more important, to make it possible to hold on to our sense of moral certainty—these divisions are, when it comes to the drug trade, rather illusory.

NACLA editor Mark Fried wrote in 1989, in the introduction of our first issue on the subject, "The moral blinders most of us wear tend to hide the essentials of the drug trade. [Drugs] are not devils, any more than they are gods. They are commodities and the drug trade is a multinational industry like any other."[1]

That is still true, and it is still helpful to examine the industry in that light, even though the drug trade has changed in some significant ways since then: In 1989, Latin American drug production meant, above all, cocaine production. In the United States, a decade-and-a-half long boom in cocaine use had culminated in "epidemic" use of crack, a particularly concentrated and dangerous form of the drug. At that time, almost all of the coca leaves used to make cocaine were grown in Peru and Bolivia, and the Colombians had recently consolidated their control over the final refining stages and over wholesale distribution to the U.S. market. Now, in the wake of successful U.S.-instigated

coca eradication campaigns in Peru and Bolivia, Colombia has replaced them as the world's largest coca producer. At the same time, Mexican traffickers have taken on a far larger role in distributing cocaine in the United States, sometimes in competition and sometimes in cooperation with the Colombians.

Until the 1990s, Mexico was the only important Latin American producer of opium poppies and the heroin that is made from them. But as heroin use has seen a recent U.S upsurge, Colombia and Peru have started producing opium poppies on a commercial scale. Meanwhile, Europe, East and West, has, over the last decade, become an important market for Latin American cocaine.

Finally, there is marijuana production, which has received less media attention than other targets of the international "war on drugs"—at least since the early 1980s, when U.S.-funded drug control efforts aimed against Mexican and Colombian producers and distributors helped encourage marijuana production in the United States to replace this supply. Large-scale marijuana production has not disappeared in either Mexico or Colombia, however; and most other American nations produce some marijuana, if only for domestic consumption. Researcher John Gettman recently calculated the value of the marijuana produce annually in North, Central and South America "conservatively" at $51.2 billion. He discusses how he reached this estimate in the expanded web edition of this issue at http://www.nacla.org

There now seems to be no nation in the Americas that has no role at all in the drug trade; most of the nations of South and Central America and the Caribbean that do not produce significant amounts of illegal drugs for export have some role as drug transit routes or money laundering centers. We detail the history and current state of some of the most important national industries below; descriptions of the others are included in our web edition.

If this is, more then ever, a truly global industry, the notion of a simple division between the "right north" as drug consumer and "poor south" as drug producer is no longer accurate—if it ever was. Though the United States undoubtedly still is the largest market for Latin American drug production; the United States, and to a lesser extent Canada, are also major producers of illegal drugs, especially marijuana but also synthetic drugs like amphetamines. The United States also benefits from the fact that by far the largest profits in the industry come at the retail end. Economists have calculated that the importers who bring illegal drugs into the United States receive only about 10% of the final retail or "street" price. The remaining 90% goes to local sellers, and most of that flows into the U.S. economy, as does the part of the importers' income that they are unable, or do not want to send home.[2] Indeed, money derived from some aspect of the drug industry has become a mainstay of many local U.S. economies, from Miami to the tiny towns of Northern California's marijuana producing "Emerald Triangle."

Despite the huge amounts of money that it generates, over the years some economists, including economists from key drug producers like Colombia, have argued that the drug industry creates negative economic effects. In the past, many of these economists pegged their argument on claims that the drug industry destroys legitimate jobs by competing for labor, or reduces food production as increasing amounts of local farmland are used for drug crops. But these claims seem rather fanciful in present day Latin America, where unemployment, not labor competition is a hugely pressing problem, and small food producers are increasingly unable to survive. A more recent, and sophisticated, criticism holds that drug production causes "Dutch disease"—an economic phenomenon, named for what happened in the Netherlands when it became awash in income from North Sea oil—whereby a commodity boom reduces local industrial production by raising the exchange rate for a country's currency, thus discouraging export sales. While the "Dutch disease" theory might have some limited application in relatively more industrialized countries like Colombia or Mexico, it seems of dubious validity in countries, like Bolivia and Peru, which have for more than a century struggled without success to establish a stable industrial sector.

Critics of the trade also argue, more convincingly, that the large amounts of drug industry-generated revenues that flow through an economy "off the books" prevent governments from setting effective fiscal and monetary policy or monitoring exchange rates. And, of course, governments are unable to tax or regulate the industry because it is illegal. But the same is true of all so-called "informal" economic activity, and other analysts have pointed out that this actually constitutes an argument in favor of legalizing the drug trade, which would allow governments to take advantage of this huge potential source of tax revenues. It's notable that the history of the legalization debate in Colombia it was precisely the Colombian business class and producer groups who favored legalization so that income generated by the drug trade could be absorbed more efficiently into the national economy.

Most often, though, the argument that drug dollars are "bad" is more closely related to their supposed social and political effects than to economics: Corruption, violence and an upsurge in other illegal activity are the most often-cited effects attributed to the drug trade. Undeniably, drug trafficking almost always occurs in an illicit nexus that includes trafficking in other illegal or stolen goods. And those involved routinely resort to violence and/or payoffs in order to stay in business. But the argument that the drug trade *causes* other illicit activity vastly over simplifies both Latin American history and present-day realities. Colombian economists Roberto Steiner and Alejandra Cochuela note, as have other writers, that Colombia has a "long history of contraband trade stretching back to colonial times."[3] Colombian traffickers were able to take advantage of pre-existing contraband networks in building their own businesses. According to Bill Weinberg's report in this issue, already corrupt and abusive Mexican *caciques* simply expanded their illicit business dealings into drugs, not the other way around.

Such arguments also overlook the social complexity of violence in the poor communities where the drug trade is often centered: In the Rio slums, residents faced with a historically corrupt police force have in some ways made their peace with violent drug traffickers who have, unlike the police, been able to reduce neighborhood crime and impose a degree of social order that did not previously exist.

In recent years—and ever more loudly since the September 11 attacks—U.S. officials have tried to sell the "war on drugs" to us by arguing that the drug trade funds terrorism and arms trafficking. While such activities are, again, undeniably linked in the real world, the officials' rhetorical sleight-of-hand encourages us to believe that it is only their enemies, designated "bad guys" like members of Al Qaeda or the Colombian guerrillas, who are involved. But many U.S. friends, or former friends, like Peru's unofficial intelligence chief Vladimiro Montesinos, now in jail, have been linked to intertwined drug, arms and money laundering schemes. And is it a case of historical amnesia that allows so many of us to forget the Iran-Contra and Contra drug scandals, in which Reagan administration officials illegally funded the Nicaraguan counterrevolutionary force with money from an illicit arms deal—and also turned a blind eye on Contra cocaine sales?

Perhaps more importantly, however, many analyses of the drug trade gloss over the important role it plays in propping up U.S.-supported free market and liberal economic programs in Latin America. Above all, the drug industry has served as a kind of social safety net, providing jobs for those displaced by liberalization: In Bolivia, many laid off miners, former employees of state-owned mines, transformed themselves into coca growers. In Mexico, many farmers are turning to opium or marijuana because their corn and other crops are unable to compete in the newly opened domestic market with cheap imported food produced by giant U.S. producers.

And the billions of dollars of successfully "laundered drug funds that now circulate within Latin America's legal economies provide at least a cushion for nations otherwise battered by the region's recurrent economic crises. Indeed, the drug trade may be Latin America's most stable and reliable export industry: While prices for other Latin American export commodities have fallen along with demand for them, drug prices and demand have stayed almost steady. Our multi-billion dollar, decades-long "war on drugs" has, in fact, had little impact on total world production and use; at best, drug interdiction and eradication efforts have caused growers, traffickers and dealers to move their operations somewhere else like inside U.S.

At the same time, as political scientist Peter Andreas has pointed out, there is a deep contradiction between U.S. drug control policy and U.S. promotion of free markets and free trade:[4] Much to the consternation of the DEA, NAFTA is making it harder to monitor cross border shipments of goods. And tightening supervision of international financial transactions in order to stem money laundering—a key goal of drug control officials—contradicts the free marketers' call to loosen financial controls.

Industry expert Han van der Veen sees the growth of the "booming drug industry and proliferating state powers to control it" as interdependent, and he says they need to be studied both as part of the trend toward globalization and in light of "intertwined symbiotic and systemic interactions of the upper and underworld, which take shape in the international political economy." He argues that there is, in the post-Cold War world, and "International Drugs Complex" analogous to the Military-Industrial Complex that drove the expansion of the military in the Cold War world.[5] Indeed, drug control, like the drug trade, is a big money pursuit: The total Federal Drug Control Budget for 2002 is, according to the White House Office of National Drug Control Policy, $18.8 billion.[6]

Van der Veen sees some form of legalization as necessary to stop the spiral: "Where drug entrepreneurial networks cannot be incorporated in local or national political and economic arrangements, their impact on society becomes much more detrimental; a situation that is only worsened as the state increasingly resorts to criminalization and repressive means to control their activities."

This is a perspective that is gaining adherents in many parts of the Americas, though the politics of this movement, defined in a traditional sense, vary widely: In the United States, a libertarian perspective, which stresses the rights of individuals to determine whether or not they will use drugs and upholds free market economic principles has been most influential.[7] In Latin America, the politics are more varied: While in Colombia legalization proponents have included both politically conservative members of the business class and left-leaning intellectuals, in the long-time coca producers Peru and Bolivia, it is the left that has most shaped the debate.[8] Since the 1970s,

socially progressive researchers in the Andes have argued that coca—the ritual and social use of which dates back thousands of years—is both culturally and economically essential. This debate took a more concrete turn in June, when coca grower leader Evo Morales, running as the candidate of the socialist MAS party, took second place in Bolivia's presidential race.

Even if, as its proponents contend, drug legalization or decriminalization would eliminate many of the negative effects of the drug trade, it would hardly be a panacea for Latin America: Abuse of power, violence, and corruption would certainly not disappear, for these have deeper roots and other causes than drug trafficking alone. And we know from Latin America's long history as a commodity provider to the world that even if drugs were to be treated, in law as well as in our analysis, as "just another commodity," this would not guarantee economic justice for the region's poor, or loosen the powerful grip that international financial institutions, led by the United States, have over the region's economies. Latin Americans, and the rest of us, would have to continue our search for alternatives to the current Washington Consensus. But abandoning drug prohibition would allow us to make a more realistic, less blinkered, appraisal of the important role the drug industry has come to play in our economies and societies and to reduce the harm that is currently being done by the drug war in the Americas, as drug producers and drug controllers clash with each other in an evermore-violent contest.[9]

PROFILE: COLOMBIA

Colombia's drug industry is the most complete and diversified in the world: Colombia is the world's only important producer of all three of the top non-synthetic illicit drugs; it provides an estimated 75% of the world's cocaine supply, 2% of the world's heroin supply, and a large amount of marijuana, including perhaps 40% of the pot imported into the United States.[1] Colombians play a major role in every stage of the industry up to the final retail level, from production of raw material, through refining, international transport and wholesale distribution.

- **Colombia's National Association of Financial Institutions (ANIF) estimated the nation's total 1999 income from the illegal drug trade to be $3.5 billion. The ANIF estimate was based on an assumption that somewhat less than 10% of total earnings from illicit drug sales are repatriated to Colombia each year, and on reported total world retail-level sales of Colombian cocaine, heroin and marijuana of $46 billion. Colombia drug trade expert Bruce Bagley notes that this calculation put Colombian drug earnings "close to the $3.75 billion made from oil—the country's top export—and more** than two and one half times the earnings from coffee exports in 1999."[2]

- As much as 3% of Colombia's work force—some 300 thousand people—are directly employed in the drug industry, according to estimates cited by Colombian economists Roberto Steiner and Alejandra Corchuelo. They cite other estimates that drug crop cultivation accounts for about 6.7% of Colombia's agricultural employment—compared to 12% for coffee farming—and they say that: "On the regional level, in centers of drug crop production like Guaviare, Putumayo and Caquetá, this percentage could reach levels close to 50%. It can be stated that—directly or indirectly—the majority of the labor force in these regions is involved" in the drug industry.[3]

The Colombian drug industry's main development has occurred within the last three decades: In the 1970s, Colombia's only large-scale illicit drug product was marijuana. In 1978, according to Juan Tokatlian, another expert on the Colombian drug industry, some 25-30,000 hectares of marijuana were being cultivated in Colombia, and 60-65% of the marijuana used in the United States was Colombian. (There was little domestic U.S. production at that time.)[4]

It was not until the late 1970s that Colombia began to play an important role in the cocaine business; Colombians then set up an efficient system for transporting cocaine in bulk to the United States, by using a network of small planes. Partly as a result, they were able to take over a large part of the wholesale trade. Groups of cocaine entrepreneurs based in the Colombian cities of Medellín and Cali became known for their efficient, if ruthlessly violent, control of the international cocaine industry. These groups were known as "cartels," though economists have debated whether the industry leaders actually coordinated their activities closely enough to merit the term.

At the time, Colombia produced almost no coca leaves; instead Colombian cocaine entrepreneurs bought finished cocaine and coca paste—semi-refined cocaine—from Peruvian and Bolivian producers. They established refining labs in Colombia to process the paste. In the mid-1990s, however, as the U.S.-funded enforcement programs began to target the so-called "air bridge" used to bring coca paste into Colombia from Peru and to eradicate significant portions of the Peruvian and Bolivian coca crop, the Colombians began to encourage coca production inside Colombian borders. Bagley notes that "by 1999 Colombia had become the premier coca-cultivating country in the world, producing more coca leaf than both Peru and Bolivia combined." In that single year Colombian coca production more than doubled. Bagley stresses that "this explosive expansion occurred in spite of a [continuous eradication] program that sprayed a record 65,000 hectares of coca" the previous year.

In the 1990s Colombia further diversified its drug industry, moving for the first time into opium poppy production. According to Bagley, Colombian poppy production "skyrocketed from zero in 1989 to 61 metric tons in 1998." This was less than 2% of world production, but Colombia soon became an important heroin supplier to the United States. In 2000, according to the U.S. State Department, about 59% of the heroin seized by federal authorities in the United States was of Colombian origin.[5] The death or jailing of many top Medellín and Cali leaders led to the near demise of these "cartels" in the 1990s. They have since been replaced by dozens of smaller organizations that observers describe as more flexible and able to respond to new law enforcement tactics by changing their own.

PROFILE: MEXICO

Mexico is according to the U.S. State Department, "a major supplier of heroin, methamphetamine and marijuana" for the U.S. market and "the transit point for more than one half of the cocaine sold in the U.S." It is also a major money laundering site: The State Department reports that "In recent years international money launderers have turned increasingly to Mexico for initial placement of drug proceeds into the global financial system."[1]

- The Mexican attorney general's office estimated that the gross annual income of Mexican traffickers was about $30 billlion dollars a year in 1994. Analyst Andrew Reding points out that "even if that sum is exaggerated by a factor of two, [it] vastly outstrip[ped] oil earnings"—some $7 billion for the same year.[2]

- The amount of money laundered in Mexico is so great—estimates range from 4% to 20% of GDP—that if laundering were successfully stopped, it could touch off an economic crisis worse than that sparked by the 1994 peso devaluation.[3]

- According to researcher Carlos Loret de Mola, the drug industry directly employs some 360,000 Mexicans. Nor should the work-generating capacity of the drug control forces be underestimated: The DEA reports that 20,000 Mexican soldiers take part in poppy eradication operations on any given day.[4]

Mexican sociologist Luis Astorga notes that "drug trafficking in Mexico began about sixty years before the Colombians got an important share of the [U.S.] drug market." Opium poppy has been cultivated in Sinaloa since the 1880s, with exports to the United States soon occurring through Mexicali and Tijuana. By the 1930s, "marijuana production could already be counted in tons in states like Puebla, Guerrero and Tlaxcala."[5] Mexico now produces about 2% of the world's opium supply, most of this converted into heroin sold in the western part of the United States. It continues to be one of the top foreign suppliers of marijuana the U.S. market.

In the 1990s, Mexican traffickers began to replace the Colombian cartels as the final importers of cocaine into the United States.[6] This represented a major expansion of the Mexican role in the Latin American drug trade, as well as a significant increase in income for the Mexican traffickers. Even more recently, the Mexicans reportedly began to work with the Colombians to import Colombian heroin into the United States, an to make use of more sophisticated Colombian refining techniques to improve Mexican heroin production.[7]

Mexicans are also increasingly involved in methamphetamine production, in the United States as well as in Mexico. According to the DEA, in 1994 Mexican groups began operating "'super labs' (laboratories capable of producing in excess of 10 pounds of methamphetamine in one 24-hour cycle) based in Mexico and in California." Many smaller meth labs also operate in Mexican border states, particularly Baja California.[8]

Astorga stresses that "since the beginning of the drug business" the best-known Mexican traffickers appear to have had close relations with high-ranking politicians and, since the Mexican Revolution, with the state party system created by the Institutional Revolutionary Party (PRI). "Controlled, tolerated, or regulated by mighty politicans in northern [Mexican] states, drug trafficking seems to have been a business that was developed from within the power structure," a situation that continues to this day. In last year's *NACLA, Report* on the Drug War in the Americas, journalist Julia Reynolds detailed the ties between Mexican banks and politicians and the drug trade in the era of PRI presidents Carlos Salinas and Ernesto Zedillo.[9] In this issue, we report on how political power is linked to the drug industry in Guerrero and other centers of poppy and marijuana production.

PROFILE: PERU/BOLIVIA

Peru and Bolivia both produce some refined cocaine for sale in international markets, but their main role in the international drug industry is as producers of coca leaf, the raw material used to make cocaine, and of coca paste, which is further processed, usually in Colombian labs, into finished cocaine hydrochloride. Until the late 1990s, when large-scale coca production began in Colombia, these two Andean nations were the world's only significant producers of coca leaf used to make cocaine.[1] Total coca production has dropped precipitously in both countries over the last few years, mostly as a result of U.S.-funded eradication programs.[2]

- **At the peak of coca production in the 1980s, Peru was reported to be earning some $600 million dollars a year from the illegal drug trade. "Coca dollars" were said to be equivalent to about 20% of the income from Peru's diverse legal exports-probably about equal to the income from what was then its most important export, oil. Bolivia's was thought to be generating about equal numbers of coca dollars, but these loomed much larger in Bolivia's smaller and far less diversified economy: Through the 1980s and into the 1990s, coca was considered to be, without a doubt, Bolivia's single most valuable product.[3]**

- **Several hundred thousand Peruvians, and a nearly equal number of Bolivians, were reported to be working as coca farmers during the boom. At least half a million more jobs were said to have been generated as coca dollars "trickled up" through the two economies.[4]**

The recent coca "bust" has apparently reduced interest in such statistics, and comparable current numbers are unavailable. Local economies are certainly suffering, however, and this has had evident political repercussions: In Bolivia, the eradication program has fueled violent protests in the Chapare coca-growing zone—and the surprise second-place finish of a coca grower leader in the recent presidential election. In Peru, it has led to a continued Shining Path presence in the main coca growing zone, the Huallaga Valley, even though the guerrillas have all but disappeared almost everywhere else.

At the same time, coca production is reported to be slowly creeping up again in both countries, with much of this increase occurring in new areas. In Peru, farmers are experimenting with opium poppy production, reportedly under the tutelage of Colombian buyers. And, as drug controllers shut down old distribution routes, like the air route between Peru's Huallaga Valley and the Colombian refining labs, traffickers have opened a plethora of new air, sea, and river routes along both countries' borders, in the process turning Brazil and other neighboring countries into new transit countries.

Coca leaves—in their raw form—have been consumed in Peru and Bolivia for millenia; they are a mild stimulant, and the leaves have deep cultural and religious meaning for many Andeans. Coca production was a mainstay of many local Andean economies even before the arrival of the Spanish; coca continued to be a staple of the colonial economy. In the nineteenth century, the governments of newly independent Peru and Bolivia depended on the crop as a source of reliable tax revenues. Cocaine was not "invented" until 1859; it was not widely used in medicine until the 1880s. At the end of the century the Peruvian government success" fully encouraged the development of an entirely legal Peruvian cocaine industry to supply the international pharmaceutical trade. In the early twentieth century, however, international drug controllers, led by the United States, began targeting Peruvian and Bolivian coca crops, and the Peruvian cocaine industry, for destruction, arguing that these were the source of the non-Andean world's cocaine "problem."

Despite continuing drug control efforts, by 1980 cocaine use was booming in the United States, and illegal industries had developed in both Peru and Bolivia to supply the demand. These were vertically integrated, with Peruvian and Bolivian nationals taking part in every stage from coca growing to refining to world distribution; In the 1980s, however, Colombian groups began to dominate the international distribution networks and they largely, if not entirely, replaced the Peruvians and Bolivians.

Already in 1980, a government official was describing Peru's cocaine income as "as kind of spring which absorbed at least some of the worst effects" of the financial crisis then underway.[5] That same year, General Luis García Meza toppled Bolivia's civilian government in what was referred to as the "Cocaine Coup," and the following year a respected newsletter on Bolivian politics and economics declared that cocaine had become the most important business in the country, accounting for more foreign exchange than the rest of the economy combined.[6]

PROFILE: CENTRAL AMERICA/CARIBBEAN

Central America and the island nations of the Caribbean primarily play a support role in the international drug industry, serving as transit points and money laundering sites. While the impact of such, ties on these small national economies is probably considerable, few relevant statistics have been compiled.

According to U.S. officials, "Central America's position as a land bridge between South America and Mexico, together with its thousands of miles of coastline, several container-handling ports, and limited enforcement capability, makes the entire region a natural conduit and transhipment area for illicit drugs bound for Mexico and the United States."[1]

The amount of cocaine shipped through Central America has reportedly increased by some 300% since 1993; journalist Ana Arana recently wrote that "the opportunities for profit and power" that this "have been rapidly exploited by many of the same groups that fought the civil wars of the Guatemalan journalist Edgar Celada noted in 1997 that the growing Central American role in the international drug trade had become "an obstacle for democratization and demilitarization" in the region.[2]

Arana says that the "cocaine trade has created a dangerous synergy between political terror and drug trafficking, and especially in Guatemala—the region's largest country—the line between criminal and violence has begun to blur." According to the State Department, "Guatemala is the preferred country Central America for shipment of cocaine to the United States."

Guatemala and Panama are the only two Central American countries Washington has currently branded as "major drug-transit countries;" according to the State Department,

"Guatemala is the preferred country in Central America for shipment of cocaine to the United States," but Costa Rica, Honduras, Nicaragua, El Salvador and Belize are also on key land and/or sea transit routes. Several large cocaine shipments were seized off the Belize coast in 2001, and the State Department reports that a highway across the Belize/Mexico border has become an important cocaine transhipment route since "Colombian drug cartels have established partnerships with the Mexican drug cartels."

Panama is still—as it was in 1989, when the United States invaded, ostensibly to capture Panamanian dictator and accused drug trafficker Manuel Noriega—a prime regional money laundering center, as well key participant in the hemispheric contraband trade.[3] Money laundered in Honduras is reported to stem from "auto theft, kidnappings, bank fraud, smuggling, prostitution and corruption," as well as drugs; while State Department says that the dollarization of the Salvadoran economy in January 2001 "increased the risk of money laundering" there.

Among the Caribbean nations, Jamaica is singled out as a money launderer, as "the foremost producer exporter of marijuana in the Caribbean" and as "a major transit country for cocaine." The Dominican republic, says the State Department, is also "a major transshipment point for narcotics moving from South America into Puerto Rico and the United States." According to the DEA, Dominicans dominate the retail level drug trade in much of the United States.[4] The State Department says currency exchanges in the D.R. play a role in laundering drug funds generated in the United States.

U.S. officials consider some nations in Central America and the Caribbean "vulnerable" to money laundering because of their large and well-developed financial sectors: These include Aruba, the Bahamas, the British Virgin Islands, the Cayman Islands, and El Salvador. Other countries are said to be attractive to money launderers because of underdeveloped financial sectors plagued by corruption or poor supervision, Haiti and Nicaragua are among these.[5]

With the exception of tiny amounts of coca reportedly produced in Panama, and a miniscule Guatemalan opium poppy crop, the only drug crop grown in Central America and the Caribbean is marijuana. Jamaica is the only major exporter to the U.S. market, though most countries in the region produce at least some pot for local consumption. The United States zealously presses for eradication of these crops: In Costa Rica, the U.S. military has participated in joint eradication under the rubric "Operation Central Skies," even though officials admit that there is no evidence that any Costa Rican marijuana is exported.

Meanwhile, the Bush administration continues to search for some evidence it can use to link Cuba to trafficking, arguing that Cuba's geographic position makes it "a logical candidate" to become a major country. Though U.S. officials admit they have so far found no such evidence, President Bush has promised that his drug controllers will "continue to keep Cuba under careful observation."[6]

PROFILE: UNITED STATES

The United States is best known as the world's largest market for drugs and, of course, as the main promoter, at home and abroad, of prohibition-based drug control policy. But the United States is also a major producer of illicit drugs, especially marijuana and amphetamines, and a major money launderer. Indeed, if one accepts that, once laundered, drug dollars function in an economy like any others, and that the largest portion, by far, of the retail value of illicit drugs sold here remains here, the United States arguably makes more than any other nation from the drug industry—even after the huge U.S. outlay for drug control is taken into account. (U.S. drug controllers, however, also subtract a long list of dubiously calculated "costs" they attribute to drug abuse; these easily swamp drug revenues.)[1] In any case, as in the rest of the world, parts of the poorest regions of the United States, including Appalachia and the south, have become dependent on drug income for their economic survival.

- **In 1998—the last year studied—people in the United States spent $66 billion on illicit drugs, including $39 billion on cocaine, $12 billion on heroin, $11 billion on marijuana, and $2.2 billion on methamphetamines, according to the White House Office on Drug Control Policy.[2]**

Although all of the cocaine and heroin consumed in the United States comes from foreign sources, U.S. producers supply much of the marijuana and amphetamines, along with changing amounts of other illicit drugs like LSD and Ecstasy (MDMA).

At least a third of the marijuana consumed in the United States is also grown here. Until the late 1970s, U.S. consumers generally disdained domestic marijuana as inferior to imported pot, but U.S.-funded crackdowns on foreign suppliers provided incentives to improve U.S. production. California and Hawaii soon sprouted large farms producing high-potency "sensimilla."[3] After forced eradication operations in both states, growers began moving production to scattered, remote sites and indoors. In 1998, the DEA reports, the leading states for indoor production were California, Florida, Oregon, Kentucky and—remarkably—Alaska. Kentucky and Tennessee have joined California and Hawaii on the list of the biggest outdoor producers.

According to researchers John Gettman and Paul Armentano, "Marijuana ranked fourth out of all United States cash crops in 1997, amassing a greater value to farmers than tobacco, wheat, or cotton. In several states —Alabama, California, Connecticut, Hawaii, Kentucky,

Maine, Rhode Island, Tennessee, Virginia, and West Virginia—marijuana stands as the largest revenue producing crop. Marijuana ranks as one of the top five cash crops in 29 others. Nationally, marijuana growers reaped an estimated $15.1 billion on the wholesale market. Only corn, soybeans, and hay rank as more profitable cash crops."[4]

Methamphetamine production is also widely scattered throughout the United States. In 2001, 46 states reported activity related to clandestine meth manufacture; production is especially concentrated in California, the midwest and the southwest. The DEA reports that U.S. law enforcers busted almost 4,600 illegal meth labs during 2000. The majority of these were "mom-and-pop" operations, run by "independent cooks" who buy the needed ingredients—including over-the-counter cold remedies—from local retail stores.[5]

It has often been noted that drug dealing is one of the few kinds of well-paid employment available to young people in the poorest urban areas of the United States.[6] Less well-reported is the fact that many of the regions of most concentrated U.S. drug production are also among the nation's poorest: In parts of the Appalachian states of Kentucky, Tennessee, and West Virginia—designated a "High Intensity Drug Trafficking Area" by U.S. officials in 1998—marijuana has "become a substantial component of the local economy, surpassing even tobacco as the largest cash crop," because, as the officials note, "in this tri-state area financial development is limited, poverty is rampant, and jobs are few." And, according to geographer Joseph Leeper, in Humboldt County, an early center of California marijuana production, "unemployment rates are usually 2 to 6% higher than those of either California or [the] United States."[7]

Better-off regions have also become key parts of the U.S. drug economy, though their role is more hidden. "Narcotics cash," *The New York Times* reported in 1982, was one of three reasons that Florida had become U.S. "banking's hottest market." At the time, Florida was the main U.S. port of entry for cocaine, and officials estimated that "drugs with a wholesale value of as much as $80 billion" a year were being imported into southern Florida. Some $6 billion in "surplus cash" was appearing on Federal Reserve books for the region, which officials took as a rough measure of the amount of drug money being "laundered" locally. Since then, U.S. drug imports—and the drug economy as a whole—have become much more widely dispersed: U.S. officials have added Houston, Los Angeles and New York to the list of important international money laundering centers.

JoAnne Kawell is editor of the *NACLA Report*.

From *Report on the Americas* by JoAnn Kawell, vol. XXXVI, no. 2, September/October 2002, pp. 8-17. Copyright © 2002 by NACLA Report on the Americas. Reprinted by permission.

Reagan's Legacy Gets Mixed Reactions In Latin America

By Kevin Sullivan and Mary Jordan

SAN SALVADOR—Gerson Martinez, a rebel leader in the 1980s, remembers Ronald Reagan as the man who funneled $1 million a day to a repressive and often brutal Salvadoran government whose thugs and death squads killed thousands of people, including the mother of his two children.

Ricardo Valdivieso, a businessman and a founder of El Salvador's main conservative political party, said Reagan "saved Central America" and was "a great ray of light and hope for civilization and liberty in a dark hour for our country."

The memory of the 40th U.S. president, who served from 1981 to 1989, is still strong in the region, and the contrasting views are passionate and polarizing.

The United States was heavily involved in wars in Nicaragua, El Salvador and Guatemala in the 1980s in what Reagan described as an effort to stem Soviet influence in the hemisphere. The United States spent more than $4 billion on economic and military aid during El Salvador's civil war, in which more than 75,000 people were killed, many of them civilians caught in the crossfire.

The United States also organized Nicaragua's contra guerrillas, who fought that country's revolutionary Sandinista government. Reagan referred to contras as "the moral equivalent of the Founding Fathers" and the United States spent $1 billion on them; the fighting in Nicaragua killed as many as 50,000 people. Honduras was a staging ground for U.S. Nicaraguan operations.

Reagan also supported the repressive military dictatorship of Guatemala, where more than 200,000 people, mostly indigenous peasants, died over 36 years of civil strife.

Reagan's support never led to a final battlefield victory in the region. Opposing sides negotiated peace in El Salvador and the Sandinistas were voted out of office in Nicaragua. But the same divisive sentiment about Reagan that existed a generation ago persists today.

Admirers credit Reagan with changing the course of Central America and helping to nurture democratic governments and free-market systems across the region. Many said Reagan's advocacy of open markets and U.S.-style capitalism sowed the earliest seeds of El Salvador's adoption of the U.S. dollar as its official currency.

"As time goes on, people are going to understand what he did for us,"

said Valdivieso, 62, a hotel owner and coffee producer. "I remember the first time I heard him speak, I thought, perhaps things will be all right, maybe we're going to be OK."

But for others, Reagan was an anti-communist zealot, whose obsession blinded him to the human rights abuses of those he supported with funding and CIA training.

"He was a butcher," said Miguel D'Escoto, who was foreign minister in Nicaragua's Sandinista government. D'Escoto, speaking by telephone from Managua, said "brutal intervention" by the United States under Reagan left "the whole country demoralized."

He said another Reagan legacy was that "Nicaragua continues to have people tied to U.S. apron strings. For some people, the lesson of the '80s is that you can do nothing without U.S. approval or you will have trouble."

In Guatemala, many remember that Reagan lent his prestige and backing to Gen. Efrain Rios Montt, who came to power in a coup and was an ardent anti-communist. Currently under house arrest for his alleged role in violent July 2003 riots, Rios Montt has been blamed by many international human rights groups for the massacre of tens of thousands of Guatemalans, including many women and children.

Carolina Escobar Sarti, a Guatemalan newspaper columnist, said many view Reagan's "interventionism" as part of a "difficult era."

"Of course," she said, "There are others, those on the ultra-right, who like Reagan," she said. "He has become a symbol of the conservatives."

In Honduras, where the little-known capital Tegucigalpa burst into the world's consciousness in the 1980s as a staging area for the U.S-funded contras, the Reagan era is viewed bleakly by many.

Reagan's critics contend that billions of U.S. dollars and U.S. arms and military intelligence inflamed and prolonged the 1980s wars because of Reagan's determination to leave no trace of communist sympathizers so close to U.S. soil.

'Lost City' YIELDING Its Secrets

There have been many theories for Machu Picchu, most wrong.

By John Noble Wilford

NEW HAVEN—Working with new evidence and a trove of re-examined relics, many of them recovered from the basement of a Yale museum here, archaeologists have revised their thinking about the significance of Machu Picchu, the most famous "lost city" of the Incas.

The new interpretation comes more than 90 years after the explorer Hiram Bingham III bushwhacked his way to a high ridge in the Andes of Peru and beheld a dreamscape out of the pre-Columbian past.

There, set against looming peaks cloaked in snow and wreathed in cloud, was Machu Picchu. Before his eyes, rising from the green undergrowth of neglect, were the imperial stones that have entranced and mystified visitors and scholars alike.

The expression "lost city," popularized by Bingham, was the magical elixir for rundown imaginations. The words evoked the romanticism of exploration and archaeology at the time, in the summer of 1911. And the lanky and vigorous Bingham seemed to personify the spirit that was driving discoveries of a forgotten past, the curiosity and courage to go seeking in remote places, as well as the hardihood to succeed.

But finding Machu Picchu proved to be easier than solving the mystery of its place in the Inca empire, arguably the richest and most powerful in the New World when Europeans arrived. The imposing architecture attested to the skill and audacity of the Incas. But who had lived at this isolated site and for what purpose?

Bingham, a historian at Yale, advanced three hypotheses—all of them dead wrong. A revival in research in recent

years, experts say, has solved the mystery and, to a large degree, demystified Machu Picchu.

The spectacular site was not, as Bingham supposed, the traditional birthplace of the Inca people or the final stronghold of the Incas in their losing struggle against Spanish conquest in the 16th century. Nor was it a sacred spiritual center occupied by chosen women, the "virgins of the sun," and presided over by priests who worshiped the sun god.

Instead, Machu Picchu was one of many private estates of the emperor and, in particular, the favored country retreat for the royal family and Inca nobility. It was, archaeologists say, the Inca equivalent of Camp David, albeit on a much grander scale.

This interpretation and other new research inform a major exhibition at the Peabody Museum of Natural History at Yale. The show, "Machu Picchu: Unveiling the Mystery of the Incas," will be here until May 3. Then it is to travel to Los Angeles, Pittsburgh, Denver, Houston and Chicago.

Dr. Richard L. Burger, the director of the Peabody and a specialist in Inca archaeology, said the show, the largest on the Incas ever assembled in the United States, would "change the way people see Machu Picchu." Dr. Burger and Dr. Lucy C. Salazar, also an archaeologist, are co-curators of the exhibit.

"Bingham's work was very important in putting Inca archaeology on the map," said Dr. Burger, who is married to Dr. Salazar. "But we can now set aside all his ideas about the meaning of the Machu Picchu site."

The new interpretation, generally supported by other experts, is based largely on a study of 16th-century Spanish legal documents and a more detailed analysis of pottery, copper and bronze jewelry, tools, dwellings, skeletal remains and other material found in the ruins.

Many of the artifacts were themselves a forgotten treasure. Shipped back by Bingham, they were stashed in the museum basement, where they remained, still in their original boxes and wrapped in pages of The New York Times from the 1920's, until renewed interest in the Incas led scientists to poke into the stash.

Until recently, there had not been much scholarly interest in Machu Picchu. Although the site has long been Peru's most popular tourist draw and a mecca for seekers of mystical and spiritual experiences, the haunting shells of temples, palaces and other structures had ceased to attract many archaeologists.

"A lot of people felt it had become so much an icon for the Inca and Peru," said Dr. Craig Morris, a specialist in Peruvian archaeology at the American Museum of Natural History in Manhattan. "They became more interested in working in places not so well known."

Bingham's long shadow may also have discouraged research. In his three expeditions to Machu Picchu from 1911 to 1915, he established himself as the "discoverer" and foremost interpreter of the lost city. His 1930 book, "Machu Picchu: A Citadel of the Incas," endured as the definitive treatise on the site. His maps and photographs of the ruins were authoritative and evocative.

But he was untrained in archaeology and he did not conduct systematic excavations and rigorous analysis. "His excavation notes," Dr. Burger said, "included more on what they were eating than what they were finding."

Bingham eventually resigned his professorship at Yale to enter politics, becoming lieutenant governor and governor of Connecticut and a senator. But his influence on Inca research remained strong, in part because of his fervid writing style.

In "Lost City of the Incas," a best seller, he wrote: "Here, concealed in a canyon of remarkable grandeur, protected by nature and by the hand of man, the 'Virgins of the Sun' one by one passed away on this beautiful mountain top and left no descendants willing to reveal the importance or explain the significance of the ruins which crown the beetling precipices of Machu Picchu."

Archaeologists today forgive some of Bingham's lapses in excavation, but they have destroyed his theories.

For example, Dr. Salazar's exhaustive examination of pottery contradicted Bingham's speculation that Machu Picchu was somehow associated with the earliest Incas. All the pottery styles were 15th century. That and other evidence suggest that construction on the site began around 1450.

That was in the reign of Pachacuti, considered the Alexander the Great of the Incas. His creation, like the empire, had a relatively brief history. From the recovered pottery and Spanish documents, scholars estimate that the site was largely abandoned after only 80 years.

Plague, brought to the New World by Spaniards, had by then left the land in turmoil, and in 1532 the Spanish conquered Peru with little resistance. The few Incan holdouts, including the last emperor, capitulated in 1572 at a tropical valley refuge that bore no resemblance in Spanish descriptions to Machu Picchu. So much for another of Bingham's suppositions.

His theory about a sanctuary for virgins and priests began to unravel in 1990 with the publication of research by Dr. John Howland Rowe, an anthropologist at the University of California at Berkeley.

In archives at Cuzco, the former Inca capital, Dr. Rowe found a 16th-century suit filed by descendants of Pachacuti. They sought the return of family lands, including a retreat called Picchu. The finding sent Dr. Burger, a onetime student of Dr. Rowe, and Dr. Salazar back to Machu Picchu.

"We then felt this was a royal estate, a country palace," Dr. Burger recalled. "All Machu Picchu is a big palace, the emperor's residence across from the temple, the dwellings and workshops, everything spread out around a great plaza."

As early as the 1960's, María Rostworowski, an ethnohistorian in Lima, pointed out that Inca rulers had established a chain of royal estates through the region. They served as occasional royal residences, but mainly as administrative centers. Many of the estates were razed by Spanish soldiers searching for gold, and some were built over and modified beyond recognition. But remote Machu Picchu, at an elevation of 6,750 feet, survived unscathed.

Dr. Susan A. Niles, an archaeologist at Lafayette College who is the author of "The Shape of Inca History," published in 1999, explained that it has long been known that the estates were peculiar to Inca royalty. Each ruler established his own and built a palace there as a monument to himself,

Machu Picchu, it seems, was an emperor's retreat— an Incan Camp David.

Each estate was the ruler's own private property, which was left to his family after death. The succeeding son could use the estates, but not own them. So he immediately began building his own monuments.

The estates, Dr. Niles said, were important centers for the economic management of agricultural lands, forests and mines in the surrounding region. That was presumably true, as well, of Machu Picchu.

Dr. Burger and Dr. Salazar agreed, but said little evidence had been found that ordinary administrative affairs were regularly conducted there. They emphasized the role of the site, 50 miles from Cuzco, as a country retreat for entertaining visiting dignitaries and for royal relaxation.

Though called a "lost city," it was not a true city. Probably no more than 750 people ever lived there at any given time, and in the rainy season the population dropped to just a few hundred. They were presumably the servants and artisans who attended to the royal family and their elite guests.

Bingham was not entirely wrong about the religious aspects of Machu Picchu. The buildings, ritual chambers, fountains and gardens, Dr. Salazar said, seemed to be arranged with Incan cosmology in mind. Rulers were believed to be descended from the sun, and wherever they went was sacred. Pachacuti, in particular, was looked upon as a creator god.

New investigations turned up bones of animals probably sacrificed in religious ceremonies. And there were dozens of obsidian pebbles, which scientific analysis showed had come from a revered volcano more than 200 miles away. The obsidian had never been modified for use as cutting tools. It is likely, Dr. Burger said, the obsidian had symbolic meaning. The Incas worshiped high mountains as the source of supernatural forces.

But Bingham had gone too far with his "virgins of the sun" hypothesis, experts say. He was misled by the findings of the party's osteologist, who reported that most of the skeletons buried at the site were those of women.

In new studies, Dr. John W. Verano, a physical anthropologist at Tulane University, determined that the ratio of female to male skeletons was comparatively even. His research also showed that many families and newborn infants lived there, not what one would expect in a community of virgins.

All the burials at the site were simple, with only modest grave goods. These were the remains of the retainers rather than royalty.

"This mortuary pattern," Dr. Burger said, "is not surprising, because if members of the Inca elite had died while residing at the country palace, they would have been transported to their principal residence in Cuzco rather than being buried at Machu Picchu."

Life at the country retreat must have been reasonably healthy. An analysis of bones showed that the workers apparently ate well. There were cases of tuberculosis and parasites, as well as considerable tooth decay from the corn diets. But nearly all the burials were of adults, including quite a few who were older than 50, an advanced age in that day.

The workers were brought from all over the empire, Dr. Verano concluded. The ethnic diversity was seen in the shapes of skulls, which had been deformed through binding in infancy. Different cultures over a wide geographic range had distinctive cranial deformations. Some came from the coast, and others from the highlands and as far away as Lake Titicaca.

Investigations by Dr. Alfredo Valencia Zegarra, a Peruvian archaeologist, and Kenneth Wright, a hydrological engineer from Boulder, Colo., have uncovered the magnitude of Machu Picchu as an engineering achievement. The Incas had not only terraced the slopes for agriculture, hauling up fine sand and topsoil from the valley and erecting stone retaining walls that have survived more than 500 years. But they had also taken an uneven ridge surface and transformed it to the flat mesalike surface seen today.

Before any of the buildings rose, the Incas leveled the site with loose rock and other fill, stabilizing it with immense walls deep beneath the surface. Mr. Wright estimates that the invisible subsurface construction constitutes some 60 percent of the effort invested in building Machu Picchu.

Whatever Pachacuti, the empire builder, had in mind, Dr. Salazar said, Machu Picchu "shows what the New World had achieved before the Spanish arrived." Some of the engineering and architecture was better than in Seville, she noted, and the Spanish "could not believe how people, people without writing, could have built something like this."

Archaeologists today may have demystified the lofty ruins, but their awe remains undiminished.

Dr. Niles of Lafayette College said the "overpowering landscape alone may be why Pachacuti chose the place for what his legacy to the world should be."

Conceding that he was biased, Dr. Morris of the Natural History Museum said that Machu Picchu "is to me the most spectacular archaeological site in the world."

Machu Picchu
- Royal estate or retreat / Royal residence — Pachacuti
- Arranged as a cosmogram ——> Sacred landscape
- Sacred landscape of mtns
  - mtns
  - rivers
  - valleys
- Agricultural terraces
- Built last half of 15th cent.
  - Econ. management
  - of resources
- Pop. approx 750
  - Entertaining visiting dignitaries

Hidden Harmony of the Q'ero

High in the Peruvian Andes, this isolated Indian community tenaciously preserves traditions that reflect its spiritual union with the natural surroundings

Aymara in Film

By Victor Englebert

A taxi left us next to a lonely house at the edge of a hairpin turn on the Puerto Maldonado road. The eight-hour eastward ride from Cuzco had been as spectacularly beautiful as it had been bumpy. It was late in August, winter in the southern hemisphere. We unloaded from the car several hundred pounds of luggage—largely rice, oatmeal, sugar, and coca leaves we were bringing to a number of Q'ero Indians whose lives my friends and I would share during the next nine days; then we bade our driver good-bye.

Five hundred years ago, the Q'ero had fled the conquistadors to hide high in steep, pathless mountains. They have remained isolated ever since, retaining better than any other Andean Indians the purity of Inca tradition, culture, and language. They claim direct descendence from Inkari, the mythical first Inca. They believe that he founded Jatun Q'eros, a village where they do not live but gather occasionally to discuss tribal concerns. What sets them apart most is their mysticism, their profound communion with the natural world.

In spite of their slight build, the Q'ero easily lifted our hundred-pound bags on their small shaggy horses

---

Our Q'ero group, five men and a twelve-year-old boy, had been waiting for us with twelve horses and a black sheepdog. A Cuzco radio station had informed them of our coming. Luckily, the single radio owner of Charcapata, the hamlet where we were headed, had heard at least one. The reception the Q'ero gave us boded well for our party. They threw themselves in our arms and patted us on our backs like long-lost friends.

In spite of their slight build, the Q'ero easily lifted our hundred-pound bags on their small shaggy horses. One man impressed us much. About five feet tall, he had a bad right leg that made his body swing widely from side to side as he walked, but a smile always illuminated his good-looking face. In spite of his 'infirmity he lifted the heaviest bags and ran around the most. Even in the following days he would work harder than anyone else. He would look after the three of us as his personal responsibility during our stay. His name was Modesto Q'espi.

Once ready, we set out to climb higher into the mountains, some of us on horseback, I walking ahead, camera in hand, and the Indians bringing up the rear on foot.

An hour and a half into the trip, the sun having set, we stopped at the house of a C'oline Indian to spend the night. Our host gave us, three visitors, a room's dirt floor to sleep on. The Q'ero stayed outside, pressing together against a stone corral under a couple of blankets. I don't think they slept. I got up twice during the night, and each time heard them joking and laughing. They may have been too cold to shut their eyes.

Next morning the ground outside was white with ice, and fluffy white alpacas in the stone corral huddled together for warmth. Leonidas, our host, gave us a breakfast of unpeeled, pockmarked, ping-pong-sized potatoes, no doubt similar to the ones the Inca ate long before the rest of the world adopted them and selectively enlarged and smoothed them down a bit. Then we helped the Q'ero load the horses, and half an hour later resumed our march inland. While the valley still lay in deep-blue shade, a bright yellow light slowly rose behind the wall of mountains to the east.

We climbed steadily, sometimes along path, most often cross-country. The sky was blue, and soon the sun shone brightly. But later clouds enveloped us in a gray mist, blowing up from the cloud forest at unfathomable depths as if from a giant pipe. The air smelled of wet stones and fragrant plants and the sounds of our caravan seemed to reach me through the ages.

Threading our way through a labyrinth of black, brown, red, yellow, green, and white mountains, of mountains in turn rocky and smooth, we reached snow, crossed Ritti C'asa (Cloud Pass) at over sixteen thousand feet, and descended into a marshy pampa. Here our group separated. Three men and the boy led the horses that carried the heavy bags of grain and sugar through a marshy shortcut, while the rest of us went on over dry ground. By one o'clock we had gone over Puka C'asa (Red Pass), named after red, rocky outcrops, and were descending into another valley.

Everywhere clear water dropped from snowy heights and gathered into rivulets and creeks.

Once in a while we passed grazing llamas and alpacas. While the llamas expressed irritation at our sight by flattening their ears back and haughtily walking away, the smaller, woollier alpacas tended to saunter towards us with curiosity.

As we descended one side of the valley, the sheepdog crossed over to the other side and hunted out of a llama herd a vicuña that was hiding there: Seeing that, the Q'ero abandoned the horses to help the dog comer the vicuña. They easily ran up the high slopes, and we soon heard the vicuñ's exhausted cries. But the animal did not exhaust itself, instead it coolly stopped, found another llama herd, and eventually escaped. We sighed with relief. The vicña is an endangered animal.

"They would not have killed it," explained Dennis, our interpreter. "They would have herded it into a corral, and there fleeced it for its fine hair. They don't kill vicuñas. They think they are the spirits of alpacas that will be reborn."

The climb got steeper and harder. Downward slopes made me feel ten years younger, but a dipping path inevitably promised another hard climb. We reached Charcapata in late afternoon, when it spread its forty or so scattered thatched stone huts and stone corrals under us. The other haft of our Indian escort had already arrived. Having walked much of the way, I was tired. In spite of my heavy clothing, I was also cold.

Outside Modesto's hut, his wife, Santusa, was squatting at a small loom anchored in the ground by four short wooden stakes. She was weaving a *lliklla* a reddish decorative small blanket that Q'ero women wear on their backs, tied around their necks. She left her weaving to prepare a hot infusion of coca leaves to warm us up. Meanwhile, we sat down together and chewed coca leaves. They soon made us feel rested and warm and serene.

The ground on which Charcapata stood fell away quickly to a narrow stream far below. Across from the hamlet, on a nearly vertical slope on the other side of the river, vast potato fields stretched upwards. I asked Modesto how the Q'ero torrid grow potatoes at such cold heights.

"We grow potatoes at four different levels. This gives us four harvests at different times. We grow corn much further down the mountain ... a nine-hour walk from here"

"We grow potatoes at four different levels," he replied. "This gives us four harvests at different times. Here, at the highest level, frost and rain sometimes spoil them, leaving them inedible except to llamas. It takes us three to four hours to walk down to the lowest potato-growing level, where the climate is milder. There we also grow *oca* and *ulluco* [other tubers]. We grow corn much farther down the mountain, at the edge of the cloud forest, a nine-hour walk from here. You just missed that harvest, which ended a week ago."

I would miss a lot more than that, for nothing is less routine than a Q'ero's life. Contrary to other Andean Indians, who get what they need form other ecological zones by trading, the Q'ero handle the whole mountain by themselves. Every month brings them other tasks in other places: planting, weeding, and harvesting different staples at different times and altitudes; guarding the corn fields against predators; herding; shearing; spinning and weaving; rethatching roofs; fixing bridges and trails; and celebrating Carnival as well as other religious festivals and rituals.

"These days, as you can see, Charcapata is nearly empty. Pasture has been overgrazed and our neighbors moved their herds to other high valleys, where they have more houses. They will return in three or four months. While a few families haven't gone yet, including ours, some of their members have already taken away their animals. At harvest time, only a few people stay with the herds, and everyone else meets down at the corn. It's a time of great rejoicing."

Sudden darkness brought more intense cold, and Q'ero families gathered inside. For half an hour we saw the light of their fireplaces flicker under their doors, and heard them talk and laugh. Then nothing. Silence had fallen as abruptly as the night. The bright stars were too icy for comfort, and our trio made its own beds over stiff alpaca skins on the floor of a vacated stone hut. It was only seven thirty, but too cold even to read at the light of a candle.

Like all Q'ero houses, ours had neither window nor chimney over the clay stove facing the low door. It had no furniture other than a squat, crude wooden stool. Straw covered the dirt floor, alpaca skins served as beds, and rough shelves and protruding sticks helped storage. Seed corn hung from the rafters.

Next morning a new sun brought the temperature back above freezing. Everybody was out. The huts are so dark that one can't do anything inside except near the fire, when one is burning. The Q'ero are careful with their firewood, as they have to walk three hours each way to get it.

Santusa was back at her weaving. She would stop only to peel and cook potatoes later on. Orelia, her eighteen-month-old daughter, sat next to her.

I followed an eleven-year-old girl and her llama and alpaca herd up the mountain. She stopped on a plateau and let her animals disperse while she pulled a loom from the carrying cloth on her back. Using a stone, she hammered its four stakes into the ground and went immediately to work on a coca bag she had already started. She interrupted her weaving only to check that no puma, fox, or condor was nearby or to reunite straying young with their mother. She told me how she had learned to weave from watching her mother and other women in her family. Every girl did.

When I returned, I found everyone sitting outside on the sunlit grass. Modesto was weaving the trimming of a skirt he had already woven for his wife. Q'ero men traditionally weave their wives' skirts, I was told, though I would not see another man weaving during my stay. They spun all the time, however, even as they climbed steep hills behind their herds. Once I even saw a man knit a *ch'ullu*, a woolen cap, while traveling on foot.

"You obviously enjoy weaving," I told Modesto. "To judge by the number of women I see weaving everywhere, you all love to weave."

"We do, and thank God for it," he said. "For we would have to weave even if we didn't like it. So much of what we need depends on our weaving, and we have to keep at it because it's such a slow process, though not as slow as spinning. We weave soft alpaca hair to make our clothes, and coarser llama hair into blankets, carrying clothes, and bags. We also braid llama hair into ropes and slings for keeping predators away from our herds. All these things wear out and must be replaced constantly."

Though every woman weaves, a few can reach the level of virtuosity that lifts their craft to art. Their peers recognize those skills and can identify the creators by their distinctive pieces.

Silence had fallen as abruptly as the night ... our trio made its own beds, over stiff alpaca skins on the floor of a vacated stone hut

---

A man came to sit with us. Dennis introduced him as José Q'espi. With a dignified, serene air, he was a *pampa mesayoq*, as was Modesto himself, we would learn. He had been struck by lightning as a young child, which had impaired his right leg but also marked him as predestined for that role. Other shamans, the *alto mesayoqs*, propitiate the *apus*, the mountain spirits.

The Q'ero believe that the world around them is charged with living energy, the *kamsay pacha*. They see the earth, the mountains, the sun, the moon, the stars, and even the stones as permeated with spirits that are as conscious of the

Q'ero as they are of them. And they won't start on a journey or an important activity unless they have first burned to them a *despacho* (an offering composed of sugar, alcohol, seeds, coca leaves, or anything else available at the time, wrapped inside a small piece of woven fabric or paper).

"I just went up the mountain to burn two *despachos*," José said, "one to the *apus* and one to Pachamama. I begged them to keep my llamas in good health. This afternoon we will have a *llama ch'allay*, a thanksgiving ceremony. We will give our male llamas *chicha* [a sweet corn beverage] to drink in gratitude for their strong backs in carrying our corn harvest up the mountains. You are welcome to attend." We gladly accepted.

When we arrived at José's house, he and two other men were sitting inside on the ground against the wall, facing their seated wives. Apparently intoxicated, or else mentally removed from this world, they seemed to look through us as if we were transparent. They alternately played the flute, said incantations, and drank the *chicha* that once in a while the women handed them in small gourds. On the ground, in front of José lay a big bundle—a *mesa* as the Q'ero call it (using the Spanish word for table, in this case, an altar table). It was filled with sacred objects charged with animated energy.

The men and women then went out, José carrying his mesa and the women buckets of *chicha*, to sit inside the corral and there repeat the same ritual. Higher up, in another corral, another family was doing the same. The women filled the small gourds, passed them on to the men, and drank as well. Once in a while the drinkers poured some *chicha* on the ground for Pachamama. Having emptied the small vessels, the men hurled them towards the llamas that pressed against each other on the other side of the corral. The women went to check whether they had fallen right side up or down, and brought them back. They fell mostly right side up, which was a sign of good things to come.

Meanwhile, their eyes lost in space, the men played the flutes and recited more incantations. Once in a while they pulled from their *ch'uspas*, or small coca bags, more leaves to chew. The women sang repetitively a monotonous song.

Finally, José opened the *mesa*, in which I saw a coiled rope, a horse bridle, a small bell such as Andean Indians use to tie around the neck of their lead llamas to cheer up the herd, some *khuyas* (stones that hold powerful living energy and help shamans to heal and to communicate with spirits), and a few things he used to make a new *despacho*. One of the women brought the *despacho* to a place in the corral wall, removed a stone at its foot, and burned it in the hole left by the stone.

Now the men got up and went to work, one forcing large plastic bottles of *chicha* down the throats of the llamas, while another held the animals' heads by their ears. By then Modesto had also arrived and was helping the third man to do the same. A woman then rang the little bell that was in the *mesa* to clean the animal's aura, and after that went for more *chicha*. Some of the animals seemed to enjoy the beverage, while others spat it out violently. After all the llamas had drunk, the men proceeded to change the faded red wool tassels in their ears for new ones, threading the wool through small holes made originally with the points of knives. The tassels help signal male llamas from a distance.

One day Modesto guided us back to the road, where we would wait for the passage of a Cuzco-bound truck. Ever the gentle friend, he hugged us a last time, and said something in my ear.

"What did he say?" I asked Dennis.

"He said that you will come back."

It had been my wish all along. The authority with which this *pampa misayoq* spoke made it a certainty.

Victor Englebert is a widely published photographer and a previous contributor to Américas.

Quero
- Andean Indians of Peru
  - Inca Tradition, Culture, Language
  - Potato Farmers => plant on 4 diff. elevations/climates
    allow for 4 harvests
  - Corn
  - Oca
- Alpaca Herders => Alpaca Hair => Textiles
- Shamans
- Apus => Mtn Spirits / Sacred Landscape
- Pachamama => Earth Mother
- Profound Communion with the Natural World

Mexico—The Sick Man of NAFTA

Christian Stracke

In the memorable debate between then-Vice President Al Gore and failed presidential candidate Ross Perot over the North American Free Trade Agreement (NAFTA) in November 1993, Perot offered up perhaps the most famous quip of the night. Mexico, he argued, was a poor choice for a trading partner, because "people who don't make anything can't buy anything."[1] In the nearly ten years since that debate, many of Perot's criticisms of NAFTA have been discredited: trade has soared between Mexico and North America (although not to the degree that some NAFTA proponents had hoped) on a net basis few jobs have been lost in the United States because of the free trade pact; and U.S. companies have not decamped to Mexico so as to avoid stricter U.S. labor and environmental regulations. But that callous remark about Mexicans not making anything and therefore not buying anything from the United States has proven depressingly correct. Most NAFTA boosters at the time dismissed Perot's comments, promising that the agreement would usher in a wave of economic modernization in Mexico that would improve the productivity of Mexico's workers, raise living standards, and create a much wealthier nation of consumers ready and able to buy American-made goods. Since NAFTA went into effect on January 1, 1994, however, economic growth in Mexico has averaged just 2.7 percent per year, exactly the same rate of economic growth as in the decade prior to NAFTA's birth. Moreover, the number of Mexican workers has increased by roughly 2.8 percent a year over the last ten years, meaning that workers' incomes have actually fallen since the agreement went into effect. At least on this one count, Ross Perot was correct: Mexicans have not become the prosperous consumers of North American goods that so many NAFTA supporters expected.

The failure of NAFTA to improve the lives and incomes of ordinary Mexicans presents a serious challenge to policymakers on both sides of the U.S.-Mexico border. The perceived shortcomings of neoliberal economic policy have already taken their toll on bilateral relations, as a sluggish Mexican economy has put President Vincente Fox Quesada—a vocal proponent of NAFTA and neoliberal reforms, including privatizing state enterprises, creating a more flexible labor market, and signing more free trade agreements—on the defensive, which in turn has helped create an environment in which he must burnish his nationalist credentials for domestic consumption. These pressures became all too apparent in the debate over U.N. approval for the war in Iraq, when the normally supportive Fox refused to back the United States in the Security Council, despite President Bush's entreaties. The failure of NAFTA to fulfill its promise of widespread Mexican prosperity has also meant continued illegal immigration into the United States, of both unskilled and skilled workers. This has contributed to tense bilateral relations, as Washington's concerns about terrorists crossing the porous U.S.-Mexico border arose just as President Fox began pressing for more relaxed border controls. Economic sluggishness in Mexico has also contributed to the trade in illegal drugs, one more point of friction between the two countries.

The Productivity Morass

Perhaps the greatest disappointment to the proponents of NAFTA after more than a decade of deregulation, privatization, and lower tariff barriers has been the stubbornly low level of productivity gains in Mexico. Simply put, labor productivity is in many ways the overwhelming determinant of improvements in living standards. But the hoped-for productivity miracle in Mexico has not materialized. Mexican labor productivity, by some measures, has actually declined since the beginning of the 1990s, whereas productivity in the United States— already a developed nation with presumably fewer opportunities for easy gains in this area—has risen by a third since 1990. The widening of the productivity gap, when all predictions were that the gap would narrow, is probably the most fundamental issue that policymakers in both Mexico and the United States must address, as the failure of Mexico's labor force to post productivity improvements undercuts the very rationale for NAFTA, especially from the perspective of Mexican workers.

To understand the productivity problem, one first has to grasp the magnitude of Mexico's dependence on growth in the labor force to drive growth in the overall economy. In 1970, 43.6 percent of Mexico's pool of potential workers (technically, all Mexicans 12 years old and older) were actually in the labor force, either working or actively seeking work. The vast majority of working age individuals not in the labor force were women. (While these women may well have been working long hours, household labor did not technically contribute to Mexico's GDP.) Over the next three decades, however, the work force participation rates of Mexican women soared, with most of the shift out of the home and into the work force occurring in the 1990s, when the Peso Crisis of 1994–95 (a crippling episode of currency devaluation, followed by a 7 percent decline in GDP and a 20 percent decline in workers' wages) forced economically inactive Mexicans to seek work. In 1991, women represented just 26 percent of the total labor force; by 2000, they represented a full 45 percent of all workers. This surge in the number of women seeking and finding work pushed the overall labor force participation rate to 56.3 percent of the population of working-age Mexicans by 2000.

While Mexican women were entering the labor force in unprecedented numbers in the late twentieth century, Mexico was also enjoying an unprecedented demographic boom in the overall population and, more importantly, in the number of working-age people. Only 50.8 percent of the population was 15 years old or older in 1970, but by 2000, a full 67.4 percent of Mexicans were 15 or older. During this period, Mexico's population also doubled, rising from 50.7 million in 1970 to 110.3 million in 2000. Thus, with women entering the work force, the surge in the working-age population, and the increase in the overall population, the size of the labor force has more than tripled over the last three decades, from 11.2 million workers, to 37.3 million workers at the end of 2000.

Comparing the growth in the labor force to the growth in the Mexican economy over the last three decades—the labor force rose by over 230 percent, GDP rose by just 220 percent—it is clear that productivity growth has been depressingly low. Thus, Mexican workers appear to be less productive today than they were in 1970.[2] This calculation could possibly overstate the problem to some degree; it is possible that the average number of hours each Mexican worker spent on the job declined somewhat over the last three decades (which would mean that workers were able to produce more per hour worked), but the lack of reliable data on the work week leaves this hypothesis difficult to test. Even assuming that the work week has grown shorter for the average Mexican worker over the last few decades, there would have to have been a 45 percent reduction— from, say, 40 hours a week to just 22 hours a week—for Mexican productivity to have kept pace with productivity growth in the United States. To put that hypothesis into perspective, changes in labor patterns in the United States since 1970, including a significant rise in female work force participation, have only decreased the average work week by about 2.5 hours, from 37 hours in 1970 to 34.5 hours in 2000. The more probable conclusion is that Mexican labor productivity has significantly lagged productivity in the United States.

The fact that productivity growth has remained so stubbornly low in Mexico comes as a surprise to those NAFTA boosters, both in Mexico and in the United States, who argued that a common market would invite a wave of foreign investment in Mexico and with that investment would come a transfer of technology and skills that would propel the Mexican economy—and the productivity of its work force— into the twenty-first century. To be sure, NAFTA has at least partially fulfilled its promise in certain geographical regions and in select sectors of the economy. The manufacturing sectors around Monterrey, in the northeast, and Puebla, a few hours south of Mexico City, for instance, have become relatively sophisticated over the last 15 years, due in large part to foreign investment and loans, and productivity in these regions has risen as a result. However, much of the rest of the Mexican economy has stalled or even regressed in terms of worker productivity. The state-controlled petroleum and electricity sectors, major components of the Mexican economy, are notorious for their inefficiency. Tourism, a labor-intensive sector where productivity gains are difficult to achieve, has grown in relation to the overall economy, dragging down productivity growth as more and more workers enter the low-value-added hotel and restaurant industries. The *maqui ladora* (manufactured goods assembly) industry is another sector where growth has been relatively high but productivity gains have been low. Built along the U.S.-Mexico border to rapidly assemble parts—often U.S. or Asian-made—into finished products like household appliances or consumer electronics, these factories are by definition low-value added operations that depend on cheap labor, not skilled workers.

By far the worst culprit in Mexico's sluggish labor productivity picture, however, has been the dramatic increase in the number of marginal workers in the informal sector. This increase is difficult to quantify, but most Mexicans will agree that the number of irregular street vendors, for example, is substantially higher today than it was 20 years ago. While these workers may be marginal, they are still members of the labor force, and their chronic lack of skills dilutes the productivity gains experienced in the dynamic sectors of the economy.

Yet another driver of low productivity growth has been Mexico's proximity to the United States, the very factor that NAFTA boosters argued would usher in an era of high productivity growth and rising living standards, as North American companies sought to benefit from Mexico's lower labor costs. As NAFTA's promises have failed to materialize, however, this proximity has led to a brain drain of relatively high-skilled Mexican workers who cross the border in search of better labor opportunities in the United States. To be sure, the porous border has also let lower-skilled Mexican workers seek work in the United States: the majority of Mexican immigrants to the United States have not completed high school. Still, roughly 1.2 million Mexican immigrants in the United States have completed some education after high school, a staggering figure when one considers that only 5.2 million Mexicans have completed any postsecondary education. This means that Mexico has lost roughly 20 percent of its most educated workers to the U.S. job market.

This last point presents a serious dilemma for policymakers looking for solutions to the problem of sluggish productivity growth. The most obvious public policy solution would be to increase Mexico's stock of human capital through increased funding for education. While increased investment in education could well result in higher levels of average educational attainment, it is by no means clear that those better educated graduates would not take their skills to the United States in search of more attractive labor opportunities. This is not to argue that investment in education would not yield some positive returns in terms of productivity growth, but so long as other conditions (poor infrastructure, labor market rigidities, institutional and regulatory deficiencies, etc.) militate against an improved economic picture, Mexican policymakers cannot be sure that better educated Mexicans will mean a better educated and more productive labor force.

Mexico's Structural Challenges

Mexico's inability to generate significant labor productivity gains—and its corollary inability to generate substantial improvement in income levels and overall development—cannot be blamed entirely on the brain drain into the United States. Mexico faces a number of mounting structural problems, including a growing social security crisis, a poor long-term outlook for the petroleum sector, chronic under-investment in the electricity sector, and an overwhelmed health care system, that will bring further economic disappointment—to Mexicans as well as to U.S. policymakers and foreign investors— unless they can be addressed. In fact, given the array of pressing structural problems facing Mexico, the productivity and income growth outlook over the next decade could be even more disappointing than that of the last three decades.

The first, and potentially most explosive, of these structural issues is Mexico's looming social security crisis, which has been growing on two fronts, with rising deficits in public-sector workers' pension plans and growing demand for health care spending. Mexico faces chronic and rising deficits in the State Workers' Institute for Social Security and Services (ISSSTE), the pension system that covers most of Mexico's public-sector workers. Largely a product of the oil-rich boom years of the 1970s, when the long-ruling PRI party extended generous benefits to government workers in order to secure their electoral support, the ISSSTE is fast becoming one of the most expensive items in an otherwise under-funded government budget. Under current ISSSTE rules, government pensioners have the right to draw 100 percent of their last 12 months of salary upon retirement. With the average worker retiring at age 56 and the average life expectancy beyond retirement at about 22 years, this guarantee imposes a significant burden on the government budget. The pay-as-you-go system's generous benefits were set up at a time when the age pyramid of Mexico's government workers was heavy at the bottom, with many young public-sector workers supporting relatively few pensioners. Now, however, the system covers only 2 million active public-sector workers while providing benefits for 400,000 retirees: a dependency ratio of 5:1. The dependency ratio, which was 20:1 just 20 years ago, 3 is projected to shrink to 2:1 by 2020, 4 meaning the pension contributions of just two government workers will have to be stretched to cover the generous pension benefits of one retiree. Because the federal government is committed to covering any deficits in the ISSSTE pension system, the worsening dependency ratio will mean rapidly increasing costs for Mexico's taxpayers (and its lenders).

Just a few years ago, the prospects for reform of the pension system—and a reduction in the long-term burden the system would impose on Mexico's fiscal accounts—appeared much brighter. Mexico, under former president Ernesto Zedillo, successfully enacted important reforms to the pay-as-you go private-sector pension system, transforming it from a defined benefit plan with serious actuarial deficits to a defined contribution plan with greatly reduced actuarial deficits. However, President Zedillo did not follow through with similar reforms with respect to ISSSTE, largely for fear of alienating the powerful public-sector unions. The Zedillo government dropped the issue when it realized that the PRI would face stiff competition in the 2000 presidential election.

Zedillo's decision to soft-pedal ISSSTE reform was not enough to help the PRI hold on to the presidency, however, and in 2000 Vicente Fox Quesada, who ran on a platform of wholesale reform, was elected. As the leader of the center-right PAN party, President Fox is much less beholden to the interests of Mexico's public-sector workers than his predecessors, although he has apparently decided that the political costs of tackling ISSSTE reform would be too great. Fox would prefer to stress other structural reforms—particularly the partial privatization and deregulation of the electricity sector that would benefit constituents in his electoral base in the more heavily industrialized north. In truth, reforming ISSSTE would be extremely difficult for any Mexican president since Mexico's three-party system virtually guarantees that no president can muster an outright electoral majority. Reform of the ISSSTE would require congressional approval, but no one party is likely to control an outright majority in either chamber of Congress, even after this summer's elections. Reform would have to be initiated by President Fox's party, but the centrist and the leftist parties, looking ahead to the 2006 presidential elections, could easily form an alliance to block reform. What this means is that Mexico's taxpayers, or the foreign and domestic lenders who cover Mexico's deficits, will be forced to finance a pension fund deficit that is projected to rise from roughly 0.3 percent of GDP (approximately $2 billion) today to 1.4 percent of projected GDP 20 years from now. This will drain funds away from the investments in education and infrastructure that Mexico needs to make in order to jump-start worker productivity.

The second half of Mexico's social security crisis derives from the government's role as the provider of most health care services. The problems in the Mexican Social Security Institute (IMSS) stem from two sources: the fact that the costs for health care services will continue to rise above the inflation rate, and the dismal state of IMSS facilities and equipment, and the corresponding need for significant investment to ward off further deterioration of services. Rapidly increasing health care costs are a problem everywhere, but in Mexico's case the problem has been compounded by a sharp rise in those covered under the IMSS health plan. Since 1974, the number of persons eligible for IMSS health services has risen by 350 percent; however, during this same period the number of hospital beds in the system increased by only 20 percent.[5] The physical condition of IMSS facilities has also deteriorated, with 40 percent of all of Mexico's primary care hospitals judged to be either in immediate need of significant repairs so as not to compromise services, or so dilapidated that they require constant and expensive maintenance. The problem of unreliable or seriously outdated medical equipment in IMSS facilities has also grown more acute in recent years, with the IMSS health service estimating that it needs some $600 million (0.1 percent of GDP) to replace broken equipment, and considerably more to modernize its equipment. All of this comes against the backdrop of Fox administration budgets that have seen drastic cuts in health care investment spending in order to offset rising expenditures elsewhere. The 2001 budget cut spending for building new hospitals, upgrading existing facilities, and purchasing new equipment by 70 percent, to just 0.02 percent of GDP, and the 2002 budget slashed them to zero, a level not seen even during the worst days of the Peso Crisis, when a steep recession forced broad cuts in government spending.

The political costs of allowing the health system to continue to deteriorate as the demands of an aging population for services increase will be hard to ignore, and this is likely to be an important issue in the 2006 presidential elections. The fact remains, however, that rehabilitating the system will require considerable government expenditures, and there are no obvious new revenue sources that can be tapped. If the federal government diverts resources that might otherwise go toward other pressing long-term needs, like investing in education and infrastructure to the health care system, it will undercut productivity growth. And so the vicious cycle continues.

If only Mexico could produce more oil, or produce it more efficiently, it is often argued, there would be plenty of money to finance the investments Mexico needs to make in order to modernize. Unfortunately, PEMEX, the government-owned oil and gas monopoly, is no longer the goose that lays the golden eggs. Mexico's long-term oil outlook is grim; proven and probable hydrocarbon reserves will be depleted in 35 years at current production levels. Oil and natural gas reserves have been steadily declining since 1984. After exponential growth in the 1970s and early 1980s, oil production was flat from 1984 to 1989, and has grown at an average annual rate of just 1.8 percent since 1990. This picture could change if there were major new discoveries, but the Fox administration, which is hamstrung by the lack of private investment in the hydrocarbon sector,

has called only for enough exploration to maintain hydrocarbon reserves at current levels. Therefore, unless there is a permanent increase in oil prices in real dollar terms, oil revenues will inevitably fall, as a percentage of GDP, as Mexico's economy grows but production remains more or less stable.

To offset stagnant oil revenues, the Mexican government will have to find a way to broaden its tax base, but there is no political appetite for tax reform that would sufficiently offset sluggish oil revenues. One of President Fox's top priorities upon entering office was to broaden the sales tax base. But Mexico's divided Congress could not agree to the politically unpopular measure, and the long-term prospects for Fox's vision of fiscal reform now appear bleak.

Yet another structural deficiency Mexico must address is its seriously outdated and inefficient electricity sector. Mexico's constitution grants the state the exclusive control of the electricity system, from generation through distribution. This state monopoly has resulted in a persistent lack of investment in generation capacity because the government has either been unwilling or unable to mount the borrowing campaign necessary to bring the system up to date. By the government's own account, the electricity sector will require $100 billion in new investment over the next decade in order to meet growing demand. President Fox's reform initiative to amend the constitution to allow the private sector to participate in the generation and distribution of electricity has been stalled in Congress since last year, however, and unless Fox's party is able to gain a significant number of seats in this summer's congressional elections, it is unlikely that Congress will agree to liberalize the electricity sector before the 2006 general elections.

If Congress does not approve the liberalization of the sector, the federal government will have to devote scarce public-sector resources to maintaining and upgrading the system. It is unlikely to be able to come up with the required $100 billion, but even partial funding will drain resources from other critical areas. And the country will still be left with an inadequate electricity grid, yet another disincentive to investment.

A Recipe for Growth

The recipe for rapid productivity growth in Mexico is not especially complicated: a mixture of increased (and more efficient) spending on education and physical infrastructure, deregulation and legal reform, and an intensified battle against corruption.

But this is all much easier said than done. Perhaps Mexico's greatest triumph in the aftermath of the Peso Crisis of 1994–95 was to have put its fiscal house in order by reducing spending in real terms, and a long record of fiscal restraint has earned Mexico investment grade ratings of BAA2 from Moody's and BBB from S&P, which allow many U.S. and European pension funds and insurance companies to invest in Mexican debt. A loosening of the fiscal reins now in the form of a wave of new education and infrastructure spending could threaten those ratings, especially if those new expenditures were not offset by politically unpopular budget cuts. A loss of investment grade ratings would entail higher borrowing costs and, potentially, a host of other economic ills that could undermine the government's ability to sustain higher investment spending.

With Mexico's options for increased spending limited, the Fox administration and Congress must undertake reforms that, however politically contentious they may be, could lead to increased, and more equally distributed, economic growth down the road. The outdated electrical system—a strong disincentive to potential new investments in manufacturing—must be opened up to private investment. The public-sector pension plan must be restructured, with pensions reduced and the minimum retirement age raised for new retirees. Bureaucratic red tape should also be addressed; one recent study showed that manufacturers looking to open a facility in China must complete only 22 bureaucratic steps to begin operations, whereas in Mexico the same firm would have to clear 359 bureaucratic hurdles.[6] Finally, Mexican labor relations should be made more transparent so as to encourage companies to invest in the country. The process by which union delegates are chosen must be made more democratic so as to avoid union bosses who represent the interests of neither the workers nor the company. The specialized labor courts, which scare away potential foreign investors wary of judicial corruption, should be integrated into the existing judicial system.

Even if Congress were to approve all of these structural reforms—an unlikely scenario, given the current political divisions— Mexico would still have considerable unmet investment needs for which there will be no funds available so long as the country pursues its International Monetary Fund-endorsed fiscal policies. Therefore, the United States should provide targeted foreign aid for Mexican education and infrastructure spending. Last year, Mexico spent just $1.1 billion on school construction and only $1.3 billion on capital investment in communications and transportation. A modest annual U.S. aid package of $1.2 billion would allow the Mexican government to boost spending in these critical areas by 50 percent. To mollify the persistent critics of NAFTA and foreign aid—both from the right and the left—in the U.S. Congress, such an aid package could be tied to the approval by the Mexican Congress of the structural reforms outlined above, reforms which would have obvious benefits for U.S. companies investing in Mexico.

A sizeable aid package directed at increasing Mexican labor productivity would be in the clear self-interest of the United States, which spends billions on policing the Mexican border against illegal immigrants and drug smugglers, and on education and health care for Mexican immigrants. Helping Mexico deal with its economic problems would make far more sense than the border controls that have proven so costly, both in financial and human terms. More importantly, U.S. companies are unable to exploit a potentially vast Mexican consumer base, because a lack of economic development has stunted the growth in Mexican wages. Faster growth in Mexican wages could be a boon to

U.S. exporters at a time when sluggish or nonexistent growth in Japan and the European Union has led to tepid demand in our traditional export markets. Finally, such a foreign aid package would be aimed directly at the problem for which NAFTA is so often blamed: income inequality. Whether or not the criticism of NAFTA as an engine of inequality is fair, a U.S. aid package targeted at building and upgrading Mexico's schools and roads could address the problem of anti-American Mexican nationalism, reflected in the popular pressure on President Fox to pursue an independent foreign policy during the debate over the Iraq War, and help improve bilateral relations over the long term.

Unfortunately, the likelihood of such a happy combination of U.S. aid and Mexican reform is slim at best, given the current political climate on both sides of the border. For the United States to offer such a package—and for the divided Mexican Congress to accept its terms and approve the required structural reforms—would require a change of heart in both Washington and Mexico City. The Bush administration would have to relax its skepticism about foreign aid that is not directly tied to strategic or domestic political interests, as with aid to Turkey or the African AIDS relief package. The administration would have to sell such aid as benefiting U.S. companies, which would be able to capitalize on a richer Mexican consumer. In Mexico, the political factions that consider themselves to be the defenders of the populist revolution of 1910 would have to abandon their vision of public control of the electricity grid as a cornerstone of national sovereignty, and in cooperating on social security reform they would have to be willing to alienate public-sector workers. Such a change of heart is not unthinkable—similar political realities existed when Mexico began to liberalize its economy in the late 1980s, and again when the private-sector social security regime was restructured following the Peso Crisis—but President Fox, or his successor, will have to be able to convince the general public that the welfare of the nation depends on reform.

Yet U.S. aid and Mexican reform could be a winning combination for both sides. In the short term, a successful structural reform agenda would improve investor sentiment, reducing the interest rates at which the government, Mexican corporations, and Mexican consumers could borrow. Over the long term, Mexico would continue to benefit from higher economic growth and rising incomes, while the United States would experience more manageable migration flows from Mexico and, more importantly, a more dynamic Mexican consumer market for U.S. exports. One thing is clear: until Mexico's problem of low productivity growth is tackled, Ross Perot's assessment of Mexico's economic potential will continue to ring true.

Notes

1. "Gore, Perot Challenge Each Other on NAFTA Money, Facts," Bloomberg News wire, November 19, 1993.
2. Mexico's official productivity data do not stretch back to 1970, and even then they are limited to only certain workers and are therefore not comparable to productivity data for the United States or other developed nations. See Christian Stracke,

"Mexico Debt Sustainability—The Bumpy Road to Single-A," November 24, 2002, at www.creditsights.com/marketing/headlines/0211/004355.htm.

3. See Interamerican Development Bank, "Health and Social Security Reform for Government Workers," mimeo, August 2002.

4. See International Monetary Fund, "Mexico: Selected Issues," Country Report No. 02/238, October 2002, p. 80.

5. See IMSS, "Informe al Ejecutivo Federal y al Congreso de la Unión sobre la Situación Financiera y los Riesgos del Instituto Mexicano del Seguro Social," mimeo, June 2002.

6. Consejo Mexicano de la Industria Maquiladora de Exportación, as reported in *Reforma*, May 23, 2003.

Christian Stracke is the head of Emerging Markets research at CreditSights, an independent financial research firm with headquarters in New York.

- Failure of NAFTA to Fulfill its Mexican Promise
  - More Jobs
  - Mexican Prosperity / Economic Growth
  - More Foreign Investment
- Low Level of Productivity
- Increasing Costs of Mexican Social Security
- Increasing Health Care Costs / 70% Decrease in Healthcare Spending
- Outdated & Inefficient Infrastructure
- Working Force (Mostly Women) Has Tripled in the Last 30 yrs → But Productivity Has Decreased → High Unemployment

- Need to Increase Funding for Education & Infrastructure
  - But Might Lead to Brain Drain
- Need for Tax Reform
Solutions → U.S. Aid Package for Education & Infrastructure Utilities / Roads

Failure for US.
- Wages so Low in Mexico, Mexican Consumer Base for US. Products/Services Never Materialized

With Little Loans, Mexican Women Overcome

By Tim Weiner

SAN MARCOS ACTEOPAN, Mexico—Guadalupe Castillo Ureña was widowed at 31, left alone with five children when her husband died trying to get to the United States from their hut here in the foothills of Mexico's southern Sierras.

She was among the poorest of the poor— scraping by, like half the people in Mexico, and half the world's six billion people, on $2 a day or less, barely surviving.

Then an organization called Finca came to the village. It asked the women there—and only the women —whether they would be interested in borrowing a little money, at the stiff interest rate of 6 percent a month, to start their own businesses.

Change came. With a loan of about $250, Ms. Ureña, now 35, started making hundreds of clay pots this winter. With Finca's help, they were sold in bulk to a wholesaler, who sells them in the city. She pocketed $15 to $20 a week in profit. That sum, the first real money she had ever earned, was enough to help feed her children and pay their school expenses.

"It's exhausting," she said, "kneading the mud, stoking the kiln. But it's something. An opportunity."

These small loans, known as microcredit or microfinance, are not a charity. They are a growing business that is producing wealth in some of the world's poorest countries. Experience has taught the lenders to make nearly all their loans—95 percent or more—to women, and preferably those like Ms. Ureña who are single mothers.

"Why only women?" said Alejandra Ayala, 36, a participant from this village 100 miles southeast of Mexico City. "Everyone knows women have the capacity to do this. Running a family, running a business, what's the difference?"

Carlos Labarte, director general of Compartamos, a larger group loaning to the poor in Mexico, had his own answer: "The men are not here. They've gone to the United States. Or maybe they're just gone. Or they've died. This is not philosophy. This is reality."

In Mexico alone, companies making these loans are reaching hundreds of thousands of women and children. Those women borrowed and repaid tens of millions of dollars in the last two years. Their default rate on these loans averages 2 percent, considered unusually low in the developed world.

The sums loaned are tiny by most standards, a few hundred dollars in most cases. But they represent a chance for the poor to acquire a little bit of wealth. The money the loans help generate can ensure that a child can attend school or that her mother can buy medicine.

That counts for something in a world where rich and poor nations are undergoing financial shocks, wealthy donors are suffering from compassion fatigue and governments everywhere have thrown up their hands at the failure of antipoverty programs and development schemes.

In Mexico, the best-run little village lenders have grown into something closely resembling banks. That is something new in Mexico. Commerical banks in this country are for rich people. They almost never serve the middle class or the poor, rarely if ever providing them with savings accounts, much less loans or credit.

"People who have nothing to eat and no table to eat on have no trust in institutions," said Serapio Reyes, who oversees 10,300 Finca clients from Cuatla, a two-hour drive southeast of Mexico City.

"They need an institution like us to have a semblance of an economic or financial life," said Mr. Reyes. "They certainly do not have access to a commercial bank. But they have us. And we are an institution based on trust."

Most of the 32 small loan institutions now operating in Mexico started in the 1990's. For many, seed money came from private American foundations or, in at least one case, the United States Agency for International Development. Many are modeled on the Grameen Bank of Bangladesh, run by an American-educated economist named Muhammad Yunis, which now lends roughly

$1.2 billion to 2.4 million poor people, overwhelmingly women.

Thousands of village banks based on Mr. Yunis's model have started around the world. But the most successful of these are now outgrowing their modest beginnings.

Compartamos, which is based in Mexico City and whose name means Let's Share, started making loans in 1990. With 140,000 clients today—98 percent of them women—it has a loan portfolio of $45 million and a default rate of just over 2 percent. Compartamos may not be Citibank, but it has the potential to offer more poor people more financial services than any commercial bank in Mexico ever has.

"People don't just need credit," said Mr. Labarte, Compartamos's director general. "They need financial services. They need savings accounts. They need insurance, life insurance, education insurance. These are things that no middle-class person, no poor person, has ever had in Mexico."

Like big banks, institutions that make these small loans can be damaged or destroyed by financial crises or by fraud. Finca's Mexico operation had to restart seven years ago, its managers say, after the peso collapsed and one local manager quit unexpectedly.

The established small loan groups like Finca, whose name means farm, and Compartamos have learned some fundamental lessons about what works and what does not work when it comes to fighting poverty.

What works, first of all, is loaning money to women.

"When you walk into a village, you do not consult the mayor and you do not consult the priest," said John Hatch, Finca's executive director. "You ask, 'Who is the woman in this community that everyone most respects?' Then you ask that woman, 'Would you be willing to convene a meeting at your house?' And at the meeting we start the bank."

The women in San Marcos Acteopan said they reacted with a rainbow of emotions after Finca came to town last year, from fear to fascination. The idea that someone was going to lend them money to start businesses was riveting.

What is also working—and this is very new—goes beyond lending money to providing poor people with the ability to save and invest. This has become possible under a new federal law in Mexico providing a legal architecture and regulations for village banks and small loans.

Those measures help combat the fiscal amateurism that had financially threatened some lending organizations, said Mary O'Keefe, who helped start Finca's operations in Mexico 10 years ago and now works as a consultant to the World Bank.

The breakthrough under Mexico's president, Vicente Fox, she said, "is the realization that poor people are deserving of financial services."

"The trick", she said, "is to create savings. The fact is that poor people can save money."

This fact qualifies as something of a revelation to economists: that given the chance, poor people would put their money into a certificate of deposit, rather than spending it, stashing it under a mattress, or fattening a pig for market.

What does not work in this kind of war on poverty, experience shows, is letting the government run things. When Mr. Fox announced that he was starting a government small-loan program back in December 2000, a week after he took office, Mr. Yunis, the godfather of microcredit, talked him out of it. "Politicians are interested in the votes of the poor," said Mr. Yunis, whose Grameen bank now has three branches in Mexico. "Politicians are not interested in getting the money back."

Mr. Reyes of Finca Mexico said demand for the loans was so great that he anticipated serving 58,000 clients by 2007, up from 10,300 today.

Among those clients is Escolástica Medina, 43, of San Marcos Acteopan, who borrowed the equivalent of $500 from Finca late last year.

With the money she bought mud and firewood to make about 1,600 clay pots. She sold them over 15 weeks for about $800, with $110 in interest going back to Finca, an effective annual rate about five times higher than any United States credit card company could legally charge.

She kept about $190—"a little money for when my children need something—enough for food, for school, for their health." Now, she said, "I feel equal to anyone."

TRADITIONALLY IN LA — WOMEN ARE UNEDUCATED OR
UNDEREDUCATED, THEY ARE PAID LESS FOR EQUAL
WORK, THEY ARE OFTEN THE HOUSEHOLD EARNERS

- BANKS ARE FOR THE RICH

- MICROCREDIT OR MICROFINANCE LOANS PROVIDE WOMEN THE
  OPPORTUNITY TO SUPPORT THEIR FAMILIES AND PROVIDE FAMILY NEEDS

Small Loan Institutions

- WOMEN MORE LIKELY SAVE, INVEST, AND PAY BACK
  THEIR LOANS

A Changing Economy

Central America continues to produce coffee and bananas, but computer chips and textiles are becoming more important.

By Alejandro Antonio Chafuen

With coasts on the Caribbean and the Pacific, a large number of well-educated people, cheap labor, and vast areas of unspoiled wilderness, Central America has a huge potential. Occupying an area about the size of Arizona and Utah combined, Central America has a regional GNP similar to that of Utah. Income per capita, at just over $2,000 per year, is lower than that of Mexico (over $3,000) but is similar to that of Colombia. Violent conflicts, especially in the 1980s, were a big deterrent to economic development. While the scars of those battles are healing slowly, civil society is learning the art of self-government.

Central America continues to produce coffee and bananas, but computer chips and textiles are each day more important, as nontraditional exports are more in demand than agricultural exports. Tourism is expanding rapidly. Apart from trade with the United States, the economies of several countries are increasingly reliant on remittances from abroad. To El Salvador alone, they rival the amount of the entire loan portfolio of the Central American Bank for Economic Integration.

Few will make more mistakes about the future of Central America than economists or other social scientists looking at the region with the blinders of their narrow disciplines. It is essential to pay attention to culture. One can't expect a similar path to economic development in all Central American countries. Guatemala, for example, has a larger Indian population than all the other countries combined. Costa Rica, on the other hand, even 50 years ago was described as a country where people "are mainly a peaceable, steady-going people, the quietest of all the Central Americans … a little country, and one which exemplifies the value of education."

The Year 2000 Human Development Index (HDI) prepared by the United Nations shows Belize, Costa Rica, and Panama with a score of 80 (out of a maximum of 100), just below the level achieved by Chile and Ireland, two recent economic success stories. El Salvador, Honduras, Guatemala, and Nicaragua score just above 60. (A nation's HDI score measures life expectancy, adult literacy, and GNP. Niger has the lowest; Japan the highest.)

Except for Honduras, all the laggard countries are those that have suffered civil strife. Pacification goes hand in hand with economic development. Guatemala and Honduras, with the largest number of indigenous people, face an extra challenge in opening the door of economic opportunity to all.

The economic indicators, as well as the economic freedom indexes, vary greatly. El Salvador has the best score in Central America, 26 out of 156 ranked countries by the Heritage—*Wall Street Journal* 2003 index. Honduras ranks the worst in the region, at 80. In the Fraser Institute index, Panama ranks 19 worldwide, and Belize, the only English-speaking country, scores the worst. The economy of Panama, which is celebrating its hundredth anniversary, has benefited from monetary stability and banking freedoms, but that did not prevent decades of corruption, nationalizations, and cronyism.

All these countries have very weak judiciary systems, a sign of fragility. Disgust with corruption can cause countries to reverse course, especially economic policies, in a very short time span. Acceptance of a "piñata" style of government, where the winners of an election feel they have the right to loot public and private treasuries, still pervades much of the region.

Most countries also have very poor protection of intellectual property rights. With human capital playing such a key role in today's economy, better protection of patents is essential to the creative energy of many researchers and small entrepreneurs.

The battle against corruption is getting more serious. The house arrest of former President Alemán of Nicaragua and the recent exposure of Guatemalan President Portillo's serious breaches of the rule of law might be indicators of more law-abiding years ahead. Increased trade should create incentives for improvements in the rule of law.

Economic reform follows education

Guatemala has almost one-third of the population and one-third of the GNP of Central America. Its frontier with Mexico's troubled region of Chiapas makes it even more strategic.

In surveys conducted during the last five years, Guatemalans consistently appear as the strongest supporters of free markets and the United States. These attitudes are the result of many decades of work. Few efforts can claim to be more important than those of the graduates of the Universidad Francisco Marroquín (UFM), which is a free-market university whose mission is to teach and disseminate the ethical, legal, and economic principles of a society of free and responsible persons.

The U.S. embassy in Guatemala has complained that business and the media too closely follow the views of free market champions F.A. Hayek and Milton Friedman, both Nobel laureates. In Central America as well as in other regions, such contradictions teach local leaders to pick and choose their U.S. friends carefully.

El Salvador is another example of good and bad U.S. involvement. FUSADES (Salvadoran Foundation for Social and Economic Development), which received U.S. Agency for International Development support for more than a decade, worked as a development bank and unofficial ministry, helping produce what today could be called a small economic miracle. Its success was built on the support it received from the local civil society and the wisdom, or luck, in choosing brilliant international economists to provide policy leadership.

Costa Rica has an economy 80 percent the size of Guatemala, within an area half as large. Its stability has allowed it to achieve the highest per capita income in the region. No other country in the Americas has a legislature more supportive of free markets. Few democracies in the world would be able to choose leaders of the caliber of Miguel Angel Rodriguez, a great champion of free markets, who was president from 1998 to 2002. Despite the role of important defenders of economic freedom in the political arena, the state maintains a big presence.

Unbridled populism and periods of totalitarianism usually accompany government interventionism. Why has intervention not led to such negative consequences in Costa Rica? While most Latin American labor union laws follow a fascist model, many workers in Costa Rica have chosen to associate with groups within the Solidarity movement, founded in 1947 by Alberto Marten. Unfortunately seldom studied and replicated, it

represents over 2,000 companies and 300,000 workers in Central America—mostly in Guatemala and Costa Rica, where it represents 24 percent of the active labor force. Based on the principle of free association, an independent board of a Solidarity union administers dues contributed in equal amounts by both owners and workers.

Like the three legs of a stool, totalitarianism in Latin America relies on the support of the military, church, and labor unions. Remove one, and totalitarianism can't stand for long. Totalitarianism can't take hold in Costa Rica, which has no armed forces but does have competitive labor and religious associations.

Free trade

Central American governments had agreed to form a customs union long before seeking an agreement with the United States. Much like other free trade pacts, however, the effort involved much more talk than action. The strategy crafted by U.S. Trade Representative Robert Zoellick gave new impetus to the effort to unite Central American economies.

During the 2001 Free Trade Area of the Americas (FTAA) meeting in Buenos Aires, it became apparent that Brazil, joined by Argentine protectionist industrialists, was going to place as many stumbling blocks as possible in the road toward a united American market. With these two countries in doubt, the United States chose a strategy that would open the doors of trade to the smallest countries of the region. Central America was the natural choice to follow the path of Chile.

The U.S. double-track strategy entails continuing the negotiations as if the FTAA were the only game in town but offering bilateral trade pacts to any country willing to move faster. It wouldn't be the first time that the leadership of small republics changed the dynamics of a continent. In the past, Venice and Genoa opened doors to European prosperity. Today, Hong Kong, Estonia, and Ireland are showing the world that smallness, lack of natural resources, security concerns, or even past poverty can't stop courageous and generous civil societies.

Approximately half of Central American exports are destined for the U.S. market. With a major port, Honduras is the country that can see its trade multiply by a large factor during the coming years. Estimating a threefold increase, authorities are already establishing better relationships with New Orleans and other port cities.

The Puebla-Panama Plan has the potential to further multiply the benefits of better integration with the U.S. economy. It is a proposal by President Vicente Fox to link Mexico to Panama with a transportation corridor. The InterAmerican Development Bank is promoting it, while antiglobalizers are vowing to block it. The project will need to respond to a nonpoliticized cost-benefit analysis. As well, it must include the participation of as many private-sector players as possible.

With marginal improvements, Central America can grow faster than South America. Unfortunately, these marginal improvements have to do with the rule of law. While a country can "dollarize" quite fast, as El Salvador has done, improving the judiciary can require generations.

Latin America and the Caribbean comprise the fastest-growing export market for the United States. President Bush has said that he believes that the twenty-first century will be known as the century of the Americas. It is up to Central Americans to move from the backyard to the vanguard.

Alejandro Antonio Chafuen is president of the Hispanic American Center for Economic Research in Fairfax, Virginia.

ECONOMIC POTENTIAL OF CENTRAL AMERICA

— SOME OF THE POOREST NATIONS IN LAC

— COFFEE & BANANA PRODUCTION PLUS Computer Chips & TEXTILES

— REMITTANCES FROM Migrant WORKER POPULATIONS

— FTAA NEGOTIATIONS

— U.S. is Cent. Amer. Biggest Customer ⟶ INTEGRATION WITH U.S. ECONOMY

— PUEBLA-PANAMA PLAN LINKING A TRANSPORTATION CORRIDOR FROM MEXICO TO PANAMA

CONDITIONS PREVENTING ECONOMIC PROGRESS

— POVERTY

— Violent Conflicts/Civil WARS

— CORRUPTION, NATIONALIZATION

— WEAK JUDICIARY SYSTEMS — LEADS TO WEAK ECON. COND.

— Military Regimes & DICTATORS / GOV'T INTERVENTION IN ECON. REFORM.

True Gold of Our Future

Costa Ricans are saying 'No!' to oil exploration companies. **Mark Engler** and **Nadia Martinez** join the growing queue in Latin America to see how it is done.

For activists engaged in seemingly impossible struggles with transnational oil interests, the scene might seem surreal. In May 2002 newly elected Costa Rican President Abel Pacheco stood on stage for his inaugural address and declared that his country would 'not be an oil enclave or land of open-pit mining'.

Pacheco promised that under new constitutional protections, such extraction would be banned. 'The true fuel and the true gold of the future,' he said, 'will be water and oxygen, our aquifers and our forests.'

The declaration delivered a firm blow to Harken Energy, a Texas-based company with close ties to US President George W Bush. The company had its sights set on precisely the kind of resource exploitation that would be outlawed, and it no doubt expected the type of governmental subservience that has allowed the removal of natural resources to continue virtually unchecked in Latin America since the early days of colonialism.

But while President Pacheco's opposition was perhaps the most public sign of defeat for the transnationals, it was only one of many.

For over three years a coalition of environmental advocates, community groups, unions and indigenous-rights organizations campaigned to end the extractions. In addition to making the mining and oil exploration into a delicate campaign issue—bolstering the new President's environmentalist resolve—they battled in courts, galvanized international support and organized local opposition in order to sink Harken's plans. In the process they secured a remarkable series of victories that reversed their country's move toward unsustainable resource exploitation.

For much of its history Costa Rica has stood out as an exception among its Central American neighbours. The country's lack of mineral resources, weak colonial institutions and early cultivation of coffee allowed stable democratic foundations to form. In 1949 a President who had recently faced a military coup took the extraordinary step of eliminating Costa Rica's armed forces. Traditions of peace and social democracy insulated the country throughout the 1970s and 1980s, when the rest of the region was mired in conflict and the CIA was at its most active.

During that time Costa Rica built the thriving tourism industry that is now the backbone of its economy. National pride in the country's extraordinary biodiversity, expansive beaches and system of parks—coupled with a limited supply of underground riches to exploit—laid necessary groundwork for an official stance against extractive resource industries.

But the Government was not always against mining and oil exploitation. Costa Rica, like most Latin American nations during the 1990s, felt the pressures of a sluggish economy, increasing unemployment and poverty. Many looked to neoliberal economics for solutions. The country began to explore new ways of attracting foreign investment. Despite its historical pro-environment policies, it passed a Law on Hydrocarbons in 1994 that divided the country into blocks for oil exploration.

Supporters sent the oil companies and the Costa Rican Government some **27,000 emails, faxes and letters of protest**

When Miguel Angel Rodriguez, an economist and entrepreneur, became President in 1998 he brokered the concession of 10 of the 22 blocks to American and Canadian companies. MKJ Xplorations, based in Louisiana, bought up interests along the Caribbean coast and soon partnered with Harken Energy to begin exploration. Isaac Rojas, board member of Comunidades Ecológicas La Ceiba, a Costa Rican affiliate of Friends of the Earth, argues that the Environment Ministry in the Rodriguez Administration 'took the country's advances in the area of environment back about 30 years'.

The location that the Harken-MKJ group selected for its activities was the Talamanca region of southeastern Costa Rica: a pristine area resting between several protected wilderness areas. Among those who lived in the area there was an immediate awareness that oil rigs could threaten the sea-life offshore—including endangered sea turtles. Locals felt stunned by the lack of consultation involved in the deal. 'We saw an announcement in the national newspaper saying that the Government had approved concessions for oil exploration,' says Enrique Joseph, a 38-year-old tour guide and restauranteur who grew up in the area. 'Within two or three days, we raised a red alert.'

Dozens of groups—ranging from farmers' organizations and the fisherpeople's union, to small-business owners, religious groups and marine biologists—came together in the following months to form *Acción de Lucha Anti-Petrolera*, or Anti-Petroleum Action (ADELA). In December 1999 they convened a meeting with about 250 people, which drafted a declaration opposing the concessions and advocating a moratorium in Talamanca.

Their declaration received national attention and the legal strategy they launched bore fruit. In September 2000 the Supreme Court ruled the oil concessions null and void because indigenous communities were not properly consulted.

But two months later, after Harken-MKJ appealed the ruling, the Court allowed the oil interests to continue their activities in the offshore blocks where no indigenous communities exist. The corporate executives were pleased with this outcome because their main targets for exploration were offshore, near Puerto Moin. As Harken Vice-President Stephen Voss explained months earlier: 'The Moin prospect is the largest structure that the company has ever tested and it offers great exposure to Harken shareholders for the discovery of significant reserves.'

Unfortunately the project's prospects for environmental sustainability were not nearly so promising as its anticipated profits. ADELA set out to amplify the message that, as Rojas says, 'local people had declared firm opposition to extractive activity. They said that this could not be done.'

Their organizing faced several difficulties. First off, they were clearly out-funded. 'We discovered that the talks we would give on the radio weren't very effective, because the oil companies bought up most of the space available,' explains Joseph. 'Or, if we drove around in a truck

with a megaphone, we'd later see the company rent three cars to do the same.'

Activists complained of the *compadrazgo*—the sense of comraderie—that seemed to exist between executives and government officials. And they saw some communities, particularly in the city of Limon, come to advocate the economic benefit of resource exploration. 'The companies promise clouds of gold,' Rojas says of his experience with extraction projects, 'and some people believe them.'

'Seeing our brothers in Limon against us, because of their false expectations,' adds Joseph, 'that was the hardest part of the campaign.'

By January 2001 groups like Natural Resources Defense Council (NRDC) and the Environmental Law Alliance Worldwide (E-LAW) began to bolster the local organizing with financial support, research, legal resources and international exposure. In what Jacob Scherr, director of NRDC's International Programme, describes as 'probably one of the largest deluges of mail they'd ever seen', supporters sent the oil companies and the Costa Rican Government some 27,000 emails, faxes and letters of protest.

That same year ADELA's campaign to raise public awareness made it politically risky for presidential hopefuls to support oil exploration. President Rodriguez wanted concessions to go forward. But he muted his enthusiasm when it appeared that the stance might hurt his party in the elections. The opposition candidate, Abel Pacheco, was outspoken on the issue and had drafted a proposal to repeal the hydrocarbons law while serving in the Legislative Assembly.

According to Enrique Joseph things shifted in Talamanca at a meeting between ADELA and Harken officials. 'People were offended to see the oil executives come in and act rudely toward residents who were well respected. Many people started to doubt the companies' promises.'

Victory came soon afterwards. In February 2002 the national technical secretariat, SETENA, rejected Harken-MKJ's plans for offshore drilling. Several months later the

Environment Ministry denied an appeal from the oil companies. Its decision noted over 50 reasons why the environmental-impact statement for the project was insufficient—a list that mirrored the meticulous documentation presented by ADELA and its international allies.

Upon his inauguration Pacheco vowed to add environmental guarantees to the Costa Rican Constitution. And on 5 June he signed a presidential decree banning open-pit mining.

For activists the fight is not over. In the past year several challenges have emerged that may determine whether Costa Rica's bans will remain intact as emblems of the country's exceptionalism, and whether it might become a model for resistance elsewhere.

According to the *National Law Journal*, Harken is not only demanding millions in payment for the money it spent on the exploration—it is trying to use the provisions of a bilateral investment treaty to win $57 *billion* from Costa Rica in lost profits that it had projected. Although the State Department denies it, ADELA believes that the US Government has lobbied on the oil companies' behalf and that it is pushing Costa Rica to ease its bans on resource extraction. 'Sources in Costa Rica have told us that the US Embassy there has been putting pressure on the new Pacheco Government to open its coast to US oil companies,' said Scherr at the NRDC.

A broader concern for activists is that treaties under negotiation, like the Central American Free Trade Agreement (CAFTA), may endanger Costa Rica's protections. Although the Pacheco Administration has defended environmental rights, it has also enthusiastically pursued a neoliberal 'free trade' agenda.

In North America, Chapter 11 of NAFTA has allowed companies to sue localities whose environmental laws interfere with their business. The challenge for Costa Rica is to guard against similar outcomes.

In mid-June ADELA struck a deal with the Government regarding the repeal of the hydrocarbons law. Activists agreed to leave in place parts of the law that enable the Environment Ministry to regulate oil refineries. Their agreement insists that a government commission will continue to review trade agreements to ensure that no provisions will invalidate local environmental laws.

But lawsuits have already affected the battle over the mining ban. The President's ban applies only to future mining projects. It allows companies that had been granted concessions prior to the ban to continue their activities. The Constitutional Chamber decided in October 2002 that Vanessa Ventures, a Canadian mining interest, was exempt from the ban on these grounds. A court decision in June ruled against the Environment Ministry for delaying approval of Vanessa's Las Crucitas project. The company is threatening to sue for $200 million under the Foreign Investment Promotion and Protection agreement if the mining does not go forward soon.

While it may prove impossible to stop this existing concession, Costa Ricans have moved to strengthen future protections. 'We need to make the presidential decree into a law, since the decree can be modified or repealed—it's not permanent,' argues Rojas.

President Pacheco sent his Environmental Bill of Rights to the legislature in September 2002, and politicians are currently negotiating the text of the legislation.

Developments on the international scene are also promising. 'I always hoped that our country would not be an exception,' says Joseph, 'that people would see it as a model of what can be done.' In several countries this has already taken place.

Nicaraguan activists
have called for their government to institute a moratorium on oil concessions

In Ecuador environmental activists are pushing to replicate the Costa Rican example. In June they presented the new government of Lucio Gutierrez with a proposed prohibition on expansion of the oil frontier. But securing a ban in Ecuador is admittedly a difficult task. Oil exploitation, which began here over 30 years ago, has polluted not only the environment but also the politics of the country. As Terry Karl, professor of political science at Stanford University, explained in an interview with the *Environmental News Network*: 'Once oil exploitation begins you get what I call the petrolization of politics, and where this is in place you would never get a Costa Rica-type situation.'

More hopeful is Nicaragua. The Nicaraguan Government recently granted exploration licenses to several US-based companies, among them MKJ Xplorations. In May Nicaraguan activists in the Oilwatch network called for their government to institute a moratorium on oil concessions.

Costa Rican activists can now take satisfaction from what has been accomplished. 'This is a step toward achieving a deeper change,' says Isaac Rojas. 'Of course it brings a certain joy. It shows that we were right.'

In Ecuadoe
environmental activists are pushing for the Costa Rican example

Enrique Joseph agrees. 'Latin Americans have endured much suffering. At least this is a guarantee that, at my age, I can continue living in a beautiful place. I want my children and their children to run on the beaches and see the trees.'

Mark Engler, is a writer based in New York City, can be reached via the website—www.DemocracyUprising.com

Nadia Martinez is a Resarch Associate with the Sustainable Energy and Economy Network, a project of the Institute for Policy Studies in Washington DC.—www.seen.org

---

The Colossus of the North

It is in U.S. political and economic interests for Central America to be stable and prosperous, as this would promote investment there and make terrorism, drugs, and immigration more manageable.

By William Ratliff

Central America has lain largely unnoticed in the shadows of its larger neighbors, the United States and Mexico. Several countries briefly assumed strategic importance and thus gained attention from Washington during the last decades of the Cold War. After the late 1980s, the region again virtually disappeared from view.

In his 2000 presidential campaign, George W. Bush promised to transform U.S. relations with Latin America as a whole, the context within which relations with Central America must be viewed. Though there have been some positive developments, on balance one may question the seriousness of Bush's promise.

Clearly, the United States is far more important to Central America than vice versa. It is reasonable to ask why, in a convulsed world, Washington should pay more than passing attention to the region. For Washington, the bottom-line issues are immigration, terrorism, and drugs.

The Drug Enforcement Administration reports that about two-thirds of the cocaine from South America passes through Central America. The CIA reports significant smuggling of aliens from Central America.

Some Central American agencies cooperate in arresting and prosecuting smugglers, particularly those that deliver their victims unwillingly into sweatshops and prostitution rings. However, law enforcement and intelligence in Central America are hampered by corruption.

Our strategic interests should be broadly interpreted. It is in the U.S. interest for Central (and Latin) America to be stable and prosperous, as this would promote investment there and make the problems of terrorism, drugs, and immigration more manageable. Finally, successful reforms would be of humanitarian value, affecting the millions of Central Americans who are living in the United States.

The legacy of colonialism

Latin America's colonial heritage remains massive, often affecting the economic development for the worse. Mexico's Nobel laureate Octavio Paz correctly observes that Spanish ideas and institutions "were built to last and not to change." Thus Peruvian journalist Alvaro Vargas Llosa writes that Latin America's main problem is not instability, as is commonly believed as it lurches through a seemingly endless cycle of political coups and economic crises, but rather excessive stability. Latin American civilization has been unable to truly change. Foremost among the lethal remnants is a mercantilist economic system that Mario Vargas Llosa (father of Alvaro Vargas Llosa), the prominent writer and Peruvian presidential candidate, has succinctly defined as "a bureaucratized and law-ridden state that regards the redistribution of national wealth as more important than the production of wealth."

This system, with its elaborate religious and legal underpinnings, was established centuries ago to serve the interests of Spain and its ruling elite in Latin America. Both crown and church deliberately prevented most people from acquiring secure title to private property or equality of political and economic opportunity under impartial law. Spain steadfastly opposed any competitive activities that could lead to representative democracy or a free, productive marketplace at home and internationally.

After independence, the colonial institutions and intellectual constructs used to justify minority manipulation of society were taken over by the victorious Creole elites and often, in time, by their successors. These successors included military forces, "nationalist revolutionary" parties like the Institutional Revolutionary Party (PRI) in Mexico, Marxist parties like the Sandinistas in Nicaragua, populist demagogues like Hugo Chávez in Venezuela, and nouveaux riches mercantilists. These groups have generally sought to defuse emerging demands by labor, the middle class, and other groups by drawing them into the fold and thus controlling them or by simply repressing opposition to their own rule. They generally thrived on influence peddling and corruption.

In the 1990s, many analysts were optimistic that reforms could tackle these problems. While the reforms often expanded private ownership, it is now clear that they did not make most economies freer. Privatizing state-dominated economies recreated what Alvaro Vargas Llosa has called the old labyrinths of power and privilege. The failures of these reforms discredited free-market ideas and contributed much to the latest round of economic breakdown and populism in Latin America.

The perspectives of the nonelites also differ from those predominant in the developed world. When the Chilean firm Latinobarometro asked Latin Americans in 2001 if democracy is the best form of government, only 48 percent said yes. Only 25 percent thought democracy was working. Moreover, two-thirds to 90 percent had little or no respect for the institutions of democracy—the presidency, congress, the judiciary, and political parties—let alone political leaders.

Exactly half of those polled said they would support a military coup if it would bring more effective government. (A poll of Hispanics in the United States in 2002 by the Pew Hispanic Center found 60 percent agreed with the statement "I'd rather pay higher taxes and support a larger government that provides more services," compared to 35 percent for non-Hispanic whites.) Serious representative democracy is not the choice of the majority in Latin America because so few have ever had it. People olden settle for corrupt leaders they believe will make the government work.

√International influences

International influences are important but in most cases secondary. Foreign economic fluctuations, ranging from the Great Depression and the U.S. slump of recent years to tariffs, are the most obvious. Even this vulnerability to foreign influences, however, must be traced directly back to the international relationships and stunted domestic economies created during the Spanish colonial period and maintained there after since independence.

Questions arose about the seriousness of President Bush's promise long before September 11. Less than a month into the president's term, he took Secretary of State Colin Powell to meet with Mexico's new president, Vicente Fox, in a loudly heralded ploy to raise Latin America's low profile. Even as they were talking with Fox, the United States resumed bombing targets inside the no-fly zone in Iraq. After that, it was impossible to find any lead media stories about the new importance of Latin America. (My phone rang constantly as radio and TV stations canceled scheduled interviews about the region.) Unassisted, the president had bombed out the grand opening of his promised "Century of the Americas."

Venezuelan-American journalist Carlos Ball has observed that U.S. officials often "act as the representatives of protectionist American corporations [and seek] to impose U.S. labor standards on local factories." To some degree, this track results from compromises in Congress to get support for trade issues. The consequence of this apparent concern for the downtrodden means that many people lose their jobs and are pushed into begging prostitu-

tion, or dependency on other family members.

Negative messages

All Latin American constitutions are littered with promises that can't be met. Pressing for more entitlements, as the ambassador to Guatemala did, simply makes things worse. Understandably, crude interference generates what Central America expert Timothy Brown calls "both obsequiousness and resentment" toward the United States in the region.

In early 2003, the United States was poised to "decertify" Guatemala for its failure to fight hard enough in our drug war, thus perpetuating an arrogant and unjust policy. As any Latin American will tell you (correctly), if any country should be "decertified" for doing more to cause than cure the drug problem, it is the United States.

Another negative message to Latin America was Bush's appointment of Otto Reich as assistant secretary of state for the Western Hemisphere Affairs. Although well qualified for the job, Reich has been haunted by leftist propagandists. His appointment predictably stoked intense partisan wrangling in Congress. Thus the president could get Reich merely an interim appointment of one year. Reich took office as a lame duck in January 2002 and left in December 2002.

At this writing, there is no new assistant secretary. One must conclude that in insisting on Reich over others who could have done the job without inciting a debilitating partisan riot, President Bush was not intent upon choosing an informed leader of hemispheric policy. With a lame duck appointment at the top State Department level, it is no wonder Bush's objective of changing the relationship has been largely illusory.

The key element of his Latin America policy has been promoting free trade, which has met with some success. The Trade Act of 2002 gave the administration *trade promotion authority* (previously called *fast track*), which immediately led to talks with the five members of the Central American Economic Community: Guatemala, Honduras, El Salvador, Nicaragua, and Costa Rica. According to U.S. Trade Representative Robert Zoellick, the objectives of U.S. Policy are to develop interim bilateral relations, encourage regional integration and competitiveness, develop a larger market for investments, and promote political cooperation.

Unfortunately, as long as the mercantilist policies of many Latin American countries remain, effective free-trade policies will be difficult or impossible to implement. Many U.S. dealings are inevitably with privileged mercantilist sectors. On the other hand, an encouraging degree of increasing political maturity and cooperation is suggested by the apparent resolutions of disputes involving the Gulf of Fonseca, the San Juan River border, and the relationship between Belize and Guatemala.

Many Central Americans have welcomed U.S. enthusiasm for combating corruption, but corruption is an inherent part of the region. Attention has been given to legal reform, which is good, though sometimes the rule of law is confused with the rule of regulations. For example, though it has benefited member countries, the NAFTA "free trade" agreement consists of many hundreds of pages. The U.S. goal seems to be as much managed trade as free trade. Arguably, agreements will be shorter as support for free trade increases.

Also, our campaign against Arnoldo Alemán's corruption in Nicaragua, however well intended, ignores the systemic corruption we had tolerated for decades. As journalist Glenn Garvin notes, it is "likely to boomerang in unexpected ways." Since much of Alemán's looted funds went into building up the National Liberal Party, Garvin says, funds that flowed into the campaigns of President Bolaños and other anti-Sandinista leaders may have been tainted. Many current politicians could be discredited, to the advantage of the last people we want to take over again in Nicaragua, the Sandinistas.

Short-term progress has been made on the immigration of Salvadorans, Hondurans, and Nicaraguans, who have suffered repeated wars and natural disasters. Several million people from these countries live in the United States, often illegally; their remittances of four billion dollars every year have become critical to their national economies. The recent Temporary Protected Status extends legal residency of people from those three countries in the United States.

Temporary Protected Status (TPS) is a temporary immigration status granted to eligible nationals of designated countries. In 1990, congress established a procedure by which the attorney general may provide TPS to aliens in the United States who are temporarily unable to return to their homeland because of ongoing armed conflict, environmental disasters, or other extraordinary and temporary conditions. Countries that are currently designated under the TPS program are Angola, Burundi, El Salvador, Honduras, Sierra Leone, Somaha, and Sudan.

If any country should be "decertified" for doing more to cause than cure the drug problem it is the United States.

In the end, President Bush's commitment is unfulfilled, though some policies have moved in the right direction. Far too often, U.S. policies fail to confront systemic problems, surely an extremely difficult challenge. Thus much of Central and Latin America remains mired in the crises bequeathed to the region by Spanish colonialism and its successors.

William Ratliff is research fellow at Stanford University's Hoover Institution and coauthor of Law and Economics in Developing Countries (2000).

Systemic Problems and Legacy of Colonial Patterns has led to a Failure of Needed Reforms
- Weak Econ. Structure
- Fragile Democracies
- Vulnerability to Foreign Economies
- Corruption
- Drug Trafficking
- U.S. Strategic Interests is a Stable & Prosperous Cent. Amer.

Discovery Pushes Back Date of 'Classic' Maya

By JOHN NOBLE WILFORD

A discovery of monumental carved masks and elaborate jade ritual objects in 2,000-year-old ruins of a city in Guatemala is raising serious questions about the chronology of the enigmatic Mayan civilization. In many respects, the city appeared to be ahead of its time.

The leader of excavations there, Dr. Francisco Estrada-Belli of Vanderbilt University in Nashville, said yesterday that the city, Cival, appeared to have been one of the earliest and largest in what is generally regarded as the preclassic period. But it has been found to have all the hallmarks of a classic Mayan city: kings, complex iconography, grand palaces, polychrome ceramics and writing.

"It's pretty clear that 'preclassic' is a misnomer," Dr. Estrada-Belli said in a telephone interview. But he added, "It may be too late to change the names" in the established framework of Maya history.

Archaeologists have long dated the start of the classic Mayan civilization at A.D. 250, which had seemed to be the time of the earliest written inscriptions in city plazas and temples. The period ended around 900 with the mysterious collapse of the largest Mayan cities in Guatemala, Honduras, Belize and parts of Mexico. The postclassic period of general decline continued until the arrival of the Spanish in the early 16th century.

The preclassic period may have begun as early as 2000 B.C. Cival reached its prime about 150 B.C. and was abandoned shortly before A.D. 100.

The new findings from Cival were announced by the National Geographic Society, which was a supporter of the research. Besides the two huge stucco masks, the discoveries included 120 pieces of polished jade, a ceremonial center that spanned a half mile and an inscribed stone slab dating to 300 B.C.

It is perhaps the earliest such monument ever found in the Mayan lowlands, Dr. Estrada-Belli said.

Other archaeologists not involved in the research said they were amazed by the size of the city but not surprised to learn that the preclassic Maya were capable of such advanced architecture, art and other classic-type culture.

Previous discoveries had already overturned the former model of the preclassic Maya as a culture of simple farming villages, Dr. David Webster, a Pennsylvania State University archaeologist, wrote in his book "The Fall of the Ancient Maya" (Thames & Hudson, 2002).

The ruins of El Mirador, also in Guatemala, have revealed a preclassic city with a highly developed culture as early as 500 B.C., a pyramid that rivaled in size those of Egypt and a population that may have reached 100,000. Cival may have had 10,000 inhabitants at its peak.

Two years ago a Harvard researcher, Dr. William Saturno, discovered a 1,900-year-old mural at San Bartolo, Guatemala, that experts hailed as a masterpiece and as fine as any wall painting ever found in Mayan ruins.

Dr. Ian Graham, a Harvard archaeologist who specializes in Maya inscriptions, said he accepted the interpretation of the Cival discovery because it seemed to corroborate other evidence of an unexpected flowering of preclassic culture.

"Extraordinary things are emerging from preclassic sites," he said. "They are simply mind-boggling."

Dr. Graham said that when he mapped the Cival site two decades ago, the jungle concealed all but some outlines of the stone buildings and pyramids that once stood there. The central plaza appeared to have been less than half the size of what has now been uncovered.

The Harvard team did not linger for extensive excavations.

Dr. Estrada-Belli's painstaking investigation began paying off with spectacular results a year ago. He was inspecting a dank tunnel in the main pyramid. Reaching into a fissure in the wall, his hand met a piece of carved stucco. Later, he saw before him the mask of an anthropomorphic face, 15 feet by 9 feet, with snake fangs in its squared mouth.

"The mask's preservation is astounding," Dr. Estrada-Belli said in a statement about the discoveries.

Last week, the archaeologist said, a second mask, apparently identical, was excavated from the same pyramid. The second mask is made of carved stone overlaid with thick plaster. Its eyes appear to be adorned with corn husks, suggesting the Mayan maize deity.

A study of ceramics associated with the mask, Dr. Estrada-Belli said, indicated that the two artifacts were part of the backdrop for elaborate rituals in about 150 B.C., plus or minus 100 years.

Other evidence suggested that Cival was occupied as early as 600 B.C. and that the broad plaza was being used for important ceremonies and ritual offerings by 500 B.C. The central axis of the main buildings and the plaza is oriented to sunrise at the equinox, presumably for solar rituals associated with the agricultural cycle.

The remains of a hastily erected defensive wall around the city attest to Cival's probable fate. Overwhelmed by an invading enemy, the city was abandoned, apparently for good.

Cival–

- 2000 years old, Maya City / Guatemala (150 BC TO A.D. 100)
- Early + longer Preclassic Period Site
- Complex iconography, grand palaces, carved stones, polychrome pottery

Even Amid Economic Woes, Argentines Live It Up

Nightlife thrives despite high unemployment and a deeply devalued currency.

By Brian Byrnes

BUENOS AIRES—Kicking back at a trendy Buenos Aires bistro, Federico With and Diego Castro look marvelous.

And looking good, whether or not they are feeling good, is an art many residents of the world's ninth-largest city take seriously. The Argentine capital boasts more plastic surgeons per capita than nearly anywhere else on the planet. Even street performers who dance the melancholy tango wear designer-label suits and skirts.

But feeling good has not been easy lately. Unemployment stands at 15 percent, personal savings have been clobbered by currency devaluation, and food shortages have led to rampant malnutrition.

Despite the bleak numbers, many Argentines are in surprisingly good spirits. On a warm Thursday night in late summer here, Messrs. With and Castro have joined thousands of other Portenos, as the Buenos Aires locals are known, in the fashionable Las Canitas neighborhood, where nightspots are packed with diners intent on dancing the night away.

"These days, it doesn't do me any good to save my money, because it could lose its value again at any time," says Castro, an accountant. "I am living for today."

This kind of attitude has always suited Argentines. It has provided them a way to forget their troubles—a military dictatorship in the 1970s, hyperinflation in the 1980s, today's economic stagnation.

But while some say a history of throwing caution to the wind has helped breed decades of government mismanagement and corruption, it has also created a rugged self-reliance that is helping the country weather its latest storm.

"We have had crises here before, with the military and others," says Juan Pedriel, a hotel manager. "But Argentines are very good at adjusting."

Argentines have been doing a lot of adjusting over the past year or so. Banking restrictions, implemented in December 2001 to limit withdrawals, have been lifted, and people once again have access to their savings—although few dare to put them back into the banks. An estimated $20 billion to $30 billion is floating around the country, secured in family lockboxes or stuffed under mattresses.

Most people cannot afford to buy new cars or apartments anymore, and trips to the US or Europe are now three times as expensive. As a result, much of the middle class now chooses to spend its disposable income by pursuing leisure.

New Nationalism

One of those benefiting from this trend is Isabel Aldao, owner of the La Bamba Estancia, a traditional Argentine ranch 75 miles northwest of Buenos Aires. In addition to a steady stream of foreigners for whom Argentina has become an inexpensive vacation spot, Ms. Aldao says the number of Argentines she has hosted in recent years has increased by 60 percent.

"Before, Argentines used to go to Punta del Este [a posh beach resort in Uruguay] for the weekend. But now they have to pay $500 for that, and they can't afford it," she says. "So they come here instead and learn more about their own country."

Aldao says she has noticed a new sense of national pride among these Argentines, who come to her ranch to ride horses with gauchos, sip herbal mate tea, and eat juicy, homegrown steaks.

This new pride has been hard won. From 1976 to 1983, Argentina was ruled by a military dictatorship that "disappeared" an estimated 30,000 people. Democracy was restored in 1983, but hyperinflation soon followed, where the price of milk would often triple in a matter of hours.

During the 1990s, Argentina enjoyed unprecedented growth and prosperity. Luxury items like perfumes and Palm Pilots were available for the first time. The telecommunications sector was privatized and upgraded, catapulting Argentina into the Internet era. Airlines began running shuttle flights between Buenos Aires and Miami, sending countless Argentines to South Beach and Disney World.

What made much of this progress possible was the peso's one-to-one peg with the dollar, an economic policy that would prove fatal for South America's second-largest economy.

The peg kept the government from being able to print money to cover costs, so instead it began unrestrained borrowing, saddling the country with unmanageable debt.

The government has defaulted on this debt and floated the currency. The peso has since lost 60 percent of its value and now trades around three-to-one to the dollar.

Self-Indulgence: The Problem

Like many Argentines, With, a lawyer, has seen his life savings slashed by two-thirds. But he is feeling better now with access to his money, and a hint of economic stability has returned.

"After all that has happened this past year, I've finally said, 'Enough with this [stuff], I want to have a good time!'" he says.

But some argue that this fondness for indulgence is exactly what has caused trouble for Argentina time and time again. Graciela Romer, a sociologist in Buenos Aires, predicts this extravagant lifestyle will continue to threaten the country.

"Argentines need to begin to see the value in saving money, not spending money. This is a weakness of the Argentine culture," she says. "Now people are spending more money because they want to gratify themselves."

Citizens go to the polls later this month to elect a new president, choosing from a field of three lifetime politicians —including former President Carlos Menem—and two relative newcomers. But with none of the candidates capturing the public's imagination—none polls above 21 percent—it seems that most people are skeptical that the politicians, some of whom had a hand in causing the current crisis, can repair the damage.

"People here can't plan for the future," says Castro, taking the last gulp from his bottle. "This is not a moment to save, this is a moment to enjoy."

From *Christian Science Monitor* by Brian Byrnes, April 2, 2003. Copyright © 2003 by Brian Byrnes.

Argentina's Economic & Political Problems

Economic — *slowed development of Democracy*

Political

- High unemployment (15%)
- Personal savings dropped as currency was devalued
  - Lack of incentive to save
- Hyperinflation in the 1980s
- Banking restrictions
- Limit on withdrawals
- Pegging peso to dollars → kept gov't from printing money to cover costs → led to unrestrained borrowing → unmanageable debt → gov't defaulted on debt → peso has lost 60% of its value

- Military Dictatorship in 1970s (1976-1983)
- Gov't Mismanagement / Corruption

New Hope for Brazil?

Stanley Gacek

On October 27, 2002, Luiz Inácio Lula da Silva—lathe operator, leader of the independent Brazilian labor movement that emerged in the late 1970s to challenge the military regime, a founder of the Brazilian Workers Party (Partido dos Trabalhadores—PT), and a former congressman from São Paulo—was elected president of Latin America's most powerful state, the world's fifth largest country, with more than 170 million inhabitants, and the ninth largest economy on the globe. His victory was unprecedented in Brazilian history: fifty-two million votes, 61 percent of the total.

> … the landless movement,… responsible for many of the rural conflicts and occupations of the last decade, expects sweeping agrarian reform to overturn a system in which less than 3 percent of the population control more than 60 percent of the arable land.

The Lula "Era," as it is being called in Brazil, has raised the hopes of leftists throughout the world. The Lula–PT platform calling for sustainable macroeconomic growth and the reduction of poverty and inequality offers a powerful and tangible alternative to the dominant paradigm of globalization. As Lula declared in his inaugural address of January 1, ending unemployment "will be my obsession" and ending hunger must be "a national endeavor." He has pledged to create at least ten million new and decent jobs by the end of his four-year term, as well as guarantee three meals a day for every Brazilian.

Lula's opposition to any version of the Free Trade Area of the Americas (FTAA) that promotes "an economic annexation" of the South to the North augurs a substantial delay if not a fatal blow to any North American Free Trade Agreement expansion in the hemisphere. There can be no FTAA without Brazil. Foreign Affairs Minister Celso Amorim made it clear to U.S. Trade Representative Robert Zoellick that his country would not be rushed into any hemispheric trade agreement—a troubling admonition for Washington, as Brazil and the United States will be co-chairing the FTAA negotiations in 2003. This new relation of forces concerning trade and the world economy led Emir Sader, a well-known University of São Paulo sociologist, to announce "the beginning of the end for the neo-liberal project in Latin America."

Lula's PT is now the largest party in the Chamber of Deputies, moving from 58 to 91 seats (out of a total of 513), as well as the third-largest party in the Senate, with an increase from 8 to 14 (out of 81). Obviously, these numbers are well short of a majority, and the PT has been compelled to maintain and consolidate alliances with other parties of both the left and the center-right.

Nevertheless, Lula's electoral coalition already included other significant parties, and a number of them are represented in his new cabinet, reinforcing the alliance he needs to govern effectively. The PT and the Lula administration have negotiated deftly with the Brazilian Congress. Executive Chief of Staff José Dirceu persuaded the center-right Brazilian Democratic Mobilization Party, which commands one of the largest legislative blocs, to elect leaders amenable to agreements with the new government. This latest negotiation guarantees a pro-Lula majority of 283 seats in the Chamber and 47 in the Senate.

Not surprisingly,... activists see a political opportunity: they want a new labor law that will weaken the posiiton of more conservative trade union leaders.

The Brazilian people decided to give Lula their mandate after two decades of military dictatorship (1964–1985), a return to democracy followed by the sudden death of the civilian president, a grueling presidential impeachment process in the early nineties, and eight years of the administration of Fernando Henrique Cardoso. Cardoso's administration succeeded in reducing inflation, following the prescription of the Washington Consensus, but it also produced growing inequality, higher unemployment, greater violence, and an astronomical increase in debt and interest rates. Much of Brazilian civil society, including business, opted for Lula because of his commitment to greater growth and social stability. Brazilian textile magnate José Alencar of the Liberal Party (PL), Lula's running mate and now vice president, represents the strong belief among national industrialists that macroeconomic stimulus is desperately needed.

Now Lula faces the challenge of managing and satisfying the soaring expectations of his supporters. For example, the landless movement, or Movimento Sem Terra (MST), responsible for many of the rural conflicts and occupations of the last decade, expects sweeping agrarian reform to overturn a system in which less than 3 percent of the population control more than 60 percent of the arable land. Gilmar Mauro, one of the top leaders of the MST, is demanding that the new government immediately expropriate "all lands belonging to owners who owe money to the state, as well as all lands left fallow." He insists that titles must be awarded now to at least a hundred thousand families currently occupying unproductive farms, and promises that the landless movement will continue to pressure the Lula administration until all its demands are satisfied.

A number of union leaders who began as young labor activists in the *novo sindicalismo* movement of the 1970s and 1980s and were founders of both the PT and the Central Única dos Trabalhadores (CUT), Brazil's largest trade union center, representing more than twenty-two million workers, have long advocated a major overhaul of the state-corporatist system of labor relations and union governance. With the presidency of the country now occupied by a metalworker who once challenged a military regime relying on state control of the union movement to repress collective worker action, these unionists say that

it is high time to end labor's dependence on the state for support of union structures. Not surprisingly, these activists see a political opportunity: they want a new labor law that will weaken the position of more conservative trade union leaders.

The Brazilian labor movement will also expect Lula to make good on his announced intention to create jobs. By the end of the 1990s, the official unemployment rate (which does not reflect chronic underemployment and the high number of Brazilians permanently discouraged from seeking work) exceeded 7 percent. In São Paulo, Brazil's largest metropolis, 20 percent of the working population is unemployed, making for two million people without jobs. And the number of unregistered, informal workers in the Brazilian economy rose from 42 percent to 58 percent of the total labor force in the last ten years, with more than 4.3 million jobs being eliminated in the formal sector.

Excessive fiscal conservatism and high interest rates will only stanch growth and erode the support from national business on the right and from unions and social movements on the left.

And the Brazilian majority expects Lula to reform the social security regime. The current public system of retirement and disability benefits actually covers only 57 percent of the eligible beneficiaries. It has generated an unsustainable deficit of more than seventy billion *reais* (well over twenty billion dollars), exacerbating Brazil's debt crisis and causing interest rates to climb. The growth of the informal sector erodes the tax base and cuts revenue. Moreover, certain public servants, including top military officers, high-ranking judges, and elite personnel in the executive branch, benefit disproportionately from the system. All told, public sector employees represent 11 percent of social security beneficiaries and receive 45 percent of the benefits.

Social Security Minister Ricardo Berzoini, former president of the São Paulo Bank Employees Union, has called for a more uniform system in which elite public servants will not receive inordinately more than other workers. Although Berzoini is not proposing to make the more egalitarian disbursement retroactive, many military officers and judges argue that any reform will interfere with their "acquired rights." And the leaders of the CUT–affiliated federal employee unions, many of whom supported Lula's electoral campaign and are among the most left-wing activists in the Brazilian labor movement, will resist tampering with the current system.

In addition to the expectations of domestic constituencies, Lula must confront the antagonistic pressures of external interests and institutions. Precisely because the new government will not be a pushover on the FTAA, multinational enterprises and global capital, along with their friends in Washington, will have every motivation to isolate and control Brazil. And with the accumulated public debt rising from 29.2 percent of gross domestic product in 1994 to well over 55 percent at present, a total of approximately 250 billion dollars, the Lula government is particularly vulnerable to the demands of international creditors and financial institutions. The International Monetary Fund's thirty billion dollar loan package negotiated at the end of last year with President Cardoso (once a left-wing sociologist who told big business during his successful 1994 electoral campaign "to disregard everything that I published") imposes high primary budget surplus requirements that may hamstring Lula's program.

In an obvious effort to show both Wall Street and the international financial institutions that the Lula administration will be fiscally responsible, Finance Minister Antônio Palocci has announced a 2003 primary surplus goal of 4.25 percent of GDP, surpassing the IMF's demand of 3.88 percent. In addition, Central Bank Governor Henrique Meirelles, a former chief executive of FleetBoston, has resisted lowering interest rates in an effort to avoid runs on Brazil's currency and an inflationary spiral. (The consumer price index for São Paulo revealed a 2.19 percent increase in January, the highest recorded for any January since 1995.)

Unfortunately, not all the financial analysts are impressed. Although the rating agency Standard and Poor's has praised Palocci and Meirelles for a "prudent and cautious policy stance," it also sent a poisonous message to the markets on January 16, asserting that the Lula administration has "limited room for policy maneuverability in a challenging global and domestic environment."

The new government's understandable attempts to avoid subversion by international finance also threaten to undo Lula's plans for macroeconomic recovery and social justice. Excessive fiscal conservatism and high interest rates will only stanch growth and erode the support from national business on the right and from unions and social movements on the left. Already, leftist militants in the PT have accused the Lula government of selling out to the IMF and the markets.

Academics and journalists have been mixed in their appraisal of Lula's capacity to deal with the formidable domestic and international challenges he faces. *Newsweek* posed the question last October, "Can Lula really lead?" and cited a Wall Street source who claimed that the new president was thoroughly "untested." Peter Hakim, from the Inter-American Dialogue, in a *Washington Post* op-ed, asked whether Lula is the "champion for Brazil's poor majority," challenging "the country's power brokers" and battling "its deep social injustices"—or is he "newly moderated," accepting "market economics" and "pluralist politics"?

They [the PT] succeeded in constructing a political force that is thoroughly open, ecumenical, non-doctrinaire, mass-based, and internally democratic.

In an article appearing in the *New York Review of Books* last December, Kenneth Maxwell of the Council on Foreign Relations claimed that Felipe Gonzalez and the Spanish Socialist Workers Party represent "one of the models for the Brazilian PT," having also managed to shake off a Marxist past and "move to the center ideologically." According to Maxwell, "perhaps 30 percent of the PT call themselves radicals... but most party members have learned to play the democratic game."

Right-wing commentators, such as Constantine Menges of the Hudson Institute, claim to know about a sinister plot being engineered by Lula and the PT that will give new life to the international communist conspiracy. Although Menges's published musings border on the psychotic, they certainly influenced Representative Henry Hyde, chair of the House International Relations Committee. Hyde wrote to President Bush on October 24, warning that Lula was a dangerous "pro-Castro radical who for electoral purposes had posed as a moderate," and likely could form "an axis of evil in the Americas" with Cuba and Hugo Chavez's Venezuela. Such a diabolical alliance would have ready access to Brazil's "30-kiliton nuclear bomb" as well as its "ballistic missiles."

The problem with most of these opinions is that they ignore the remarkable twenty-five year political development of Lula, the PT, and Brazil itself. The trade unionists, human rights advocates, progressive clergy, academics, and other opponents of the military dictatorship who founded the Workers Party more than twenty years ago were dedicated to building a new and unprecedented movement, which would break radically from the Brazilian tradition of top-down populist and *caudilho* politics. They succeeded in constructing a political force that is thoroughly open, ecumenical, non-doctrinaire, mass-based, and internally democratic. Although the PT accepted members who claimed allegiance to revolutionary ideals (many of whom are currently excoriating the government for compromising its socialist principles), it also rejected Leninist vanguardism and "democratic centralism" as defined by the orthodox communist left. PT militants did not simply "learn to play the democratic game" in Brazil; for all intents and purposes, they *created* it.

In a 1991 press interview in Mexico City, Lula made it clear that the PT never accepted the Soviet model, nor did it have to "renounce" a dogmatic past: "From its birth, the PT criticized Eastern Europe. We criticized the Berlin Wall, state bureaucracy, the absence of union freedom; we defended Walesa from the moment we were founded."

As Johns Hopkins political scientist Margaret Keck has correctly observed, the PT's socialism was never a dogmatic ideology, but rather an "ethical proposal, within which a number of alternative visions of the good society competed… " For Keck, the party's moral aspiration for socialism has been "an aspiration for democracy."

It is also misleading to imply that a "newly moderated" Brazilian Workers Party has only recently learned to deal with "pluralist politics" and "market economics" and that Lula is "untested." Over the last twenty years, the Workers Party has built a successful and corruption-free record of governing hundreds of cities, including the world's third-largest, São Paulo, as well as several important state governments. In addition, hundreds of able PT activists have served in the Brazilian Congress, including Lula himself. None of this could have happened had the PT not respected the principles of democratic pluralism and coped with the realities of a market economy.

As for the "axis of evil" theory, Menges, Hyde, and other fearmongers should be reminded that Lula has spoken out forcefully against the spread of nuclear arms and has promised Brazil's total compliance with the Non-Proliferation Treaty. Like many other Latin American leaders and statesmen, Lula maintains constructive relations with the current leadership in Cuba and Venezuela. However, he has said that Castro needs to encourage internal democracy, and he has admonished Chavez for "acting like a military officer" rather than "a civilian politician" when dealing with the Venezuelan opposition. Although the anti-Chavez forces have denounced Brazil for sending oil and humanitarian assistance to Venezuela, Lula's special secretary for foreign relations, Marco Aurelio Garcia, recently assured me that the aid is meant to encourage peaceful negotiations and an outcome that will respect Venezuela's democratic institutions and rule of law.

The Lula administration has so far demonstrated good faith in responding to Brazil's domestic constituencies and their demands. The new government has convened a National Social and Economic Council that directly involves representatives from all segments of Brazilian civil society-national business, the trade unions, the religious community, cooperatives, and civil and human rights organizations—for the purpose of negotiating an inclusive social pact. By the end of 2003, the Executive will have presented the Council's proposals to the Congress that address the country's most pressing issues, including social security, tax, and agrarian reform. In addition, the new government is convening a National Forum on Labor to review all the employment, labor rights and labor relations concerns facing Brazil. After extensive debate and negotiation among government, trade union, and business representatives, the Forum will also have its proposals delivered to the Congress by the end of the year.

This inclusion of Brazilian civil society in serious negotiations is something that Lula's predecessors were both unwilling and unable to do. Without question, the task is not an easy one. CUT President João Felicio has already made it clear that social pact discussions do not mean that labor will give in on wages and working conditions. And speaking for the interests of capital, Alencar Burti, president of the São Paulo Chamber of Commerce, announced, "I am not interested in talking about what I'm supposed to give up." Nevertheless, by involving all representative groups in the decision-making process, the new government is engaging in constructive dialogue with Brazilian society, as well as with its own current and potential opposition. Lula claims that agrarian reform will go forward during his administration precisely because he and his government will actually sit down with the MST and negotiate.

Several of the new government's ministers have proposed specific public policy initiatives. Labor minister Jaques Wagner is considering reducing the standard workweek from forty-four to forty hours in order to increase employment. Wagner has spoken of targeted tax incentives and subsidies to employers who hire more workers by reducing hours without cutting wages and benefits. Economist Paul Singer, recently named Secretary of the *Economia Solidária* (Solidarity Economy), is proposing the massive creation of service, family farm, labor, and microcredit cooperatives as a means of putting thousands of Brazilians back to work.

Although the IMF's draconian pressures and Wall Street's insatiable demand for "fiscal responsibility" could spell political and economic disaster, there are several factors working in Brazil's favor. For example, World Bank President James Wolfensohn has openly praised the new government, and has committed funding to Jaques Wagner's youth employment programs and to Lula's anti-hunger campaign. The plans to multiply domestic credit sources could help reduce the dependence on foreign investment. And the Argentine tragedy should have taught a serious lesson to the world's financial markets and institutions: if Lula is not allowed to bring a minimal amount of growth and social stability to Latin America's largest economy, the prospects for security in the hemisphere are bleak.

The international union movement can play an important role in helping to ameliorate some of the external pressures. For example, workers' pension funds are responsible for literally trillions of dollars in financial markets and can exert substantial influence when it comes to Wall Street's relationship with Brazil. In addition, the labor-friendly social development projects of the Lula administration suggest interesting partnerships with international workers' capital.

More direct contacts between unionists from the United States and Brazil, as well as exchanges involving city and state governments, small agricultural producers and family farmers, educators, musicians, artists, and progressive legislators, can only enhance North American support for Brazil's new democratic hope, as well as help to counter any revanchism on the part of the Bush administration. And the AFL-CIO and its affiliates have a promising opportunity to work with the Lula government to guarantee labor rights to the more than one million Brazilians living and working in the United States.

Lula is well aware of the importance of international union solidarity to his success, as he made clear to hundreds of admirers from the U.S. labor movement who were present at AFL-CIO headquarters to receive him during his visit to Washington last December. He deserves, and needs, many more friends. For the real question is not whether "Lula can lead" or whether "Lula can succeed." The real question is whether the world can afford for Lula to fail.

STANLEY GACEK is a labor attorney and AFL-CIO International Affairs Assistant Director, responsible for the Federation's relations with Latin America and the Caribbean. He has spoken and written extensively on Brazilian labor and politics and has been a friend and adviser to Lula and the PT for more than twenty-two years.

Lula — New Hope for Brazil
— World's 5th Largest Country (Area/Pop)
—   "   9th Largest Economy

Lula Era
① Sustainable Economic growth
② Reduce Poverty & Inequality
③ Ending unemployment
④ End Hunger

— Opposition to FTAA — Sees This as Economic Annexation of the South to the North

— Decades of Military Dictatorship (1964-1985)

Cardoso Gov't
— Inflation
— Inequality
— Violence
— Increasing Debts / Interest Rates
— Less Than 3% of Pop owns 60% of Land — Landlessness
— Vulnerable to International Creditors & Financial Institutions
— Lula's support Base; Working class, Small Farmers / Trade Unions
— 1 million Brazilians living / working in U.S. — Migration

Update — Corruption Threatening The Gov't

Chile, the Rich Kid on the Block (It Starts to Feel Lonely)

By Larry Rohter

SANTIAGO, Chile—Achieving a free trade agreement with the United States was supposed to be the magic moment that certified Chile's entry into the elite club of stable, democratic and prosperous nations. Instead, the new accord, signed in January, has reignited a sometimes anguished debate here about what it means to be Latin American and whether Chile has somehow lost those essential characteristics.

Since the beginning of the decade, all three of Chile's neighbors have suffered political and economic convulsions that have forced changes of government. In sharp contrast to Argentina, Bolivia and Peru, not to mention the rest of South America, Chile these days looks "dull but virtuous," to borrow the title of a recent report by one Wall Street brokerage house.

This is a country where most people actually pay their taxes, laws are rigidly enforced and the police only rarely seek bribes. That is unusual for Latin America and probably should be cause for celebration. Yet, it has the rest of the region looking at Chile as if there is something wrong with it because it is not what the Brazilians call "bagunça" or what the Argentines call "quilombo"—passionately messy, turbulent and chaotic.

"The image of Chile for many years has been that of a country that is 'different' and solitary," the Peruvian political commentator Álvaro Vargas Llosa, wrote in a recent essay that generated much comment here. "Curiously, although Chile has undertaken a growing trade with the world and attracted investments, it was perceived as 'isolated' in a space that is psychological more than political or economic."

Chileans were abruptly reminded of their unpopularity late last year, when an uprising in neighboring Bolivia overthrew a president who favored exporting natural gas through a Chilean port. Bolivia has been landlocked since losing a war to Chile in 1879, and Chileans were shocked to discover that many Latin Americans—led by the Venezuelan president, Hugo Chávez—supported Bolivia's historical claim to recover a piece of its coastline.

It has been this way for at least a generation, Chile as South America's odd man out. When Salvador Allende came to power in 1970, the rest of the Southern Cone was ruled by right-wing military dictatorships, and by the time those countries began to swing back toward democracy in the 1980's, Gen. Augusto Pinochet was in power.

Chileans gained sympathy during the Pinochet dictatorship, when "solidarity with Chile was a cause that people in Latin America believed in, somewhat like Spain was in the late 1930's," said the writer Ariel Dorfman. But that changed when democracy returned, he said. "Many people resented how Chileans, especially our entrepreneurs, behaved, buying up all the banks and telephone and electricity companies and acting arrogant and obnoxious."

Today, Chile is a hypercapitalist state at a time when Brazil, Argentina, Venezuela, Ecuador and Uruguay are all moving leftward and questioning free trade and open markets. Chile may also be suffering from what might be called teacher's pet syndrome. Since the 1980's, other Latin American countries have had to endure repeated lectures from the United States, Europe and Japan on the need to become more like Chile in opening their economies to the outside world and combating corruption.

"For some mysterious sociocultural reason we have been 'the serious kids in the neighborhood,'" Mario Waissbluth, a business consultant here, acknowledged recently in an essay for the newspaper La Tercera. "We are the nerd student who does all of his homework and is loathsome to the rest" of his classmates.

"Obviously it's better to be dull and virtuous than bloody and Pinochetista, but Chile has been a very gray country for many years now," said Mr. Dorfman, one of whose novels, "The Nanny and the Iceberg," deconstructs Chile's "cool and efficient" image of itself. "Modernization doesn't necessarily have to come with soullessness, and I think there is a degree of that happening."

The Free Trade Agreement with the United States has aggravated all these contradictory sentiments. "Everyone, from Peru to Colombia, is fighting tooth and nail to get the same status," Mr. Vargas Llosa wrote. "But it pains them that Chile has got there ahead of them."

Government officials here say they are conscious of their spotty image and are moving to mend fences with their neighbors. But their initiatives have been limited mainly to trade-oriented efforts like dispatching delegations to Brazil and Central America to advise on what to expect in negotiations with the United States.

"My concern is that of Chileans being seen as the new Phoenicians of Latin America, just good at trade," said María de los Ángeles Fernández, director of the political science program at Diego Portales University here. "That is not sufficient to have good relations. You need other means of communication."

Simultaneously, though, Chile is also looking to establish alliances beyond its own troubled neighborhood. A free trade agreement with the European Union went into effect last year, a similar accord with South Korea has been ratified, and in November Chile will play host to a conference of the 21-member Asia-Pacific Economic Cooperation conference—an effort to strengthen its identity as part of the equally dull but virtuous Pacific basin.

"We are trying to do everything within our reach to integrate ourselves" with the world beyond Latin America, Ricardo Lagos Weber, the main organizer of the conference and the son of the president, acknowledged in an interview. But there is also a recognition here, he added, that Chile "can't be an enclave of modernity surrounded by poverty, instability and bad vibes" without suffering the consequences.

South America's New Leaders with Muscle

Kirchner Reorients Foreign Policy

Argentina's new president, Néstor Kirchner, has formed an ambitious common cause with Brazil's President Lula da Silva. Their goals: to expand the alliance and, hence, the political and economic clout of the region's two giants and revive the Southern Common Market to act as a counterweight to U.S. influence in the region.

BY MARTÍN RODRÍGUEZ YEBRA

A SENTENCE THAT NÉSTOR KIRCHNER is fond of repeating illustrates precisely the path Argentina wants to take vis-à-vis the world: "Politics drags the economy, and not the other way around, as they had us believe for years." Last week, Kirchner outlined a new direction for foreign policy, which he considers a cornerstone of his administration. He first met Tuesday [June 10] with U.S. Secretary of State Colin Powell at the Casa Rosada [Argentina's presidential palace]. It was an hour of frank dialogue, which produced the conclusion: The government wants help to negotiate the foreign debt, but it first needs internal stability.

Powell took note and promised to relay the idea to President George W. Bush. Kirchner was impressed by the United States' chief diplomat and emphasized a concept that Powell had put forth during the conversation. "Brazil is a leader on account of its size; Chile is a conceptual leader. You should find a different formula that sets you apart," stated Powell at the meeting. That idea captured the imagination of the government's front line. Kirchner believes Argentina should end its automatic alignment with the United States and that his country can play a fundamental role in Latin America, especially if it can bring about serious integration with Brazil and make radical changes in its public sector. With those ideas in mind, Kirchner left for Brasília a couple of hours after Powell had flown to Washington.

On Wednesday, Kirchner met for the third time in a month with [Brazil's] President Luiz Inácio Lula da Silva, whom he enthusiastically calls his "new friend." They spoke alone for an hour and dedicated another three to discussing the principal themes of the bilateral agenda with their ministers. Afterward, in the presence of the entire local press, the two presidents announced that the decision to politically integrate Mercosur [the Southern Common Market, comprising Argentina, Brazil, Paraguay, and Uruguay] "was irreversible" and that the deadlines for making good on that long-standing promise would be shortened. "The union between Argentina and Brazil will encourage the other countries in the region to realize the dream of continental integration," said Lula. Kirchner echoed Lula to the word. Kirchner's visit to Brazil, the first of his term, bore the desired fruit: a strong gesture of alliance with Brazil and a handful of pronouncements of diplomatic impact, such as the will to present a united front in negotiations to enter the Free Trade Area of the Americas (FTAA) and to form a Mercosur parliament.

"This time it's for real," said [Argentina's Foreign Minister] Rafael Bielsa at the end of the summit. In his first meeting with Brazil's Foreign Minister Celso Amorim, he had used an ironic method to question the way bilateral relations had been managed during recent decades. He asked his interlocutor to listen to him for five minutes without interrupting and he read a selection of verbs culled from all the documents signed within Mercosur: to tend toward, to promote, to coordinate, to endeavor…. The list was interminable. "We must do away with rhetoric and move on to action," he said.

Lula's position seemed in line with Kirchner's foreign-policy plan: With political fortitude, economic objectives will be attained more easily. "If we tarry with commercial conflicts, we will continue to move backward," Lula maintained during the working meetings. Lula and Kirchner's synchronicity impressed some Argentine diplomats. "Before, when we would speak with Brazilian authorities, we had three litmus tests: what is said, what is not being said, and what they want to tell us. Now there is only one message," explained a presidential spokesperson. The government knows that it cannot yet count on Brazil to negotiate with international institutions while Argentina is in default. However, regional political strength will give impetus to the national position, Kirchner believes.

The same concept holds for the intention to expand Mercosur: Lula and Kirchner spoke of enticing Peru and Venezuela. And they hope Chile will finally join as a plenary member, although that nation signed a free-trade agreement with the United States this month. "The government of Chile is acting with great generosity," a source close to Bielsa pointed out.

During his meeting with Powell, Kirchner outlined his idea of a renegotiation of the multilateral debt tied to trade surplus and growth, with extended deadlines and a cut in interest rates. The United States promised concrete aid at the meeting that Economy Minister Roberto Lavagna held Thursday in Washington with the No. 2 official at the Treasury Department, John Taylor [undersecretary for international affairs].

The time for effective negotiation is dizzyingly near: In September, Argentina will again face million-dollar due dates with international lenders. Lavagna appears to be motivated by the philosophy that gives primacy to politics over economics. In fact, he was a pioneer of that strategy with the International Monetary Fund (IMF) during Eduardo Duhalde's government.

The following weeks will prove decisive for the success of that strategy. Wednesday, Brazil and Argentina will set forth in Asunción the relaunch of Mercosur (aside from what has already been negotiated). They will speak of promoting a harmonious regimen of public services, coordinating a plan of infrastructure integration between countries, and creating a system of joint purchasing, social cooperation, and significant immigration reforms.

A week later [June 23], Horst Köhler, the IMF's head, will arrive. He is one of the most stubborn opponents of the new Argentine position but at the same time the holder of a key that may open the doors to an improved financial future.

Kirchner vows to stay on his chosen path, although he is suspicious of his foreign ministry. "Half the foreign ministry plots against me," he said recently in private meetings. But he repeatedly insisted: "I cannot turn back."

— Conjoining of Political & Economic Policy Making
— Expansion of Mercosur (Venezuela, Chile, Peru)
   — Regional Economic Strength
   — Regional Political Strength
   — Continental Integration
   — United Front in FTAA Negotiations
—
— Renegotiate Debt.

Europe and South America Near Trade Accord

By TODD BENSON

SÃO PAULO, Brazil, April 19— South America's biggest economies moved a step closer on Monday to sealing a long-sought trade deal with the European Union, a shift that could complicate efforts to forge a free trade area stretching from Alaska to Argentina by the end of the year.

Officials from the European Union and Mercosur—the trade bloc that includes Argentina, Brazil, Paraguay and Uruguay as full members, and Bolivia and Chile as associates— exchanged offers over the weekend in Buenos Aires with an eye on closing the deal by October. The talks also managed to make progress in agriculture, a sector that has proved to be a deal breaker repeatedly in global trade talks.

"We improved our offer substantially, and they are complying by responding to our top priorities, primarily in agriculture," said Regis Arslanian, Brazil's chief negotiator in the talks with the European Union.

The rush to expand trade with Europe gained a new urgency for the South American bloc earlier this month after separate talks in Buenos Aires to start a 34-nation Free Trade Area of the Americas ended in an impasse for the third time in two months.

The stalemate prompted negotiators to postpone indefinitely a pan-American trade summit scheduled for later this week in Puebla, Mexico, casting doubt on the likelihood that a free trade zone, including all countries in the Western Hemisphere except Cuba, can be reached by a target date of January 2005.

"Two years ago, nobody thought that the talks with the European Union were serious—everyone said that the F.T.A.A. would come first," said Marcos Sawaya Jank, president of the Institute for International Trade Negotiations, or Icone, a research group in São Paulo."What's so surprising is that now it's the other way around."

Fearing that a trade bloc in the Americas might leave European goods and companies at a disadvantage in South America, the European Union began courting the Mercosur countries—especially regional heavyweights Brazil and Argentina—about five years ago.

Disputes over agriculture—an area in which the South American countries are most competitive and where Europe is most protectionist—have hobbled the talks from the outset. But the European Union wooed Mercosur back to the table last weekend by offering some concessions in agriculture.

European officials refused, however, to discuss farm subsidies, a sore spot also slowing the talks for the Americas. Instead, Brussels proposed a two-step approach by which it would initially offer Mercosur more generous import quotas for agricultural goods like beef, dairy products, sugar and instant coffee. The remaining quotas would later be divided up in the continuing Doha round of negotiations at the World Trade Organization.

In exchange, Mercosur offered the European Union privileged access to new investments and service sectors, particularly telecommunications and banking. The South American bloc also pledged to speed the phasing out of import tariffs on European goods and provide greater legal protections to foreign investors.

Negotiators from both sides are scheduled to meet in Brussels in the first week of May to iron out any wrinkles, and a blueprint of a deal could be introduced as early as May

28 at a European Union-Latin America summit in Guadalajara, Mexico.

The apparent shift away from a pan-American pact in favor of Europe also reflects domestic politics in Brazil and Argentina, where two left-leaning governments have sought to avoid being seen as overly friendly to Washington. Luiz Inácio Lula da Silva, president of Brazil, even warned while he was on the campaign trail that the Free Trade Area of the Americas could mean Brazil's "annexation" by the United States if negotiators were not careful, though he has since toned down his rhetoric.

And with the presidential campaign in the United States already latching on to widespread fears of jobs migrating abroad as a result in part of free trade agreements, the hemispheric pact is facing stiff winds in Washington as well.

"There's no question that there is not much fire in the belly in Washington for any further trade arrangements this year," said Riordan Roett, director of the Western Hemisphere Program at Johns Hopkins University.

"And it's no mystery that the current administration in Brasília would much prefer to close the E.U. agreement this summer or early fall," Mr. Roett said. "It will be a relatively light trade agreement, but symbolically and diplomatically it's going to be very important."

Mercosur is also seizing on the chill in the Americas trade talks to warm up to another of Washington's closest trade partners—Mexico. On a trip to Argentina last week, the foreign minister of Mexico, Luis Ernesto Derbez, said Mexico had asked to join the South American trade bloc.

Still, the United States, which is the co-leader of the pan-American negotiations with Brazil, says it is confident that a deal can be reached by next year's deadline. "We are working hard with others in the F.T.A.A. process to fulfill our mandate," Richard Mills, a spokesman for the United States trade representative, Robert B. Zoellick, said on Friday.

In Brasília, the mood is less upbeat. While Brazilian officials insist that they have not lost hope in the hemispheric talks, they say it is up to the Bush administration to break the deadlock.

After the negotiations hit a road block earlier this month, Brazil's foreign minister, Celso Amorim, sent an unusually harsh letter to Mr. Zoellick in which he suggested that the Americans had poisoned the talks. In the letter, Mr. Amorim criticized the United States for insisting on "side issues" instead of moving on to "mutually beneficial market access negotiation."

According to Adhemar Bahadian, co-chairman of the talks with the Americas and Brazil's consul-general in Buenos Aires, "If the United States doesn't take a more flexible approach, then it's obvious that there won't be an agreement."

Contraband Is Big Business in Paraguay

By TONY SMITH

ASUNCIÓN, Paraguay—Judging by this country's trade figures, Paraguayans would seem to be a nation of ferocious chain smokers.

Last year, the tobacco industry here produced about 45 billion cigarettes, worth about $600 million all told. But according to the central bank's records, only about 300 million cigarettes, or $4.3 million worth, were exported. That leaves about a pack a day to be smoked by each Paraguayan man, woman and child.

In reality, of course, most of the cigarettes really are exported—just not legally. Industry experts say that nearly 95 percent of the country's output, including counterfeit versions of American brands like Marlboro and Camel and Latin American favorites like Derby and Free, are smuggled through Paraguay's porous borders with Argentina, Bolivia and Brazil and then on to destinations as distant as the Caribbean, the United States and Mexico.

"Paraguay's tobacco exports are, almost without exception, illegal," said Milton Cabral, vice president of Souza Cruz, the Brazilian subsidiary of British American Tobacco.

The situation is typical of many industries here. About one-fifth of the Paraguayan economy has been driven for years by illicit cross-border trafficking in everything from cigarettes and pirated Nintendo games to submachine guns and stolen BMW's.

But now the country has a president-elect—Nicanor Duarte Frutos, who won handily in April and is to take office in August—who has promised to clamp down on "the chronic problems" of counterfeiting and contraband and "build a system of legality," as he put it in remarks he made to reporters on a recent visit to Brazil.

Even if he is to be believed—and there are already signs his zeal for reform might not survive long once he is in office—Mr. Duarte Frutos will have a tough job weaning the economy off contraband, even though most Paraguayans agree it is increasingly urgent that the country kick the habit.

Paraguay is one of the poorest countries in Latin America and getting poorer. Income per person has fallen by half in the last six years, to less than $900 a year, and more than one-third of the population lives in abject poverty. With the public payroll bloated with 200,000 government employees, nearly all affiliated with Mr. Frutos's Colorado Party, the country's finances are draining fast.

The government has been especially short of cash in the last 12 months, as the money from a $400 million loan from Taiwan has started to run out, obliging the central bank to print fresh stacks of increasingly worthless Guaraní notes, sending inflation spiraling up from the single digits to about 17 percent this year. Nearly one-fifth of the work force is jobless.

The lack of funds has stopped public-sector investment in its tracks, and is effectively stalling multilateral loans to agribusiness because the Paraguayan state cannot afford its part. That industry's hard-currency earnings had been one of the legitimate economy's only real bright spots.

"We are staring what could be a notorious social collapse right in the face," said Luis Campos, economist and partner at KPMG's local affiliate. "And the crisis has its epicenter in the government's fiscal problems."

Mr. Campos estimates that annual government revenues have slid from more than $1 billion in 1996 to just $700 million today, which could explain why Mr. Duarte Frutos is keen to clamp down on what is politely called the informal economy.

Putting a precise figure on the size of that economy is difficult, naturally. Most economists estimate it at one-fifth of Paraguay's $7 billion in gross domestic product, twice the proportion in the United States.

And as a source of fake goods or as a transshipment point between Asia and Latin America for gray-market computer components or for counterfeit high-fashion eyeglass frames, the country is a source of substantial losses to major companies worldwide as well.

Among those eager for a crackdown are executives at the Nintendo Company. Since 1998, when Paraguay signed a bilateral trade agreement with the United States, more than 4.4 million counterfeit Nintendo products have been seized in Paraguay, more than anywhere else in the Western Hemisphere, said Jodi Daugherty, chief piracy fighter at Nintendo's American unit.

Ms. Daugherty said she could not say exactly how much of the $649 million that Nintendo calculates it lost to piracy globally in 2002 could be attributed to Paraguay, but she called it "a serious business number."

The company has gotten scant relief from the inefficient and corruption-plagued Paraguayan legal system. Since 1997, Nintendo has brought 36 cases in Paraguayan courts over counterfeiting or smuggling of its goods. None have been resolved yet.

Neighboring Brazil also has a big informal economy— for example, two-thirds of the personal computers sold there are supplied by the gray market. The components are made mainly in Asia; they come in through Paraguay.

"We all suffer," said Reinaldo Opice, an executive at *Microsoft's* Brazilian unit. "If we managed to cut piracy by just 10 percent, the industry would grow by about $17 billion in four years."

So far, Mr. Duarte Frutos's pledges to crack down on cronyism and smuggling, and to "build a much more respected judicial system," have caused more raised eyebrows than optimism. After all, his Colorado Party has been in power since 1947, longer even than the Communists in North Korea.

Among the first people to congratulate Mr. Duarte Frutos the morning after the election at his heavily guarded home in Asunción was Senator Julio Dominguez Dibb. Mr. Dominguez Dibb's father, Oswaldo, is Paraguay's richest tobacco baron and is under investigation for trying to copy and register as his own trademark the Brazilian customs seal that must be placed on every pack of cigarettes sold in that country. He has denied any wrongdoing.

Soon after, Alejandro Nissen, one of the attorney general's most dogged investigators, was suspended without pay. Mr. Nissen revealed in 2001 that the current president, Luis Gonzalez Macchi, was driving around in a BMW limousine stolen from *Johnson & Johnson* in Brazil.

"So far I have seen no change in attitude despite the new incoming government," said Mr. Nissen, who said he believed that he was now paying the price for accusing Oscar Gonzalez Daher, another Colorado senator and an important ally of the new president, of also having a stolen car— a shiny red Mercedes Benz.

According to many analysts here, as long as Paraguay's shady business culture continues, it will prevent the country from focusing on what it does best: agriculture.

Thanks largely to new techniques and investments brought by the 450,000 "Brasiguayos"—Brazilian farmers and their Paraguayan-born descendants who have populated regions along the border—Paraguay has become the world's fourth-largest soy exporter and a major regional beef producer.

Brazil has other interests in Paraguay as well. It is by far the leading source of foreign investment in the country, with $1 billion in agribusiness, banking and a handful of "maquila" assembly plants that export legally to Brazil.

"We have to try and give Paraguay an alternative" to smuggling and counterfeiting, said Luiz Augusto de Castro Neves, Brazil's ambassador in Asunción. "Otherwise it could become a colony of evil, a paradise for drug dealers, money launderers and gun runners."

Indigenous Bolivians Are Rising Up And Taking Back Power

By Hector Tobar

SORATA, Bolivia—The police won't return to this village in the Andes unless the peasants promise not to throw rocks at them.

The peasants rose up and chased the police out months ago, along with the local representative of the provincial government, the judges and even the army. The authorities fled Sept. 20 in the face of a crowd of Aymara Indians armed with little more than sticks and stones, enraged by an insult uttered by an army general hours earlier, and moved by centuries of pent-up frustration.

Since the uprising, this corner of Bolivia—where the dry Altiplano, a high plateau, around Lake Titicaca meets lush tropical mountains—has become a kind of an Indian liberated zone.

"Before, they were the bosses. They made us work, they would run everything," said Felix Puna Mamani, a resident of the neighboring village of Viacha, referring to the people of European descent who have dominated Bolivian society since the 16th century Spanish conquest.

As many as 1.5 million people—almost a fifth of Bolivia's population—live in areas where indigenous authorities have replaced at least some government functions, said Alvaro Garcia Linera, a university professor in La Paz who has studied the popular movements of Bolivia's two main indigenous groups, the Aymara and the Quechua.

"Since 2000, we have seen an enormous, continual uprising of indigenous people, with a strong element of Indian nationalism," Garcia Linera said. "In many places, the institutions of the Republic of Bolivia have begun to fade away."

Bolivia's new president, Carlos Mesa, is attempting to lead a sharply divided country and a democracy teetering on the brink of collapse a generation after the last in a se-

ries of military dictators stepped down. His government has shown little inclination to confront the peasants.

Guido Arandia, the chief of police for La Paz department, says his officers won't go back to Sorata and the other towns in rebellion unless they are welcomed. "It's not that we don't want to return," Arandia said. "But as long as there are no guarantees from the community and its leaders, we cannot place our people at risk."

In Sorata as in other towns of the region, the locally elected mayor and City Council remain in office—most of them are Aymara speakers and appear to have the support of the town's non-Indian minority. But Sorata's connection to the federal and provincial governments in La Paz remains tenuous at best. The City Council recently considered a hunger strike to force provincial authorities to free up education funds.

> *"In many places, the institutions of the Republic of Bolivia have begun to fade away."*
>
> **PROF. ALVARO GARCIA LINERA**

With the police gone, "peasant union police" are the only forces of order. They wear the tasseled chicote staff that is an ancient Aymara symbol of village authority.

The Aymara are redistributing land in communal assemblies called Open Councils that issue edicts in the mode of government pronouncements. At some public schools, the rainbow-colored Indian wipata flag flies in place of the Bolivian flag.

People in villages such as Sorata feel that Bolivia's highly centralized government has failed them. Even before the uprising, the long, slow decline of that government—which seems more cash-strapped and corrupt with each passing year—had led the Aymara to rely more on their own, pre-Columbian forms of communal rule.

In the face of the central government's broken promises, City Council members and Indian leaders in Sorata and elsewhere routinely organize "methods of pressure," such as blocking roads. Since 2000, such tactics have become commonplace throughout a wide swath of South America, from Lima, Peru, to the northern Argentine provinces of Jujuy and Salta.

In the Aymara villages of the Altiplano, the most vocal leader of the rebellion is Felipe Quispe, a heavyset former professor and president of the peasants union. Known to the Aymara as "el Malku," or the Condor, he has promised to set off a guerrilla war if his demands on behalf of Bolivia's indigenous people are not met.

The Los Angeles Times is a Tribune Co. newspaper.

AYMARA — TAKING MATTERS INTO THEIR OWN HANDS
- GROWING INDIAN NATIONALISM
- LAND REDIST.
- Local SELF-govt
- Rejection of Colonial Legacy

China Fuels Brazil's Dream of Being a Steel Power

By TODD BENSON

(handwritten: Brazil's Dependence on overseas markets)

SÃO PAULO, Brazil, May 20—Blessed with abundant raw materials, cheap labor and top-notch technology, Brazil is looking to become a major player in the global steel industry with the help of some deep-pocketed foreign friends.

Despite recent efforts in China to cool its red-hot economy, Brazil's steel industry is racing to keep up with the demand there, investing billions of dollars to increase its production capacity by more than 30 percent in the next four years. While the bulk of the financing is expected to come from Brazil's own steel industry and local banks, international steel companies are also getting in on the action to guarantee a steady flow of resources back home.

Leading the way are global heavyweights like the Baosteel Shanghai Group of China and Arcelor of Luxembourg, both of which are still planning to build blast furnaces to churn out steel in Brazil. The Dongkuk Steel Mill Company of South Korea and the Italian metals group Danieli are also helping to finance the construction of a $700 million steel slab plant in the northeastern Brazilian state of Ceará.

Brazil, the largest economy in South America, has long had big dreams about steel. But years of sluggish economic growth and the high cost of credit kept the industry from expanding faster. With China's voracious appetite for raw materials having breathed life into steel companies across the globe, Brazil is among the main beneficiaries.

"The plans this time for these facilities look as though they are well founded, mainly because they are on the basis of the need for steel from a growing economy in China and will partially be funded by outside sources with a lot of money," said Karlis Kirsis, a managing partner at World Steel Dynamics, a steel research group in New Jersey.

To be sure, now that the Chinese government is taking steps to stop the economy from overheating, many analysts fear that China may not need as much foreign steel as initially thought. Indeed, when Chinese authorities recently said they were considering canceling some construction projects and even temporarily forbidding banks from making loans, share prices in Brazilian steel companies sank on the São Paulo Stock Exchange on fears of diminished demand.

Others, however, say those fears are overblown.

"Even if China's economy starts to cool off," said Germano Mendes De Paula, a steel specialist at the Federal University of Uberlândia in Minas Gerais, "millions of Chinese are leaving the countryside each year to move to cities. It's not hard to imagine the demand for housing, and thus steel, that migration like that generates."

"Demand from China may shrink a bit, but it's not going to dry up anytime soon," he added.

The prospect of steady demand from China has steel makers just about everywhere looking to increase production, stoking concerns about a possible glut of steel that could drive down prices. Even so, a low-cost producer like Brazil appears better positioned than most to weather the storm, if there is one. *(handwritten: low wages — more competitive)*

"If anyone is going to be able keep raising production capacity, it's Brazil," said Luciana Machado, a steel analyst at Fator Corretora, a São Paulo brokerage firm. "It's simply not feasible for other countries, even the big steel consumers, to keep raising output because they don't have all the conditions that Brazil does."

Drawn by Brazil's vast reserves of iron ore, an essential ingredient of steel, Baosteel is joining forces with the Brazilian mining giant *Companhia Vale do Rio Doce*, or C.V.R.D., as it is widely known, to build a $1.5 billion steel mill in São Luís, a coastal city in the poor northeastern state of Maranhão. Instead of shipping iron ore back to China, Baosteel hopes to save money by shipping lighter, ready-to-go steel slabs from São Luís, which is only a few days by ship from the Panama Canal.

The plant is expected to produce about 3.5 million metric tons of steel a year by 2007 and eventually expand output to 7.5 million tons. Arcelor, which is the world's largest steel maker, and Brazil's state-run National Bank of Economic and Social Development are also considering investing in the project.

In addition, the bank recently signed a letter of intent to offer $110 million in financing for a separate steel slab plant further to the east in Ceará, where the state government is trying to woo foreign investors by guaranteeing natural gas supplies for the venture. Dongkuk Steel and Danieli have already signed up for the project, as has C.V.R.D., the world's biggest producer and exporter of iron ore. The plant is expected to produce 1.5 million tons of steel for export by the middle of 2006, bringing in about $600 million a year in revenue.

Brazil is not relying only on outsiders to bankroll its steel boom. The Brazilian Steel Institute, a trade association, announced late last month that the industry planned to invest $7.4 billion through the end of 2008 to raise production by 10 million tons a year, to 44 million. If the plan is a success, Brazil would jump from the world's ninth-largest steel producer to No. 6, displacing Germany.

"What we're seeing today is a great window of opportunity for the Brazilian steel industry to grow, and in a big way," said Gabriel Stoliar, executive director of strategic planning at C.V.R.D., which brought in a record $1.55 billion in profit last year. "But we can't waste time. The time to invest is now so we can guarantee our place in the world."

While some of the extra capacity will be consumed domestically, most is likely to be exported to faster-growing economies abroad, especially China. The Chinese gobbled up 2.4 million tons of Brazilian steel in 2003, or $730 million worth, making it the country's top export market ahead of the United States. China is on track to import even more from Brazil this year.

Betting that China will continue to play a crucial role in Brazil's economic future, President Luiz Inácio Lula da Silva is heading to Beijing on Friday in a bid to expand trade. It will be the first visit to China by a Brazilian head of state in nine years.

Almost 400 Brazilian business leaders have signed up to join Mr. da Silva on the six-day trip, some hoping to persuade the Chinese to invest in infrastructure projects in Brazil to speed up the export of raw materials like iron ore and other goods. Brazil's foreign minister, Celso Amorim, already traveled to China in March to lobby the Chinese to help finance as much as $5 billion in railways and roads in Brazil.

"Without a doubt, China has been a huge stimulus for steel makers everywhere, but we're particularly important to them because of our iron ore," said Jorge Gerdau Johannpeter, president of Grupo Gerdau, the Brazilian steel company.

— Brazil's Steel Industry and the Chinese Market
— Domestic Consumption but greater Export quotas to faster-growing economies like China
— Hope for China to invest in infrastructure of Brazil → to speed up export of raw materials

Brazil's Dependence on overseas markets
Export Oriented

Between Hard and Soft

The Dilemma of the Guambianos

Poppy Cultivation

By Victor Englebert

In the Andes' central cordillera, in a southwestern corner of Colombia, live some seventeen thousand Guambiano Indians. Since the Spanish conquest they have been gradually pushed out of the best parts of their original land. Today they eke out a difficult existence in the colder, rainy, and more abrupt parts of Colombia's mountains.

Guambianos weave their own clothes, grow their own food, and build their own adobe houses. Like most Andean Indians, they practice minga, an Incan custom of community support. They help each other plant, harvest, build houses, and create new rural paths. Guambianos are also very young: only 13 percent of their population is over fifty.

Though they have interacted with whites and mestizos for around five hundred years, Guambianos have retained their language, traditional customs, and typical dress. That attire consists, for men, of a knee-length blue *anaco* (a skirt-like wraparound garment), a short blue or gray poncho, a red scarf, and a narrow-brimmed felt hat. (To an outsider, the anaco contrasts almost comically with their army-type boots and bare legs.) Women dress similarly but wear a blue cape rather than the poncho, a true skirt, and heavy white bead necklaces instead of the scarf.

Floro Tunubala, the Guambiano governor of Cauca State, goes about his official duties wearing the traditional dress. Before he took office, people asked the governor elect if he was going to govern in a *falda* (skirt). He corrected them by explaining the Guambiano term: he would govern in an anaco. Indeed, he had worn it while attending university, even in Mexico while studying on a grant for three years. Another member of the state congress, also a Guambiano, wears traditional attire when attending the assembly.

Most Guambianos must eke out a living based on subsistence agriculture. Many live in the Páramo de las Delicias, a moor above ten thousand feet, roughly between Purace volcano and snowcapped Huila Mountain. Because of lack of roads, most of them find it difficult—if not impossible—to bring their produce to the market at Silvia, a pleasant small town of about four thousand mestizos. Those who live close enough walk, or ride horses or bicycles, to reach the market. Others make it by traveling in a *chiva*, a colorfully painted wooden bus built over a truck bed. The chiva has to struggle over steep, sinuous dirt roads, if roads exist at all.

The people get so little money for their potatoes, manioc, onions, beans, wheat, corn, and rice that they are hardly better off than those who cannot come to the market. Guambianos who lived closer to Silvia used to benefit from regional tourism. The town was a weekend destination for the people of Cali, Colombia's second city, a winding, two-hour car ride to the north. The tourists came to escape Cali's heat, walk, ride horses, and fish for trout in the mountain streams.

Many bought vegetables, woolen blankets and sweaters, and other handmade articles from the Guambianos. Now, even that small income is gone: in recent years guerrilla and paramilitary forces have been shooting it out all over the region. So Guambianos were pleased when, a decade ago, drug traffickers from Cali asked them to grow poppies.

To the Guambianos, this was an easy crop to produce. They saw the poppy as a beautiful red or purple flower but had little understanding of its role as the basis for a drug. Besides, poppies are weeds and grow without help. The crop gives two harvests a year, and the gum that must be extracted from the flowers could be gathered right outside their doors, wherever they lived.

In fact, two or three acres of poppy cultivation can bring a family several hundred dollars twice a year. This is a considerably greater cash return than for any other crop. For a while, the poppies brought the Guambianos some relief from poverty. A few young people even started buying motorcycles. But Guambiano society, otherwise highly respectful of traditions and the *cabildo*, or council of yearly elected leaders, started to break apart. Their communal system of agriculture turned to a more individual one. Alcoholism, divorce, school dropouts, and loss of respect for the elders changed a hard but honest and self-respecting way of life more than the average Guambiano could accept.

The situation worsened as the guerrillas increased their presence for the purpose of taxing the poppy production. This, in turn, brought more paramilitaries in to fight them. The Colombian government sprayed their fields from the air with herbicides, destroying their poppies, and other crops. The chemicals tainted the streams that gave them drinking water and even poisoned a few cows. Spraying also did a good deal of collateral ecological damage. It killed the trout some people had started cultivating—in large outdoor tanks—as a possible alternative to the poppies.

The Colombian government offered the Guambianos land, credit, and technical assistance if they would agree to alternative crop programs. One of those programs was the trout hatchery, which would supply regional restaurants and markets. Though the plan was long on promise, it was short on delivery. This angered the Indians, many of whom returned to their illegal activity after seeing how little they were getting in return for pulling up 90 percent of their poppies. The mafia was paying them double what the government promised, and unlike the government, it fulfilled its pledges.

The Guambianos' cosmogony, however, told them that continued poppy cultivation and drug trafficking would affect the balance between "the world's hard and soft elements." They realized that if they stuck to the easy path, the acculturation pressure they had successfully resisted for five hundred years would soon become impossible to hold back. The people hope to find a successful cash crop or product as an alternative to the red flowers.

Indeed, since 1997, the Colombian government and the Guambianos have agreed several times on the eradication of poppies in exchange for roads, schools, and alternative schemes. So far, failure to make any headway has rested on the government's nonfulfillment of the deals. Colombia's new president, Andres Pastrana, is the first one in thirty-five years of guerrilla activity to make a real effort to face that challenge and has achieved, partly with American aid, the first real military successes. All hope that he will be able to find a way to bring legal prosperity, and a little peace and happiness, to the Guambianos.

Victor Englebert is a freelance photojournalist.

GUAMBIANO INDIANS OF COLOMBIA

— MINGA → INCAN CUSTOM OF COMMUNITY SUPPORT
— PLANTING, HARVESTING, CONSTRUCTING HOUSES

— RETAINED THEIR TRADITIONAL CUSTOMS, DRESS, LANGUAGE

— SUBSISTENCE Ag.
— Poppy Cultivation FOR INCOME → RELIEF FROM POVERTY
↓
LED TO EROSION OF THEIR Culture

— PARAMILITARIES / GUERRILLAS
⌐→ SPRAYING FIELDS
— NEED TO Replace Poppy PRODUCTION WITH A Successful CASH Crop.
legal
— Gov't HAS NOT FulFilled ITS Promises

Venezuela's Chávez as Everyman

By Gary Payne

He looks like he could be the driver of the most decrepit taxi on the streets of Caracas. He could be a streetsweeper, a waiter, a shoe shiner. The tiny upper and middle classes of Venezuela think he is uncivilized. But to the three-quarters of Venezuelanos living in poverty, he is a mirror image of themselves. He is Everyman.

When he addresses the nation on the popular Sunday broadcast, "Hello Mr. President," *his* somewhat darker-skinned crowds gather in Plaza Bolivar to listen to him carry on over government-installed television screens and radio speakers. For five hours. He answers questions from everyday citizens in an engaging conversational tone, without notes, without hesitation, but with an air of calm informality that sets him apart from the pretentious-ness of his competition.

And then, without any warning, albeit at regular inter-vals in the broadcast, he will bathe his detractors—the richest and most powerful people in the history of the world—with a shower of insults that confirm his peculiar position in global politics. For Hugo Chávez is not merely the President of this poor majority, but the long-stifled ex-pression of its collective historical frustration and the em-bodiment of its hopes. Hopes that would have seemed terribly naïve only a few years ago.

Some would say that those hopes are still naïve. Like many populist-left leaders before him, Chávez has be-come an absolutely intolerable barrier to business as usual in this hemisphere. Venezuela's oligarchs and free marketeers everywhere both hate and fear him, for noth-ing succeeds like success. And if his administration is al-lowed to succeed in South America, a new cadre of followers might rise across the continent, threatening his-torical privilege and economic privatization like never before.

Not even Cuba poses such a threat, for as an island na-tion, it has been more easily isolated. What Fidel Castro has accomplished may be miraculous considering the challenges he has faced—attacks, embargoes, harass-ment, covert destabilization efforts—but Cuba's potential for democracy was subverted by this history. Fidel seems to have recognized this himself, having reportedly told his friend Hugo that he has an historic opportunity in front of him, and he should not waste it.

For by contrast, Venezuela is a vast territory, and Hugo Chávez was twice elected by overwhelming majorities in this oldest democracy on the continent. By all accounts, the nation's energy resources are legendary. Chávez is raising the cost of oil royalties paid by Exxon-Mobil, Conoco-Phillips, and supporting OPEC production limits instead of overpumping as his predecessors have done. This income could be diverted from its customary path to the pockets of the oligarchy and distant stockholders. It could build a modern egalitarian society.

The untold story in Venezuela is that this new society is sprouting legs and moving off the drawing board. Chávez has not yet turned the tide on poverty—not by any means—but he has set the stage for a fundamental shift in economic and educational opportunity. He banned school entrance registration fees for students which previously served as a barrier to much of the child population. "Bolivarian" schools have opened in poor neighborhoods, often maintained and run by parents and volunteers, but supported by the government. Literacy is increasing rapidly as millions of new students have en-tered school.

Chávez's "Inside the Barrio" health plan is setting up clinics in the poorest communities, often staffed by re-spected Cuban doctors and nurses who are on loan to a society that in return provides cheap oil to the island na-tion. Some of the better new Venezuelan students, previ-ously unable to even dream of college, have found themselves enrolled in Cuban medical schools.

His land reform legislation limits individual owner-ship to 5,000 hectares (12,350 acres), and allows idle land to be redistributed to peasant cooperatives, which will likely lead to much greater fairness in a nation where 2% of the people own 60% of the territory.

The Venezuelan oligarchy and various international entities connected to Venezuela's natural resource food chain have taken notice, and are coordinating actions to bring Chávez down. Their task is formidable, because of his popularity and elected status. But they maintain sev-eral overwhelming advantages.

Most useful is information control. Almost all of the private television and radio stations and all but one of the major newspapers in the country are owned and operated by those who loathe the Chávez administration. The feeds that go to the mainstream international media come almost exclusively from these sources, and the hype and spin against Chávez is spectacular, even by today's cynical standards.

The International Monetary Fund has indicated that it supports a transitional government and was reported by Caracas' right-wing *El Nacional* to be willing to bankroll those who would replace Chávez. The National Endowment for Democracy, long used as a cover for the CIA projects, brought several opposition leaders to Washington for consultations in the months preceding the attempted coup in 2002. Fedecarmaras, an unnatural alliance of upper-middle class trade unionists and business owners that called for a strike two days before the coup attempt in order to promote the impression of chaos, is waiting backstage. Coke, along with Venezuelan-owned Polar breweries, recently provided bus service for members of the opposition, and/or help with blockades during both violent and non-violent protests by the opposition.

And now the two major party candidates for the U.S. Presidency—both rich, both Yale graduates, and both tightly bound to the oil industry—have begun a verbal rivalry to see which is the most aggressive anti-Chavista.

The list of privileged rogues in this alliance will continue to disrupt everyday life and the economy of Venezuela. The present referendum fiasco is merely its latest tactic in a disinformation campaign. What remains to be seen is whether a steady diet of it can exhaust the will of Chávez supporters or lead Chávez himself into making a serious mistake with the military forces at his disposal.

In the old English passion play, Everyman asks Death to give him more time. Death complies, although Everyman eventually must succumb, taking only his good deeds with him to the afterlife. Chávez is asking for more time. But in the Venezuelan version of the play, his good deeds may never be fully implemented. And Death, in some form, may be forced upon him prematurely.

Gary Payne is a professor of sociology who teaches at Central Lakes College in Minnesota. He recently visited Venezuela for a few weeks.

From Maid to Rio Governor, and Still Fighting

By LARRY ROHTER

RIO DE JANEIRO—Just before Benedita da Silva was sworn in as the new governor of the State of Rio de Janeiro a few months ago, a political rival accused her of being excessively fond of the luxuries of office. As Brazil's first black woman governor, he argued, she had an obligation to remain true to her origins in the slums here and shun all pomp and privilege.

"I have no problem whatsoever in walking on red carpets," she immediately retorted, "because I've certainly washed enough of them in my life."

As chief executive of a state of 14 million people, Ms. da Silva insists on exercising her right to live in the governor's mansion. But she makes a point of reminding Brazil's almost exclusively white political establishment that her long and arduous journey to power began in maid's quarters and that she speaks for millions of black Brazilians who remain poor and disenfranchised.

"We may be a majority, but blacks are invisible in this country, and I want to make them visible," Ms. da Silva said recently. "But at the same time, I want my mistakes and my achievements to be attributed to the person I am, and not to the color of my skin or my gender. That's what I am fighting for."

Officially, less than half of Brazil's 175 million people are classified as black. But in a nation that likes to consider itself a racial democracy, 70 percent of those living below the poverty line are black, as are 80 percent of those who are illiterate, and some studies indicate that on average, whites live longer than blacks and earn twice as much.

As a political leader, Ms. da Silva owes much of her popularity and her credibility among voters here to the powerful symbolism of her rise to power from poverty, which kindles hope that others can do the same.

But on a personal level, that background remains extremely painful to her.

"To experience misery, discrimination, prejudice and social exclusion in one's own skin, the truth is that's a memory I don't like to dwell on very much," she said.

Born in one of Rio's innumerable favelas, or squatter slums, Ms. da Silva, now 60, was one of 13 children, only 8 of whom survived to adulthood. Her mother, a priestess in the Afro-Brazilian religion of Candomblé, was a washerwoman for the family of Juscelino Kubitschek, who was Brazil's president from 1956 through 1960.

Her first toys, she recalls, were hand-me-downs from the future president's daughter Márcia, with whom Ms. da Silva would later serve in Congress.

Eventually, Ms. da Silva would also work for years as a maid and cleaning lady. But even as a child, she shined shoes, sewed and sold fruits and candy because her father, who washed cars and worked in construction, could not make enough money to support his family.

The family was so poor that Ms. da Silva had only one set of clothes to wear to school, and the temptations and dangers of the street were many. A sister drifted into prostitution, one of her brothers became a pickpocket, and Ms. da Silva herself was molested during her childhood by a boarder her parents had taken in to earn a little extra money.

"One reason I don't like to remember what I went through is that it takes me to one of the things that has most marked my life, which is sexual violence," she said. "I was raped at the age of 7, and it took decades and decades of my life for me to recompose myself."

At 16, she married a heavy-drinking house painter 10 years her elder, and her life became even more difficult. Two of her four children died as infants, one of whom was buried in a pauper's grave. After being widowed at the age of 38 she married again, only to have her second husband die six years later.

All the while, though, Ms. da Silva, or Bené, as she is widely known here, was working in community groups in the favela, where she had lived since she was an infant, and was continuing to study, eventually earning a degree as a social worker. When the left-wing Workers' Party was founded two decades ago and came looking for candidates with grass-roots credentials, she was an obvious choice.

"I will never forget that first election, when I ran for the City Council in 1982," she said. "I was trembling with fear from all the misery and atrocities that I had been through, but I knew that I had to crack the code, that if I didn't figure out the code, I would be eternally subjugated. And there is nothing worse than that, than feeling you are imprisoned."

Initially, she was the only Workers' Party representative on the Council here, but other firsts followed quickly: the first black woman to serve as a congressional deputy, the first black woman to be a senator. Elected lieutenant governor here in 1998 on a multiparty slate, she became governor in April when her predecessor, Anthony Garotinho, stepped down nine months early to run for president.

But Mr. Garotinho, anticipating an early departure, had already spent nearly the entire budget for the year, she complains, making it almost impossible for her to strengthen the social programs that are her trademark. There has also been a surge in violent crime and attacks on police officers, as the drug lords who now control the favelas openly challenge her authority.

In the most recent polls leading up to October's election for a full four-year term, Ms. da Silva is running a strong third in a field of eight. Two of her strongest rivals are also women, one of them Mr. Garotinho's wife, but both are white, and Ms. da Silva argues that she continues to face an uphill climb because of the color of her skin.

"Just because I am the governor of the state of Rio de Janeiro doesn't mean that racial prejudice has been done away with," she said. "The favelas are still there, their residents still encounter the same problems of racism and class prejudice, and only a black person who isn't awake doesn't see what is going on around him."

Unusually for a left-wing politician in Brazil, the world's most populous Roman Catholic nation, Ms. da Silva is also a member of an evangelical Protestant religious group, which she joined in her mid-20's. She admits that her colleagues in the Workers' Party find that odd, but she argues that "Christ was the greatest revolutionary" and that the Gospel and the fellowship of believers has helped make her whole and has brought her peace.

"I've been through my ugly duckling phase, of not liking my kinky hair and big feet," she said. "But the church taught me that I need not be ashamed of that."

For the past decade, Ms. da Silva has been married to Antônio Pitanga, a popular television soap opera star whom she met when both were campaigning for office. Three years ago, the couple left the Chapéu Mangueira favela, where Ms. da Silva had spent virtually her entire life and reluctantly moved into a middle-class neighborhood close to the studio where Mr. Pitanga often works.

Ms. da Silva continues to worry about those left behind in the favelas. Everywhere she goes, she is reminded that hers is an exceptional case.

"I go into a five-star hotel and they address me in English," she said. "That's because a well-dressed black woman simply isn't within the standard model for black Brazilians, and they assume she has to be a foreigner. So I have to tell them, 'No. I am a Brazilian.'"

— Racial Prejudice / Class Prejudice
— Gender Disadvantage

Khipu — Bookkeeping
Record Census
Keeping Narratives of Events

String, and Knot, Theory of Inca Writing

By JOHN NOBLE WILFORD

Of all the major Bronze Age civilizations, only the Inca of South America appeared to lack a written language, an exception embarrassing to anthropologists who habitually include writing as a defining attribute of a vibrant, complex culture deserving to be ranked a civilization.

The Inca left ample evidence of the other attributes: monumental architecture, technology, urbanization and political and social structures to mobilize people and resources. Mesopotamia, Egypt, China and the Maya of Mexico and Central America had all these and writing too.

The only possible Incan example of encoding and recording information could have been cryptic knotted strings known as khipu.

The knots are unlike anything sailors or Eagle Scouts tie. In the conventional view of scholars, most khipu (or quipu, in the Hispanic spelling) were arranged as knotted strings hanging from horizontal cords in such a way as to represent numbers for bookkeeping and census purposes. The khipu were presumably textile abacuses, hardly written documents.

But a more searching analysis of some 450 of the 600 surviving khipu has called into question this interpretation. Although they were probably mainly accounting tools, a growing number of researchers now think that some khipu were nonnumerical and may have been an early form of writing.

A reading of the knotted string devices, if deciphered, could perhaps reveal narratives of the Inca Empire, the most extensive in America in its glory days before the Spanish conquest in 1532.

If khipu is indeed the medium of a writing system, Dr. Gary Urton of Harvard says, this is entirely different from any of the known ancient scripts, beginning with the cuneiform of Mesopotamia more than 5,000 years ago. The khipu did not record information in graphic signs for words, but rather a kind of three-dimensional binary code similar to the language of today's computers.

Dr. Urton, an anthropologist and a MacArthur fellow, suggests that the Inca manipulated strings and knots to convey certain meanings. By an accumulation of binary choices, khipu makers encoded and stored information in a shared system of record keeping that could be read throughout the Inca domain.

In his book "Signs of the Inka Khipu," being published next month by the University of Texas Press, Dr. Urton said he had for the first time identified the constituent khipu elements. The knots appeared to be arranged in coded sequences analogous, he said, to "the process of writing binary number (1/0) coded programs for computers."

When someone types e-mail messages, they exist inside the computer in the form of eight-digit sequences of 1's and 0's. The binary coded message is

sent to another computer, which translates it back into the more familiar script typed by the sender. The Inca information, Dr. Urton said, appeared to be coded in seven-bit sequences.

Each sequence could have been a name, an identity or an activity. With the possible variations afforded by string colors and weaves, Dr. Urton estimated, the khipu makers could have had at their command more than 1,500 separate units of information. By comparison, the Sumerians worked with fewer than 1,500 cuneiform signs, and Egyptian hieroglyphs numbered under 800.

Dr. Urton concedes that his interpretation of a khipu writing system may be hard to prove. No narrative khipu has been deciphered. Spanish conquerors, who suspected the knotted strings might contain accounts of Inca history and religion, destroyed those they came across as idolatrous objects. The few existing descriptions of the khipu by explorers and missionaries lack enough detail for an understanding of the way the Inca made and "read" them.

Other Inca scholars generally agree that the khipu may have served as more than accounting devices or memory aids, and may have been a medium for recording historical information. But they reserved judgment on Dr. Urton's binary code hypothesis.

"Most serious scholars of khipu today believe that they were more than mnemonic devices, and probably much more," said Dr. Galen Brokaw, a specialist in ancient Andean texts at the State University of New York at Buffalo. He was quoted in an article about the khipu in the June 13 issue of the journal Science.

Dr. Patricia J. Lyon of the Institute of Andean Studies in Berkeley, Calif., was unmoved from her position that the khipu were mnemonic devices, personalized visual and tactile cues for the recall of the information retained in the memory of the maker. If that was the case, the khipu would not be a form of writing because they would have been understood only by their makers, or someone familiar with the same memorized accounts or narrative.

"People feel this great need to pump up the Inca by indicating that the khipu were writing," Dr. Lyon said.

Dr. Urton said in an interview that others would soon be able to test his theory and possibly find other patterns and clues in the khipu he studied. A detailed khipu database, financed by the National Science Foundation and prepared with the help of Dr. Carrie Brezine, a Harvard mathematician and weaver, is expected to be ready this fall and will eventually be available online.

Experts in the culture of early Peru think it understandable that textiles would have been the chosen medium for writing. The Sumerians and Babylonians wrote on clay, the Egyptians on stone and papyrus. The Inca may have used cloth, though, to store and communicate knowledge because to them cloth was a widely used marker of status, wealth and political authority.

Dr. Heather Lechtman, an archaeologist at the Massachusetts Institute of Technology who specializes in early Andean technology, said that "fibers were the heart of Andean technologies of all kinds, even long before the Inca, and so it doesn't surprise me that people would have thought of using khipu perhaps for some sort of writing system."

Early Spanish colonists gave conflicting accounts of the practice. A drawing of a khipu maker in an Inca storehouse seemed to reflect the view that the knotted strings involved record keeping. A Jesuit chronicler said the khipu were like ledgers or notebooks that overseers and accountants used "to remember what had been received and consumed."

Another account tells of Spanish travelers who came upon an old Indian man who tried to hide the khipu he was carrying. Under questioning, the Indian claimed the khipu recorded the activities of the conquerors, "both the good and the evil." The Spanish burned the khipu and punished the Indian.

Not until the 1920's did scholars seem to reach a consensus on what the khipu were. From studies of a collection of knotted strings at the American Museum of Natural History in New York City, L. Leland Locke, a science historian, concluded that they did not represent a conventional scheme of writing but signs recording columns of numbers. Khipu makers must have been bookkeeping bureaucrats.

This remained the prevailing opinion until the last two decades. Husband and wife researchers, Dr. Robert Ascher, a retired Cornell archaeologist, and Dr. Marcia Ascher, a mathematician at Ithaca College, reopened debate by pointing out that khipu seemed to use numbers as both numbers and labels. They estimated that about 20 percent of existing khipu were "clearly nonnumerical" and could have been examples of an early form of writing.

Dr. Urton has carried the idea further. A creator of khipu, he posits, made a series of choices involving the type and color of string and each knot. Each choice contributed to creating a binary signature. A certain string configuration could represent signs that stood for a value, object or event, much as graphic signs do in familiar forms of writing.

Emboldened by this insight, Dr. Urton said in his book that the Inca "may

well have been recording full subject-object-verb notations in the khipu."

Dr. Urton based his research primarily on khipu specimens at museums in the United States, Germany and Peru. A discovery in 1997 in northern Peru, at a burial site of the Chachapoya culture, yielded 32 khipu with exceptionally elaborate and varied types of string patterns. Strands hanging from the horizontal cord had their own secondary and tertiary pendants.

These complex pendant attachments, he wrote, "must have been an important mode of binary coding in the khipu."

A close examination of Dr. Urton's new database of khipu elements by other scholars, including linguists and pattern-recognition experts, may win wider support for the writing hypothesis.

"It's much too early to say anything about how this will all come out," said Dr. Lechtman of M.I.T.

More definitive would be the discovery of an Inca "Rosetta stone." It was such a trilingual inscription that finally enabled scholars to decipher Egyptian hieroglyphics.

A colonial governor had khipu makers "read" some strings and scribes record the accounts in Spanish. This could have been a start toward decipherment, if only the khipu had been preserved.

A prospective Rosetta stone was announced in 1996 by an Italian amateur historian, who claimed to have found a translation into Spanish of a song encoded in a khipu. But other researchers have not been allowed to examine the material, and Dr. Urton said that many questions had been raised about its authenticity.

Dr. Urton holds out more hope of making a breakthrough discovery in the Chachapoya material. Most of the khipu there appear to be from the early colonial period. For that reason their encoded messages are more likely to have been transcribed in Spanish documents as the sought-after Rosetta stone of Inca writing. If, that is, the Inca wrote with strings and knots.

Castro Attacks Enrope for Meddling

By Richard Lapper

Fidel Castro has marked the anniversary of the failed guerrilla attack that sparked the Cuban Revolution by launching a ferocious assault on some of the countries who have helped the island's battered economy survive the collapse of the Soviet Union.

Stung by recent criticism of human rights abuses, the Cuban president lambasted the European Union on issues ranging from rules on foreign aid and sugar subsidies to their historical involvement in the slave trade and the plunder and extermination of entire peoples.

In the low-key atmosphere at Saturday's rally in Santiago, 10,000 specially invited supporters waved small paper Cuban flags in unison as Mr Castro delivered his tirade. All were wearing red or black T-shirts specially printed with the slogan "carry in the heart the doctrines of the master"—a reference to the writings of José Marti, the 19th century Cuban nationalist hero who inspired Mr Castro's political thinking.

Mr Castro was speaking in front of the former Moncado barracks, the mustard yellow building that he together with a few dozen nationalist rebels—had unsuccessfully attacked in July 1953, kicking off the revolution that was to make Cuba communist. His guerrilla fighters finally overthrew the dictatorship of Fulgencio Batista in 1959.

His speech was intensely ideological, and seems to mark an intention to dig in and weather the political storms that are buffetting the island, rather than reaching out to the wider world.

Cuba had welcomed the creation of the EU, Mr Castro said, because it counterbalanced the military and economic dominance of the US. "But now the EU adopts this arrogant and calculated attitude in hope of reconciliation with the masters of the world," he added, referring to the US.

Over the past 10 years, European trade and investment—especially in the tourism sector—have cushioned Cuba from the loss of billions of dollars a year in Soviet subsidies. In addition, European links have helped the island counter the impact of the US economic embargo which has been in place for four decades.

But Cuba's decision in April to execute three hijackers who were trying to reach the US by ferry and to impose draconian prison sentences on 75 opposition activists led to a crisis in the relationship with the EU, including fierce diplomatic protests and the suspension of humanitarian aid.

The European Commission froze Havana's request to join its Cotonou accord, which would have given Cuba access to substantial development and aid assistance, but was conditional on respect for human rights. During his speech, Mr Castro said he would in future only accept aid from organisations that did not "impose political conditions on Cuba".

Despite Mr Castro's outburst, the EC said on Sunday it remained committed to supporting the Cuban people.

But if Mr Castro is serious in rejecting dialogue, he appears to be embracing isolation and a further period of socialist austerity at a time of increasing economic difficulty. A decline in revenues from tourism and the closure of 71 of 156 state-owned sugar mills have hit the economy hard.

In spite of a gradual recovery since 1994, economic output has still not fully recovered to the levels of 1990.

Britain, Spain, and the former communist countries of eastern Europe—allies of the United States in the Iraq war—came under especially heavy fire from Mr Castro. He challenged the British government to tell the world how David Kelly, the biological weapons expert, who died 10 days ago "was brutally murdered, or how he was led to commit suicide".

The Cuban leader criticized José Maria Aznar for holding "fascist" ideas and said Spain's education system was equivalent to that of a "banana republic" and "an embarrassment for Europe".

The leaders of eastern Europe were opportunists and their countries US Trojan horses in the heart of the EU. They were full of hatred for Cuba, he said. Eastern Europe could not forgive the communist island "for having demonstrated that socialism is capable of achieving a society a thousand times more just and humane than the rotten [capitalist] system they were adopting", he claimed.

Fidel Castro is now 77 and dogged by rumours of ill health, a far cry from the 26-year-old revolutionary who stormed the barracks here. But on Saturday, his rhetoric was a reminder that the last communist state in the western hemisphere still has no intention of accepting criticism from outsiders.

Sources for Statistical Reports

U.S. State Department, *Background Notes* (2000–2001).
The World Factbook (2001).
World Statistics in Brief (2001).
World Almanac (2001).

The Statesman's Yearbook (2001).
Demographic Yearbook (2001).
Statistical Yearbook (2001).
World Bank, World Development Report (2000–2001).

Glossary of Terms and Abbreviations

Agrarian Relating to the land; the cultivation and ownership of land.

Amerindian A general term for any Indian from America.

Andean Pact (Cartagena Agreement) Established on October 16, 1969, to end trade barriers among member nations and to create a common market. Members: Bolivia, Colombia, Ecuador, Peru, and Venezuela.

Antilles A geographical region in the Caribbean made up of the Greater Antilles: Cuba, Hispaniola (Haiti and the Dominican Republic), Jamaica, the Cayman Islands, Puerto Rico, and the Virgin Islands; and the Lesser Antilles: Antigua and Barbuda, Dominica, St. Lucia, St. Vincent and the Grenadines, St. Kitts– Nevis, as well as various French departments and Dutch territories.

Araucanians An Indian people of south-central Chile and adjacent areas of Argentina.

Arawak An Indian people originally found on certain Caribbean islands, who now live chiefly along the coast of Guyana. Also, their language.

Aymara An Indian people and language of Bolivia and Peru.

Bicameral A government made up of two legislative branches.

CACM (Central American Common Market) Established on June 3, 1961, to form a common market in Central America. Members: Costa Rica, El Salvador, Guatemala, and Nicaragua.

Campesino A Spanish word meaning "peasant."

Caudillo Literally, "man on horseback." A term that has come to mean "leader."

Carib An Indian people and their language native to several islands in the Caribbean and some countries in Central America and South America.

CARICOM (Caribbean Community and Common Market) Established on August 1, 1973, to coordinate economic and foreign policies.

CDB (Caribbean Development Bank) Established on October 18, 1969, to promote economic growth and development of member countries in the Caribbean.

The Commonwealth (Originally the British Commonwealth of Nations) An association of nations and dependencies loosely joined by the common tie of having been part of the British Empire.

Compadrazgo The Mexican word meaning "cogodparenthood" or "sponsorship."

Compadres Literally, "friends"; but in Mexico, the term includes neighbors, relatives, fellow migrants, coworkers, and employers.

Contadora Process A Latin American intiative developed by Venezuela, Colombia, Panama, and Mexico to search for a negotiated solution that would secure borders and reduce the foreign military presence in Central America.

Contras A guerrilla army opposed to the Sandinista government of Nicaragua. They were armed and supplied by the United States.

Costeños Coastal dwellers in Central America.

Creole The term has several meanings: a native-born person of European descent or a person of mixed French and black or Spanish and black descent speaking a dialect of French or Spanish.

ECCA (Eastern Caribbean Currency Authority) A regional organization that monitors the integrity of the monetary unit for the area and sets policies for revaluation and devaluation.

ECLA (Economic Commission for Latin America) Established on February 28, 1948, to develop and strengthen economic relations among Latin American countries.

FAO (Food and Agricultural Organization of the United Nations) Established on October 16, 1945, to oversee good nutrition and agricultural development.

FSLN (Frente Sandinista de Liberación Nacionál) Organized in the early 1960s with the object of ousting the Somoza family from its control of Nicaragua. After 1979 it assumed control of the government. The election of Violeta Chamorro in 1990 marked the end of the FSLN.

FTAA (Free Trade Area of the Americas) An effort to integrate the economies of the Western Hemisphere into a single free trade arrangement.

GATT (General Agreement on Tariffs and Trade) Established on January 1, 1948, to provide international trade and tariff standards.

GDP (Gross Domestic Product) The value of production attributable to the factors of production in a given country, regardless of their ownership. GDP equals GNP minus the product of a country's residents originating in the rest of the world.

GNP (Gross National Product) The sum of the values of all goods and services produced by a country's residents in any given year.

Group of 77 Established in 1964 by 77 developing countries. It functions as a caucus on economic matters for the developing countries.

Guerrilla Any member of a small force of "irregular" soldiers. Generally, guerrilla forces are made up of volunteers who make surprise raids against the incumbent military or political force.

IADB (Inter-American Defense Board) Established in 1942 at Rio de Janeiro to coordinate the efforts of all American countries in World War II. It is now an advisory defense committee on problems of military cooperation for the OAS.

IADB (Inter-American Development Bank) Established in 1959 to help accelerate economic and social development in Latin America.

IBA (International Bauxite Association) Established in 1974 to promote orderly and rational development of the bauxite industry. Membership is worldwide, with a number of Latin American members.

IBRD (International Bank for Reconstruction and Development) Established on December 27, 1945, to make loans to governments at conventional rates of interest for high-priority productive projects. There are many Latin American members.

ICAO (International Civil Aviation Organization) Established on December 7, 1944, to develop techniques of international air navigation and to ensure safe and orderly growth of international civil aviation. Membership is worldwide, with many Latin American members.

Glossary of Terms and Abbreviations

ICO (International Coffee Organization) Established in August 1963 to maintain cooperation between coffee producers and to control the world market prices. Membership is worldwide, with a number of Latin American members.

IDA (International Development Association) Established on September 24, 1960, to promote better and more flexible financing arrangements; it supplements the World Bank's activities.

ILO (International Labor Organization) Established on April 11, 1919, to improve labor conditions and living standards through international action.

IMCO (Inter-Governmental Maritime Consultative Organization) Established in 1948 to provide cooperation among governments on technical matters of international merchant shipping as well as to set safety standards. Membership is worldwide, with more than a dozen Latin American members.

IMF (International Monetary Fund) Established on December 27, 1945 to promote international monetary cooperation.

IPU (Inter-Parliamentary Union) Established on June 30, 1889, as a forum for personal contacts between members of the world parliamentary governments. Membership is worldwide, with the following Latin American members: Argentina, Brazil, Colombia, Costa Rica, Haiti, Mexico, Nicaragua, Paraguay, and Venezuela.

ISO (International Sugar Organization) Established on January 1, 1969, to administer the international sugar agreement and to compile data on the industry. Membership is worldwide, with the following Latin American members: Argentina, Brazil, Colombia, Cuba, Ecuador, Mexico, Uruguay, and Venezuela.

ITU (International Telecommunications Union) Established on May 17, 1895, to develop international regulations for telegraph, telephone, and radio services.

Junta A Spanish word meaning "assembly" or "council"; the legislative body of a country.

Ladino A Westernized Spanish-speaking Latin American, often of mixed Spanish and Indian blood.

LAFTA (Latin American Free Trade Association) Established on June 2, 1961, with headquarters in Montevideo, Uruguay.

Machismo Manliness. The male sense of honor; connotes the showy power of a "knight in shining armor."

Marianismo The feminine counterpart of machismo; the sense of strength that comes from controlling the family and the male.

Mennonite A strict Protestant denomination that derived from a sixteenth-century religious movement.

Mercosur Comprised of Argentina, Brazil, Paraguay, and Uruguay, this southern common market is the world's fourth largest integrated market. It was established in 1991.

Mestizo The offspring of a Spaniard or Portuguese and an American Indian.

Mulatto A person of mixed Caucasian and black ancestry.

Nahuatl The language of an Amerindian people of southern Mexico and Central America who are descended from the Aztec.

NAFTA (North American Free Trade Agreement) Established in 1993 between Mexico, Canada, and the United States, NAFTA went into effect January 1, 1994.

NAM (Non-Aligned Movement) A group of nations that chose not to be politically or militarily associated with either the West or the former Communist Bloc.

OAS (Organization of American States) (Formerly the Pan American Union) Established on December 31, 1951, with headquarters in Washington, DC.

ODECA (Central American Defense Organization) Established on October 14, 1951, to strengthen bonds among the Central American countries and to promote their economic, social, and cultural development through cooperation. Members: Costa Rica, El Salvador, Guatemala, Honduras, and Nicaragua.

OECS (Organization of Eastern Caribbean States) A Caribbean organization established on June 18, 1981, and headquartered in Castries, St. Lucia.

PAHO (Pan American Health Organization) Established in 1902 to promote and coordinate Western Hemisphere efforts to combat disease. All Latin American countries are members.

Patois A dialect other than the standard or literary dialect, such as some of the languages used in the Caribbean that are offshoots of French.

Peon Historically, a person forced to work off a debt or to perform penal servitude. It has come to mean a member of the working class.

PRI (Institutional Revolutionary Party) The dominant political party in Mexico.

Quechua The language of the Inca. It is still widely spoken in Peru.

Rastafarian A religious sect in the West Indies whose members believe in the deity of Haile Selassie, the deposed emperor of Ethiopia who died in 1975.

Rio Pact (Inter-American Treaty of Reciprocal Assistance) Established in 1947 at the Rio Conference to set up a policy of joint defense of Western Hemisphere countries. In case of aggression against any American state, all member countries will come to its aid.

Sandinistas The popular name for the government of Nicaragua from 1979 to 1990, following the ouster of President Anastasio Somoza. The name derives from César Augusto Sandino, a Nicaraguan guerrilla fighter of the 1920s.

SELA (Latin American Economic System) Established on October 18, 1975, as an economic forum for all Latin American countries.

Suffrage The right to vote in political matters.

UN (United Nations) Established on June 26, 1945, through official approval of the charter by delegates of 50 nations at an international conference in San Francisco. The charter went into effect on October 24, 1945.

UNESCO (United Nations Educational, Scientific, and Cultural Organization) Established on November 4, 1946, to promote international collaboration in education, science, and culture.

Unicameral A political structure with a single legislative branch.

UPU (Universal Postal Union) Established on July 1, 1875, to promote cooperation in international postal services.

World Bank A closely integrated group of international institutions providing financial and technical assistance to developing countries.

Bibliography

GENERAL WORKS

Mark A. Burkholder and Lyman L. Johnson, *Colonial Latin America*, 4th ed. (New York: Oxford University Press, 2000).

E. Bradford Burns, *Latin America: A Concise Interpretive History*, 6th ed. (New Brunswick, NJ: Prentice-Hall, 1994).

David Bushnell and Neill Macaulay, *The Emergence of Latin America in the Nineteenth Century*, 2nd ed. (New York: Oxford University Press, 1994).

Franklin W. Knight, *Race, Ethnicity and Class: Forging the Plural Society in Latin America and the Caribbean* (Waco, TX: Baylor University Press, 1998).

Thomas E. Skidmore and Peter Smith, *Modern Latin America*, 5th ed. (New York: Oxford University Press, 2000).

Barbara A. Tenenbaum, ed., *Encyclopedia of Latin American History*, 5 vols. (New York: Charles Scribner's Sons, 1996).

Claudio Veliz, *The Centralist Tradition of Latin America* (Princeton, NJ: Princeton University Press, 1980).

NATIONAL HISTORIES

The following studies provide keen insights into the particular characteristics of individual Latin American nations.

Argentina

Leslie Bethell, *Argentina Since Independence* (New York: Cambridge University Press, 1994).

Nicholas Shumway, *The Invention of Argentina* (Berkeley, CA: University of California Press, 1991).

Bolivia

Herbert S. Klein, *Bolivia: The Evolution of a Multi-Ethnic Society*, 2nd ed. (New York: Oxford University Press, 1992).

Brazil

Thomas Skidmore, *Brazil: Five Centuries of Change*, (New York: Oxford University Press, 1999).

Caribbean Nations

Franklin W. Knight, *The Caribbean: The Genesis of a Fragmented Nationalism*, 2nd ed. (New York: Oxford University Press, 1990).

David Lowenthal, *West Indian Societies* (New York: Oxford University Press, 1972).

Louis A. Perez Jr., *Cuba: Between Reform and Revolution*, 2nd ed. (New York: Oxford University Press, 1995).

Central America

Ralph Lee Woodward Jr., *Central America: A Nation Divided*, 3rd ed. (New York: Oxford University Press, 1999).

Chile

Brian Loveman, *Chile: The Legacy of Hispanic Capitalism*, 3rd ed. (New York: Oxford University Press, 2001).

Colombia Frank Safford, Macro Palacios, *Colombia: Fragmented Land, Divided Society*, (New York: Oxford University Press, 2001).

Mexico

Michael C. Meyer and William L. Sherman, *The Course of Mexican History*, 5th ed. (New York: Oxford University Press, 1995).

Eric Wolf, *Sons of the Shaking Earth: The Peoples of Mexico and Guatemala; Their Land, History, and Culture* (Chicago: University of Chicago Press, 1970).

Ricardo Pozas Arciniega, *Juan Chamula: An Ethnolographical Recreation of the Life of a Mexican Indian* (Berkeley, CA: University of California Press, 1962).

Peru

Peter Klaren, *Peru: Society and Nationhood in the Andes*, (New York: Oxford University Press, 1999)

David P. Werlich, *Peru: A Short History* (Carbondale, IL: Southern Illinois University Press, 1978).

Venezuela

John V. Lombardi, *Venezuela: The Search for Order, The Dream of Progress* (New York: Oxford University Press, 1982).

NOVELS IN TRANSLATION

The Latin American novel is perhaps one of the best windows on the cultures of the region. The following are just a few of many highly recommended novels.

Jorge Amado, *Clove and Cinnamon* (Avon, 1988).

Manlio Argueta, *One Day of Life* (Vintage, 1990).

Miguel Ángel Asturias, *El Señor Presidenté* (Macmillan, 1975).

Mariano Azuela, *The Underdogs* (Buccaneer Books, 1986).

Alejo Carpentier, *Reasons of State* (Writers & Readers, 1981).

Carlos Fuentes, *The Death of Artemio Cruz* (FS&G, 1964).

Jorge Icaza, *Huasipungo: The Villagers* (Arcturus Books, 1973).

Gabriel García Márquez, *One Hundred Years of Solitude* (Penguin, 1971).

Mario Vargas Llosa, *The Green House* (FS&G, 1985).

Victor Montejo, *Testimony: Death of a Guatemalan Village* (Curbstone Press, 1987).

Rachel de Queiroz, *The Three Marias* (University of Texas Press, 1991).

Graham Greene's novels about Latin America, such as *The Comedians* (1966), and V. S. Naipaul's study of Trinidad, *The Loss of El Dorado: A History* (1969), offer profound insights into the region.

Index

Index